T0407981

POTENT ANTICANCER MEDICINAL PLANTS

Secondary Metabolite Profiling,
Active Ingredients, and Pharmacological Outcomes

POTENT ANTICANCER MEDICINAL PLANTS

Secondary Metabolite Profiling, Active Ingredients, and Pharmacological Outcomes

Edited by
Deepu Pandita
Anu Pandita

AAP APPLE ACADEMIC PRESS

First edition published 2024

Apple Academic Press Inc.
1265 Goldenrod Circle, NE,
Palm Bay, FL 32905 USA

760 Laurentian Drive, Unit 19,
Burlington, ON L7N 0A4, CANADA

CRC Press
2385 NW Executive Center Drive,
Suite 320, Boca Raton FL 33431

4 Park Square, Milton Park,
Abingdon, Oxon, OX14 4RN UK

Library and Archives Canada Cataloguing in Publication

Title: Potent anticancer medicinal plants : secondary metabolite profiling, active ingredients, and pharmacological outcomes/ edited by Deepu Pandita, Anu Pandita.
Names: Pandita, Deepu, editor. | Pandita, Anu, editor.
Description: First edition. | Includes bibliographical references and index.
Identifiers: Canadiana (print) 20230445446 | Canadiana (ebook) 20230445497 | ISBN 9781774913116 (hardcover) | ISBN 9781774913123 (softcover) | ISBN 9781003431190 (ebook)
Subjects: LCSH: Medicinal plants. | LCSH: Antineoplastic agents.
Classification: LCC QK99.A1 P68 2024 | DDC 581.6/34—dc23

Library of Congress Cataloging-in-Publication Data

....

CIP data on file with US Library of Congress

....

ISBN: 978-1-77491-311-6 (hbk)
ISBN: 978-1-77491-312-3 (pbk)
ISBN: 978-1-00343-119-0 (ebk)

About the Editors

Deepu Pandita

Senior Lecturer, Government Department of School Education, Jammu, Union Territory of Jammu and Kashmir, India

Deepu Pandita is working as a Senior Lecturer in the Government Department of School Education, Jammu, Union Territory of Jammu and Kashmir, India. Deepu Pandita has more than 20 years of teaching experience and has done her Masters in Botany (MSc) from the University of Kashmir, Jammu and Kashmir, India and Master of Philosophy (MPhil) in Biotechnology from the University of Jammu, Jammu and Kashmir, India. Deepu Pandita has a number of international and national courses to her credit. She has qualified for fellowships, such as JRF-NET and SRF from the Council of Scientific and Industrial Research (CSIR), New Delhi India; Biotechnology Fellowship, Government Department of Science and Technology, Jammu & Kashmir, India; and IAS-INSA-NASI Summer Research Teacher Fellowship, India. Deepu Pandita has presented her research work at both national and international conferences and was awarded Best Oral Presentation Award at an International Conference on Biotechnology for Better Tomorrow, Avtar Krishan Award of Cytometry in 12th Indo-US Cytometry Workshop on Clinical Research; a Women Researcher Award; and a Research Excellence Award from two professional associations in India. She is a life member of various scientific societies. Deepu Pandita is also a reviewer, associate editor, and editor of a number of journals of international repute. She has published a number of editorials, book chapters (Springer, CRC, and Elsevier), reviews, and research articles in various journals of national and international repute, including *Cells, Frontiers in Plant Sciences, Journal of Fungi, Frontiers in Sustainable Food Systems,* and *Frontiers in Physiology.* Deepu Pandita has several books currently under production.

Anu Pandita

Senior Dietician, Vatsalya Clinic,
New Delhi, India

Anu Pandita is working as a Dietician at Vatsalya Clinic, E-5/11, Krishna Nagar, New Delhi, India. Previously, she worked as a Lecturer at Bee Enn College of Nursing, Talab Tillo, Jammu, India, and Dietician at Ahinsa Dham Bhagwan Mahavir Charitable Health Centre, New Delhi, India. Anu Pandita has done her MSc internship and a course in the Dietetics Department of Post Graduate Institute of Medical Education & Research (PGI), Chandigarh, India. She conducted a case study at the Pediatric Gastroenterology Ward at Nehru Hospital, PGI, Chandigarh, India, on a patient suffering from chronic liver disease. She has done a Certificate in Food & Nutrition as well. Anu Pandita has presented her research work at both national and international conferences. She has a number of trainings, refresher courses, and workshops to her credit. She is a life-time member of the Indian Dietetic Association and Indian Science Congress Association, Kolkata, India. She has published several book chapters in Springer and CRC journals and a number of reviews and research articles in various journals of national and international repute, including *Cells, Journal of Fungi, Frontiers in Sustainable Food Systems,* and *Frontiers in Physiology*. She has several books under production currently.

Contents

Contributors

Madhusudhana Reddy A.
Department of Botany, Yogi Vemana University, Kadapa, Andhra Pradesh, India

Iffat Zareen Ahmad
Department of Bioengineering, Integral University, Dasauli, Lucknow, Uttar Pradesh, India

Rohit Sam Ajee
Independent Researcher and Alumni Amity Institute of Biotechnology, Amity University, Madhya Pradesh, India

Ambedkar
Department of Biochemistry, Yogi Vemana University, Kadapa, Andhra Pradesh, India
Department of Biotechnology, Yogi Vemana University, Kadapa, Andhra Pradesh, India

Sharmistha Banerjee
Biomedical Engineering and Bioinformatics, University Teaching Department, Chhattisgarh Swami Vivekanand Technical University, Newai, Bhilai, India

Sagar Barge
Chemical Biology Lab I, Institute of Advanced Study in Science and Technology, Paschim Boragaon, Guwahati, Assam, India

Lepakshi M. D. Bhakshu
Department of Botany, PVKN Government College (Autonomous), Chittoor, India

Jagat C. Borah
Chemical Biology Lab I, Institute of Advanced Study in Science and Technology, Paschim Boragaon, Guwahati, Assam, India

Shalini Gurumayum
Chemical Biology Lab I, Institute of Advanced Study in Science and Technology, Paschim Boragaon, Guwahati, Assam, India

Adhikesavan Harikrishnan
Department of Chemistry, School of Arts and Science, Vinayaka Mission Research Foundation-AV Campus, Chennai, India

Bhoomika Inamdar
JSS Medical College (Deemed to be University), Mysuru, Karnataka, India

Dhananjay Jade
Biomedical sciences, University of Leeds, Leeds LS2 9JT, UK

Shuchi Kaushik
State Forensic Science Laboratory, Madhya Pradesh, India

K. M. Srinivasa Murthy
Department of Microbiology and Biotechnology, Jnanabharathi Campus Bangalore University, Bengaluru, India

Manohar M. V.
JSS Medical College (Deemed to be University), Mysuru, Karnataka, India

Chandramathi Shankar P.
Department of Biotechnology, Yogi Vemana University, Kadapa, Andhra Pradesh, India

Amogha G. Paladhi
Christ (Deemed to be University), Bengaluru, Karnataka, India

Ramachandra Reddy Pamuru
Department of Biochemistry, Yogi Vemana University, Kadapa, Andhra Pradesh, India

Anu Pandita
Vatsalya Clinic, Krishna Nagar, New Delhi, India

Deepu Pandita
Government Department of School Education, Jammu, Jammu and Kashmir, India

K. Venkata Ratnam
Department of Botany, Rayalaseema University, Kurnool, India

B. Uma Reddy
Department of Studies in Botany, Vijayanagara Sri Krishnadevaraya University, Ballari, Karnataka, India

Rajagopal Reddy S.
Department of Botany, Yogi Vemana University, Kadapa, Andhra Pradesh, India

Ramasamy Shanmugavalli
Department of Chemistry, School of Arts and Science, Vinayaka Mission Research Foundation-AV Campus, Chennai, India

Chandrasekhar T.
Department of Environmental Sciences, Yogi Vemana University, Kadapa, Andhra Pradesh, India

Narayan Chandra Talukdar
Chemical Biology Lab I, Institute of Advanced Study in Science and Technology, Paschim Boragaon, Guwahati, Assam, India
Assam Down Town University, Panikhaiti, Guwahati, Assam, India

Heena Tabassum
Dr. D. Y. Patil Biotechnology and Bioinformatics Institute, Pune, Maharashtra, India
Department of Bioengineering, Integral University, Lucknow, Uttar Pradesh, India

Tewin Tencomnao
Natural Products for Neuroprotection and Anti-ageing Research Unit, Chulalongkorn University, Bangkok, Thailand
Department of Clinical Chemistry, Faculty of Allied Health Sciences, Chulalongkorn University, Bangkok, Thailand

Rajesh Singh Tomar
Amity Institute of Biotechnology, Amity University Madhya Pradesh, Maharajpura Dang, India

Sugumari Vallinayagam
Department of Biotechnology, Mepco Schlenk Engineering College, Sivakasi, Tamil Nadu, India

Vijay Kumar Veena
Department of Biotechnology, School of Applied Sciences, REVA University, Bangalore, India

Pulala Raghuveer Yadav
Department of Biotechnology, Indian Institute of Technology Hyderabad, Kandi, Telangana State, India

Abbreviations

AA	asiatic acid
ACh	acetylcholin
ADA	D-Adenosine deaminase
AM	acrylamide
ALP	alkaline phosphatase
ALT	alanine aminotransferase
AST	aspartate aminotransferase
ATM	ataxia telangiectasia mutated
BA	boswellic acid
BAP	Benzylaminopurine
BSA	bovine serum albumin
CA	*Centella asiatica*
CCA	Cholangiocarcinoma
CAT	catalase
CPT	camptothecin
DR	death receptors
DTQ	dihydrothymoquinone
EAC	Ehrlich ascites carcinoma
EGFR	epidermal growth factor receptor
ELSD	evaporative light scattering detection
EMT	epithelial to mesenchymal transition
ER	endoplasmic reticulum
ERK	extracellular signal-regulated kinase
FAK	focal adhesion kinase
FasL	factor associated-suicide ligand
FCM	flow cytometry
FGF	fibroblast growth factor
FPS	farnesyl pyrophosphate synthase
FT-IR	Fourier transforms infrared spectroscopy
GBD	glabridin
GC	gas chromatography
GGPP	geranylgeranyl diphosphate
GI	*Garcinia indica*
GPS	geranyl diphosphate synthase

GPS	ginseng polysaccharide
GST	glutathione-S-transferase
HCC	hepatocellular carcinoma
HGF	hepatocyte growth factor
HPLC	high-performance liquid chromatography
HPTLC	high-performance thin layer chromatography
HRMS	high-resolution mass spectrometry
HSCCC	high-speed counter-current chromatographic
HSP	heat shock proteins
IAA	iso-angustone A
IAP	inhibitor of apoptosis protein
IFP	interstitial fluid pressure
IL	interleukin
LC	liquid chromatography
LCM	licocoumarone
LG	Liquiritigenin
MAA	methacrylic acid
MAP	microtubule associated protein
MAPKs	mitogen-activated protein kinases
MDR	multidrug resistance
MeJA	methyl jasmonate
MIP	molecular imprinting
MMPs	matrix metalloproteinases
MS	mass spectrometry
MS	multiple sclerosis
MVA	mevalonate
NAA	naphthalene acetic acid
NAC	N-acetylcysteine
NCAM	neural cell adhesion molecule
NMR	nuclear magnetic resonance
NMR	nuclear resonance spectroscopy
NO	nitric oxide
NOS	nitric oxide synthases
NPC	nasopharyngeal carcinoma
OSCs	oxidosqualene cyclases
PARP	poly ADP ribosyl polymerase
PBD	polo-box domain
PBMC	peripheral blood mononuclear cells
PCNA	proliferative cell nuclear antigen

PDA	pancreatic ductal adenocarcinoma
PDGF	platelet derived growth factor
PMA	phorbol myristate acetate
PMK	phosphomevalonate kinase
PSA	prostate-specific antigen
RNAi	RNA interference
ROS	reactive oxygen species
RSM	response surface methodology
SAR	structure-activity relationships
SLE	systemic lupus erythematosus
SOD	superoxide dismutase
SRE	serum response factor
SS	squalene synthase
THQ	thymohydroquinone
THY	thymol
TME	tumor microenvironment
TNBC	triple negative breast cancer
TLC	thin-layer chromatography
TQ	thymoquinone
UV	ultraviolet
VEGF	vascular endothelial growth factor
WGOS	water-soluble ginseng oligosaccharides
WHO	World Health Organization

Preface

Cancer is the leading cause of death worldwide, and the quest for effective treatments has been ongoing for decades. While conventional treatments like chemotherapy, radiation therapy, and surgery have improved survival rates, these often come with severe side effects that can reduce the quality of life for cancer patients.

In recent years, there is an increasing interest in using medicinal plants as a complementary or alternative treatment for cancer. Many studies have shown that certain plant compounds can inhibit the growth and spread of cancer cells, induce apoptosis (cell death) in cancer cells, and even enhance the effectiveness of chemotherapy and radiation therapy. Herbal medicines play a critical role in the prevention as well as the treatment of various forms of cancer which is the foremost cause of death worldwide. Synthetic drugs have prolonged toxic side effects, and substitute panacea is the economical and natural medicinal plants.

This book aims to provide an in-depth compilation of most vital plant genera and species, namely *Garcinia indica, Centella asiatica* Linn, *Nigella sativa, Ocimum sanctum, Boswellia serrata* Roxb, *Catharanthus roseus, Withania somnifera* (Linn.) Dunal, *Camptotheca acuminata* Decne, *Taxus baccata* L., *Panax ginseng, Tinospora cordifolia* (Wild) Miers, *Taxus brevifolia,* and *Glycyrrhiza glabra,* which have potent anticancer activity. These anticancer medicinal plants are bestowed with novel and essential cradle of chemotherapeutic complexes, biologically active molecules, and secondary metabolites like taxol, vinblastine, vincristine, camptothecin, topotecan, etc. with promising properties to cure cancer and have the potential for new drug discovery in the era of modern medicine to fight against cancer. This book also covers the evidence for their effectiveness in preclinical and clinical studies, as well as any potential side effects or interactions with other medications.

We hope that the information in this book will prove to be a treasured resource for researchers, healthcare professionals and residents in pharmacy and medicine, oncologists, biotechnologists, medicinal chemists, pharmacists, pharmacologists, phyto-chemists and members of biomedical and pharmaceutical sciences working in the areas of cancer treatment for the welfare of human society and anyone interested in using natural remedies for cancer prevention and treatment and those seeking a comprehensive understanding of the potential of medicinal plants in the fight against cancer. We hope that this book will contribute to the growing body of knowledge on medicinal plants and inspire further research and development in this field.

It is important to note that while medicinal plants may hold promise as a complementary or alternative cancer treatment, they should never be used as a substitute for conventional medical treatment. Cancer patients should always consult their healthcare providers before using any herbal remedies or supplements.

Few Lines on Anticancer Medicinal Plants

In gardens, fields and forests, they quietly thrive and grow,
Nature's wonders, to help fight the foe, with a healing flow.
Potent anticancer plants with healing power and strength so rare,
To ward off the cancer, hour by hour and power to spare.
With healing powers to mend and renew,
Anticancer properties to see us through.
With ancient knowledge and modern research,
Their benefits for health, we can now perch.
In ancient wisdom, they were known,
To aid in healing, to strengthen and hone.
And now science has shown their worth,
As potent allies in the fight on earth.
From *Boswellia's* boswellic acid to *Catharanthus's* vincristine and
vinblastine a potent blend,
They hold the key to health, longevity and cancer prevention,
on their help we depend.
These plants of life with rainbow of color, we must treasure,
A natural cancer-fighting crew, we cannot measure.
The rose periwinkle, a wondrous flower, with its vibrant hue,
Holds a secret medicinal power to cure cancer, that could save me and you.
The *Withania somnifera*, a beacon of light,
In the darkness of cancer's plight.
Panax Ginseng, a medicinal wonder, its roots hold the secret key,
Its benefits we cannot ponder, to fight cancer, a remedy free.
Tinospora cordifolia, a herb so pure,
Its anticancer compounds we can't ignore.
Taxus, a plant with healing power,
Its anticancer compounds we can't devour.

Glycyrrhiza glabra, a herb so pure,
Its anticancer compounds we cannot ignore.
Ocimum sanctum, a herb so divine,
Its anticancer compounds we must incline.
Garcinia indica, a plant so rare,
Its anticancer compounds we must share.
Centella asiatica, a herb so bright,
Its anticancer compounds shine in the light.
Nigella sativa, a herb so divine,
Its anticancer compounds we must incline.
Camptotheca acuminata, its name,
Holds the key to a cancer-fighting game.
Their bark, leaves and roots, a source of great hope,
Contain alkaloids, flavonoids and novel compounds that help us cope.
With leukemia and lymphoma, they can fight,
And slow the spread of cancer's might.
Against cancer cells, they can prevail,
And slow their growth without fail.
And the list goes on and on,
A multitude of plants to call upon.
May we continue to learn and explore,
And find new ways to fight cancer's core.
Let us honor these plants with a heart of gold, for all what they provide,
Their healing properties are a source of hope for those who have cried.
With compounds that can heal and cure,
And offer hope in times unsure.
In the face of cancer's deadly grip,
These plant offers a chance to slip.
May these greens continue to heal and inspire,
Until cancer is conquered, we never tire, and all can aspire.

CHAPTER 1

Garcinia indica

TEWIN TENCOMNAO[1,2]

[1]*Natural Products for Neuroprotection and Anti-Ageing Research Unit, Chulalongkorn University, Bangkok, Thailand*

[2]*Department of Clinical Chemistry, Faculty of Allied Health Sciences, Chulalongkorn University, Bangkok, Thailand*

ABSTRACT

Garcinia indica (GI), a plant native to India commonly known as kokum, belongs to the Clusiaceae family. GI has been famous for a long time due to its various benefits such as culinary, pharmaceutical, and industrial applications. GI is recognized as a natural antioxidant, thus making the GI extract so valuable for both research and application. Interestingly, GI was found to possess numerous activities such as anti-inflammatory, anti-infectious, antineurodegenerative, and anticancer effects. Emerging roles of garcinol, the main bioactive compound of GI, have been grown since it exhibits various pharmacological effects, particularly anticancer. Anticancer mechanisms of garcinol have been documented vastly depending on the cancer cell types. These mechanistic targets of garcinol include cell cycle, proliferation, migration, invasion, metastasis, apoptosis, angiogenesis, stem cell-like phenotype, epigenetics, miRNAs, and so forth. Therefore, garcinol may represent an attractive lead compound for further anticancer drug development.

Potent Anticancer Medicinal Plants: Secondary Metabolite Profiling, Active Ingredients, and Pharmacological Outcomes. Deepu Pandita and Anu Pandita (Eds.)

1.1 BOTANICAL DESCRIPTION AND TRADITIONAL USES OF *GARCINIA INDICA*

Commonly recognized as kokum, *Garcinia indica* (GI), a plant in the mangosteen family (Clusiaceae), is a slender, evergreen, and fruit-bearing tree with drooping branches, which can grow approximately 15 m tall with a dense canopy of green leaves and red-tinged tender emerging leaves (Singh, 1993). The fruits of GI are spherical with diameters of 2.5–5.0 cm (Yella Reddy and Prabhakar, 1994). This plant is found widely in tropical regions, especially in Asia and Africa. GI grows widely on the western coast of India (Padhye et al., 2009).

To date, GI has been known for numerous benefits to humans including culinary, pharmaceutical, and industrial purposes. As revealed in many reports, this particular plant species has long been known for culinary uses (Chate et al., 2019) and industrial uses (Maheshwari and Reddy, 2005). In addition, the health benefits of GI include medicinal and cosmetics properties (Baliga et al., 2011).

1.2 MEDICINAL PROPERTIES OF *GARCINIA INDICA*

Furthermore, certain studies on the functional properties of GI demonstrated its antioxidant effects (Mishra et al., 2006; Panda et al., 2012; Jayakar et al., 2021), anti-inflammatory effects (Khatib et al., 2010; Panda et al., 2013) and anti-infectious effects (Lakshmi et al., 2011; Varalakshmi et al., 2011; Sutar et al., 2012; Tharachand et al., 2015).

GI extract is well-classified as a natural antioxidant. Since oxidative stress is contributed to the development of a vast array of diseases such as age-related, inflammatory, neurodegenerative, and noncommunicable diseases, antioxidants play a critical role in alleviating these diseases. For example, numerous investigations have revealed that antioxidants play a crucial role in neurodegenerative diseases such as Alzheimer's disease, and Parkinson's disease. As depicted in Figure 1.1, the neuroprotective effect against 6-hydroxydopamine (6-OHDA) neurotoxicity for striatal dopaminergic neurons in the rat model was demonstrated using GI extract, thus implicating its potential benefit for Parkinson's disease (Antala et al., 2012). Furthermore, GI was found tofight against both brain ischemic reperfusion in the rat model (Ahmed, 2020) and depression via monoaminergic pathway (Dhamija et al., 2017).

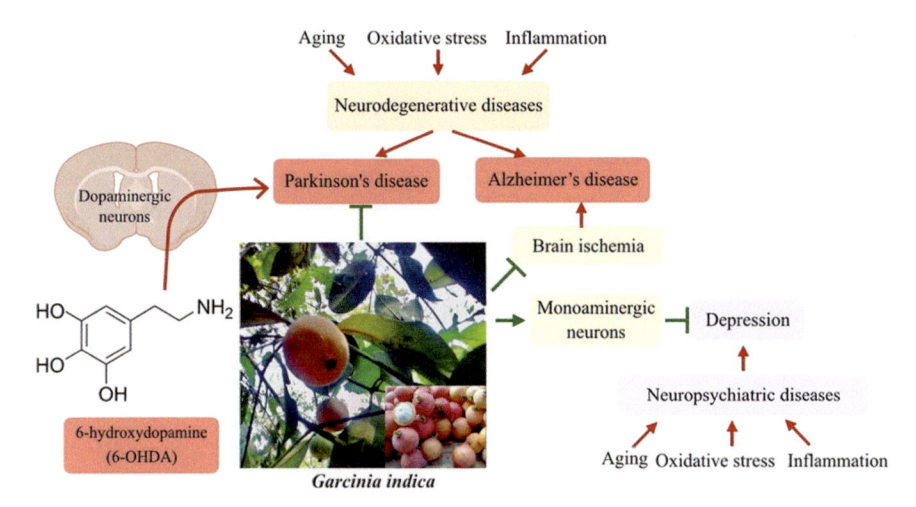

FIGURE 1.1 The protective effects of GI extract on Parkinson's disease, Alzheimer's disease, and depression.

1.3 CHEMICAL COMPOSITION OF *GARCINIA INDICA*

Various medicinal and cosmetics propertiesof GI fruitrinds and leaves were uncovered to link with the chemical constituents of GI (Baliga et al., 2011). GI was found to possess protein, tannin, pectin, sugars, fat, organic acids including (−)-hydroxycitric acid, hydroxycitric acid lactone, and citric acid; the anthocyanins, cyanidin-3-glucoside, and cyanidin-3-sambubioside; and the polyisoprenylated phenolics garcinol and isogarcinol (Krishnamurthy et al., 1981; Baliga et al., 2011; Kaur et al., 2012; Biersack, 2016; Jayakar et al., 2020). Secondary metabolites, garcinol and isogarcinol, are found in both rinds and leaves of GI. The chemical structures of both natural polyisoprenylated benzophenone derivatives garcinol and isogarcinol are shown in Figure 1.2. In line with the biological activity spectra of GI extract, the activity spectra of garcinol and isogarcinol include anticancer, antibiotic, antioxidant, and anti-inflammatory effects (Hemshekhar et al., 2011; Schobert and Biersack, 2019).

1.4 ANTI-CANCER ACTIVITIES OF GI OR ITS CHEMICAL COMPOUNDS

Herein, anti-cancer effects exhibited by GI and its chemical constituents are presented and discussed according to numerous previous reports. GI leaf extract was demonstrated to exert a cytotoxic effect to A498 human renal carcinoma cells, but the finding was based on the MTT assay only (Jayakar et al., 2021). As far as the molecular mechanisms of cancer are concerned, the association between endoplasmic reticulum stress signaling and cancer has been firmly established (Oakes, 2020). Interestingly, there was a study demonstrating thatGI extract standardized for 20% garcinol could reduce endoplasmic reticulum stress in both cultured cells and adipose tissue, thus suggesting the medical effects of GI for not only diabetes and obesity, but cancer as well (Majeed et al., 2020). Furthermore, GI extract was shown to possess an inhibitory effect on cultured 3T3 mouse fibroblasts (Varalakshmi et al., 2011).

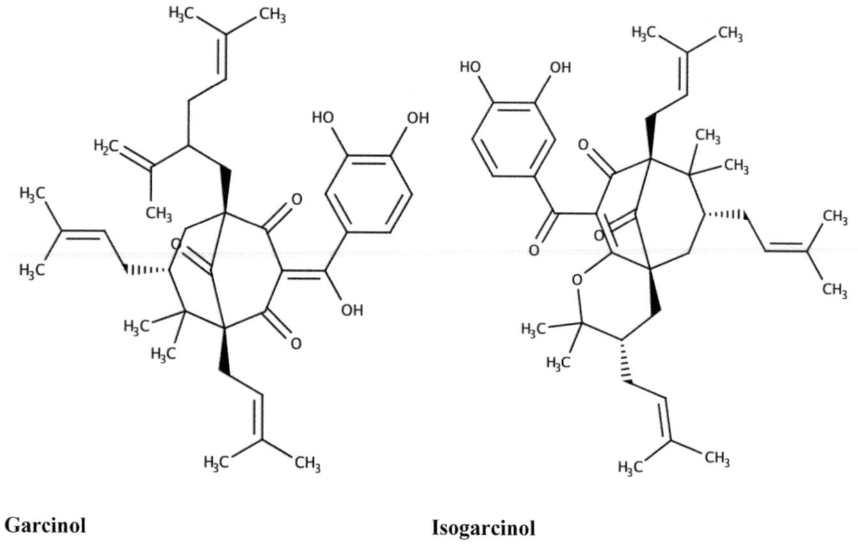

Garcinol   Isogarcinol

FIGURE 1.2   Structures of the secondary plant metabolites garcinol and isogarcinol.

In fact, the activities of cancer suppression were mainly found due to the main active substance, garcinol, but not due to the GI extract. Aggarwal et al. (2020) discussed a variety of anticancer mechanisms of garcinol, but a few interesting examples are addressed in this book chapter. Table 1.1 summarizes the anticancer mechanisms of garcinol as reported in numerous publications.

Antiproliferative and anti-invasive effects of garcinol on gallbladder carcinoma cells and the inhibitory mechanism of garcinol were associated with the suppression of Stat3 and Akt signaling pathways, which might

TABLE 1.1 Anti-Cancer Mechanisms of Garcinol in Various Cancer Types.

Types of cancer cells	Proposed signaling pathway involved (mechanisms of action)	References
ISH and HEC-1B cells (endometrial cancer cells)	Garcinol exhibited cell cycle arrest and activated JNK/c-JUN signaling pathway to cause apoptosis	Zhang et al. (2021)
U-87 MG and GBM8401 cells (glioblastoma cells)	Garcinol showed antiproliferative effect by inhibiting the activation of STAT-3, 5 by inducing the expression of Hsa-miR181d	Liu et al. (2019)
HGC-27 cells (gastric cancer cells)	Garcinol inhibited the invasion of cancer cells through the downregulation of PI3K/AKT signaling pathway	Zheng et al. (2020)
OVCAR-3 cells (ovarian cancer cells)	Garcinol alone or in combination with cisplatin showed apoptotic effect by inhibiting the activation of PI3K/AKT, NF-κB pathway	Zhang et al. (2020)
KYSE150 and KYSE450 (Esophageal cancer cells)	Garcinol inhibited metastasis by downregulating p300 and p-Smad2/3	Wang et al. (2020)
CAL-27, SCC-15 cells (oral squamous cell carcinoma cells)	Garcinol inhibits ATP production, mitochondrial respiration, and basal respiration and subsequently affected the energy-producing pathway in cancer cells. It also reflexively boosted glycolysis apart from the upregulation of glucose transporter 1 and 4, and HIF-1α, AKT, and PTEN	Zhang et al. (2019)
A549 and H1299 cells (nonsmall cell lung carcinoma cells)	Garcinol affects epithelial to mesenchymal transition by modulating miR-200b to sensitize the carcinoma cells to chemotherapy	Farhan et al. (2019)
Hela and SiHa cells (human cervical cancer cells)	Garcinol showed antiproliferative activity by attenuating PI#/AKT pathway and inhibited cell migration and invasion by suppressing MMP-2,9 and inducing T-cadherin	Zhao et al. (2018)
Caki cells (renal carcinoma cells)	Garcinol upregulated the expression of DR5 and downregulated the expression of c-FLIP, thereby inducing apoptosis of the TRAIL-negative cells	Kim et al. (2018)
A549 and H441 cells (nonsmall cell lung carcinoma cells)	Garcinol diminished the ability of the NSCLC cells to form spheres and form colonies, by impairing phosphorylation of LRP6, dysregulating the Wnt/β-catenin pathway, and reducing the Dvl2, Axin2, and cyclin D1 expressions	Huang et al. (2018)

TABLE 1.1 *(Continued)*

Types of cancer cells	Proposed signaling pathway involved (mechanisms of action)	References
A549 cells (nonsmall cell lung carcinoma cells)	Garcinol enriched DNA damage-inducible transcript 3 (DDIT3), altered DDIT3-CCAAT-enhancer-binding proteins β (C/EBPβ) interaction resulting in the attenuation of the prognostic cancer cell marker aldehyde dehydrogenase 1 family member A1 (ALDH1A1) expression	Wang et al. (2017)
4T1 cells (breast cancer cells) and female Balb/c mice	Garcinol sensitized the cells toward Taxol treatment and improved the efficacy of the drug in mice through the suppression of caspase-3/iPLA2 and NF-κB/Twist1 signaling pathways.	Tu et al. (2017)
HT-29 cells (colorectal cancer cells)	Garcinol exhibited an antiproliferative effect and antiangiogenic effect by suppressing of HIF-1α and VEGF expression	Ranjbarnejad et al. (2017)
SCC-4 cells (oral squamous cell carcinoma cells)	Garcinol exerted antiproliferative, pro-apoptotic, cell-cycle regulatory, and anti-angiogenic effects by inhibiting the expression of STAT-3, c-Src, JAK1, JAK2, NF-κB, COX-2 and VEGF	Aggarwal and Das (2016)
PC-3 cells (prostate cancer cells)	Garcinol inhibited autophagy and exhibited apoptosis through the activation of p-mTOR and p-PI3K/AKT	Wang et al. (2015)
PANC-1 SP cells (Pancreatic cancer cells)	Garcinol suppressed the stem-like properties and metastasis by attenuating the Notch1 signaling pathway through the upregulation of miR-200c	Huang et al. (2017)
UMSCC1 cells (head and neck squamous cell carcinoma cells) and HNSCC Xenograft mouse model	Garcinol enhanced the apoptotic effect of cisplatin by attenuating surviving, Bcl-2 and suppressed tumor growth through the downregulation of Ki-67 and CD-31	Li et al. (2015)
MCF-7 cells (breast cancer cells)	Garcinol exhibited anti-proliferative activity against 17β-estradiol and promoted cell cycle arrest and apoptosis by suppressing ac-p65/NFκB pathway	Ye et al. (2014)
HepG2, HUH-7 cells (hepatocarcinoma cells) and athymic nu/nu female mice	Garcinol exhibited apoptotic and anti-tumor growth effects by inhibiting dimerization, phosphorylation, and acetylation of STAT3	Sethi et al. (2014)
p53 wild type H460 and p53-null H1299 cells (nonsmall cell lung carcinoma cells)	Garcinol induced cell cycle arrest by inducing CDK inhibitors p21[Waf1/Cip1] and exhibited apoptosis through p38-MAPK pathway	Yu et al. (2014)

TABLE 1.1 *(Continued)*

Types of cancer cells	Proposed signaling pathway involved (mechanisms of action)	References
Panc-1 (Pancreatic cancer cells)	Garcinol sensitized the cells toward gemcitabine treatment and induced apoptosis by altering the miRNA profile involved in the oncogenic signaling pathway	Parasramka et al. (2013)
MDA-MB-231, BT-549 cells (Breast cancer cells), and xenograft mice model	Garcinol inhibited epithelial to mesenchymal transition through the upregulation of miR200b, miR-200c and regulating NFκB, Wnt signaling pathways	Ahmad et al. (2012)
Hep3B cells (hepatocellular carcinoma cells)	Garcinol elevated ROS levels and caused caspase-3-mediated apoptosis	Cheng et al. (2010)
HT-29 cells (colorectal carcinoma cells)	Garcinol inhibited cell invasion, decreased tyrosine phosphorylation, and prevented activation of the Src, MAPK/ERK, and PI3K/Akt signaling pathways with an increase in caspase-3 activity to promote apoptosis	Liao et al. (2005)
HeLa cells (cervical cancer cells)	Garcinol inhibited HAT activity, thereby repressing chromatin transcription and inducing apoptosis by altering global gene expression involved in oncogenic signaling	Balasubramanyam et al. (2004)

contribute to the downregulation of the molecular targets such as matrix metalloproteinase 2 (MMP2) and MMP9 (Duan et al., 2018).

Garcinol was found to suppress cancer stem cell-like phenotype via inhibition of the Wnt/β-catenin/STAT3 axis signaling pathway in human nonsmall cell lung carcinomas, thus suggesting the roles of garcinol in blocking cancer recurrence and metastasis (Huang et al., 2018).

In addition to the modulation of an increased hsa-miR-181d/STAT3 and hsa-miR-181d/5A ratio by garcinol, Liu et al. (2019) reported that garcinol treatment significantly suppressed the proliferative, invasive, and migratory potential of U87MG or GBM8401 glioblastoma cell lines in a dose-dependent fashion and the anticancer effect by garcinol significantly attenuated the glioblastoma stem cell-like phenotypes, as reflected by the reduced capability of both cell lines to form colonies and tumorspheres, and suppressed expression of OCT4 and SOX2.

Using the human colorectal cancer cell line, HT-29, as a cell model, garcinol was revealed to modulate tyrosine phosphorylation of focal adhesion kinase (FAK) and subsequently induce apoptosis through downregulation of Src, ERK, and Akt survival signaling (Liao et al., 2005). Also, they demonstrated that garcinol significantly inhibited the expression of MMP-7 in interleukin 1β (IL-1β)-induced HT-29 cells, thus suggesting that garcinol reduced cell invasion and survival through the inhibition of FAK's downstream signaling. Garcinol was also reported to exhibit anti-proliferative activities by targeting microsomal prostaglandin E synthase-1 in human colon cancer cells (Ranjbarnejad et al., 2017).

Regarding the effects of garcinol on breast cancer cells, antiproliferative and apoptosis-inducing actions were reported and the apoptosis-inducing effect of garcinol was mediated by the NF-κB signaling pathway (Ahmad et al., 2010). Consistently, the same might be true for the antiapoptotic mechanism of garcinol in other cancer types. In particular, garcinol was proven to inhibit tumor cell proliferation, angiogenesis, and cell cycle progression and induce apoptosis via NF-κB inhibition in oral cancer (Aggarwal and Das, 2016). As an acetyltransferase inhibitor, garcinol was demonstrated to suppress the proliferation of breast cancer cell line MCF-7 stimulated by 17β-estradiol, thus indicating the epigenetics mechanism of the action exerted by garcinol (Ye et al., 2014). Although the data pertaining to precise toxicity, dosages, routes of administration, and safety for clinical applications are still limited, garcinol seems to serve as a promising anti-cancer epigenetic drug (Kopytko et al., 2021).

Garcinol was shown to dose-dependently suppress cell viability, colony formation, invasion, migration, and cell cycle progression, and promoted cell apoptosis in cervical cancer cell lines, as well as inhibited tumor growth in a xenograft model, and its mechanism of action was found to be associated with activating T-cadherin/P13 K/AKT signaling pathway (Zhao et al., 2018). Garcinol was shown to inhibit the proliferation of endometrial cancer cells by inducing cell cycle arrest, and following the garcinol treatment, the expression levels of p53 and p21 were increased, while the expression levels of CDK2, CDK4, cyclin D1, and cyclin B1 were gradually decreased in a dose-dependent manner in endometrial cancer cell lines (Zhang et al., 2021).

In each cancer cell type, garcinol may participate to fight against cancer in more than one signaling pathway. For example, a single strategy for successful cancer treatment in certain cancer types may be very rare. Integrative therapeutic approaches may shed light on the success of cancer management. For example, pancreatic cancer represents one of the most aggressive types of malignancy due to its high resistance toward most clinically available treatments. Therefore, research in this field has been continued and made natural product research significant and intriguing. It has been reported that garcinol modulates specific MicroRNA (miRNA) biomarkers associated with the sensitization of pancreatic cancer cells to gemcitabine treatment, thus attenuating the drug-resistance phenotype (Parasramka et al., 2013). More recently, garcinol was found to suppress the stem-like properties of human pancreatic cancer cell line and inhibit the metastatic potential by downregulating the expression of Mcl-1, EZH2, ABCG2, Gli-1, and Notch1 (Huang et al., 2017). Remarkably, garcinol treatment led to the upregulation of several tumor suppressor microRNAs, and downregulated notch1 via modulating mir-200c by garcinol was observed.

1.5   ISOGARCINOL AS AN IMMUNOSUPPRESSANT

Unlike garcinol, isogarcinol extracted from the *Garcinia* species has been shown to exert many attractive activities, predominantly as an immunosuppressive agent. Searching for immunomodulatory drugs from natural sources is very critical in the pharmaceutical industry because immunosuppressants are used for post-transplant organ rejection. It has been reported that isogarcinol could inhibit calcineurin, a ubiquitous protein phosphatase that is involved in several physiological roles such as apoptosis, T-cell activation, and cell cycle control as well as in other pathological conditions including neurodegenerative disease, thus serving as an immunosuppressant (Cen et al., 2013). Enzyme kinetic analysis demonstrated that inhibition of calcineurin by isogarcinol was competitive, and it bound directly to calcineurin in vitro (Cen et al., 2015).

Psoriasis is an incurable, long-term (chronic) inflammatory skin condition featured by keratinocyte hyperproliferation of epidermis. HaCaT keratinocyte cell line has been a proper cell model for antipsoriatic drug research. Certain Thai medicinal extracts were shown to possess an antipsoriatic effect via the modulation of NF-B signaling markers in HaCaT cells (Saelee et al., 2011). Isogarcinol was reported to improve imiquimod-stimulated psoriasis-like skin lesions in mice and suppress inflammatory factor expression in lipopolysaccharide-induced HaCaT cells, thus considered an antipsoriatic agent (Chen et al., 2017).

Using collagen-induced arthritis mice as a model, the oral administration of isogarcinol was revealed to reduce not only serum inflammatory cytokines, but the erosion of cartilage and bone as well, thus making it an antiarthritic agent (Fu et al., 2014). Furthermore, the in vitro study of Fu et al. demonstrated that isogarcinol downregulated iNOS and COX-2 mRNA expression and NO content by suppressing the NF-B signaling pathway.

A previous study has shown isogarcinol as a promising therapeutic agent due to its ability to reduce proteinuria, corrected the abnormal serum biochemical parameters, lowered the levels of serum antibodies, alleviated the abnormal activation of CD4 T cells, and decreased the inflammation in a murine model of systemic lupus erythematosus (SLE)-like disease (Li et al., 2015).

Using a murine model of multiple sclerosis (MS), isogarcinol improved clinical scores, reduced inflammation, lowered demyelination of the spinal cord, and decreased intracranial lesions, and isogarcinol appeared to inhibit T helper cell differentiation (Wang et al., 2016). Therefore, isogarcinol was established as a promising candidate for treating MS.

1.6   ANTI-CANCER ACTIVITIES OF *GARCINIA INDICA* AND GARCINOL

Herein, anti-cancer effects exhibited by GI and its chemical constituents are presented and discussed according to numerous previous reports. GI leaf extract was demonstrated to exert a cytotoxic effect on human renal A498 carcinoma cells, but the finding was based on the MTT assay only (Jayakar et al., 2021). As far as the molecular mechanisms of cancer are concerned, the association between endoplasmic reticulum stress signaling and cancer has been firmly established (Oakes, 2020). Interestingly, there was a study demonstrating that GI extract standardized for 20% garcinol could reduce endoplasmic reticulum stress in both cultured cells and adipose tissue, thus suggesting the medical effects of GI for not only diabetes and obesity, but cancer as well (Majeed et al., 2020). Furthermore, GI extract was shown to possess an inhibitory effect on cultured 3T3 mouse fibroblasts (Varalakshmi et al., 2011).

In fact, the activities of cancer suppression were mainly found due to the main active substance, garcinol, but not the GI extract. Aggarwal et al. (2020) discussed a variety of anti-cancer mechanisms of garcinol, but a few interesting examples are addressed in this book chapter. Table 1.1 summarizes the anti-cancer mechanisms of garcinol as reported in numerous publications.

Garcinol was shown to suppress the Stat3 and Akt signaling pathways as well as downregulated the expression of matrix metalloproteinases 2 and 9, thereby exerting antiproliferative and anti-invasive activities against gallbladder carcinoma cells (Duan et al., 2018).

Garcinol was found to suppress cancer stem cell-like phenotype via inhibition of the Wnt/-catenin/STAT3 axis signaling pathway in human nonsmall cell lung carcinomas, thus suggesting the roles of garcinol in blocking cancer recurrence and metastasis (Huang et al., 2018).

In addition to the modulation of an increased hsa-miR-181d/STAT3 and hsa-miR-181d/5A ratio by garcinol, Liu et al. (2019) demonstrated the anti-proliferative, anti-invasive, and anticancer effects of garcinol against glioblastoma cell lines (U87MG or GBM8401) in a dose-dependent fashion garcinol was revealed to modulate tyrosine phosphorylation of FAK and subsequently induced apoptosis of human colorectal HT-29 cancer cell line through the downregulation of Src, ERK, and Akt signaling (Liao et al., 2005). Also, they demonstrated that garcinol markedly mitigated MMP-7 expression in HT-29 cells, suggesting that garcinol reduced cell invasion and survival through the prevention of FAK's downstream signaling. Garcinol was also reported to exhibit anti-proliferative activities via the modulation of microsomal prostaglandin E synthase-1 in human colon cancer cells (Ranjbarnejad et al., 2017).

Regarding the effects of garcinol on breast cancer cells, antiproliferative and apoptosis-inducing action were reported and the apoptosis-inducing effect of garcinol was mediated by NF-B signaling pathway (Ahmad et al., 2010). Consistently, the same might be true for the anti-apoptotic mechanism of garcinol in other cancer types. In particular, garcinol was proven to inhibit tumor cell proliferation, angiogenesis, and cell cycle progression, thereby inducing apoptosis via NF-B inhibition in oral cancer (Aggarwal and Das, 2016). As an acetyltransferase inhibitor, garcinol was demonstrated to suppress the proliferation of breast cancer MCF-7 cell line stimulated by 17-estradiol, thus indicating the epigenetics mechanism of the action exerted by garcinol (Ye et al., 2014). Although the data pertaining to precise toxicity, dosages, routes of administration, and safety for clinical applications are still limited, garcinol seems to serve as a promising anti-cancer epigenetic drug (Kopytko et al., 2021).

Garcinol was shown to promote cell apoptosis in cervical cancer cell lines, as well as prevented tumor growth in a xenograft model, and its mechanism of action was found to be associated with activating T-cadherin/P13K/AKT signaling pathway (Zhao et al., 2018). Garcinol treatment was shown to increase the expression levels of p53 and p21, while decreased the expression levels of CDK2, CDK4, cyclin D1 and cyclin B1, resulting in the inhibition of endometrial cancer cells in a dose-dependent manner (Zhang et al., 2021).

In each cancer cell type, garcinol may participate to fight against cancer in more than one signaling pathway. For example, a single strategy for successful cancer treatment in certain cancer types may be very rare. Integrative therapeutic approaches may shed light on the success of cancer management. For example, pancreatic cancer is one of the most aggressive types of malignancy because it shows increased resistance to most of the cancer treatments available today. Therefore, research in this field has been continued and made natural product research significant and intriguing. It has been demonstrated that garcinol modulates specific miRNA biomarkers associated with the sensitization of pancreatic cancer cells to gemcitabine treatment, thus attenuating the drug-resistance phenotype (Parasramka et al., 2013). More recently, garcinol was found to suppress the stem-like properties of human pancreatic cancer cell line and inhibit the metastatic potential by the downregulation of Mcl-1, EZH2, ABCG2, Gli-1, and Notch1 (Huang et al., 2017). Remarkably, garcinol treatment led to the upregulation of several tumor suppressor microRNAs, and downregulated notch1 via modulating mir-200c by garcinol was observed.

To search for the mechanisms of natural products for anticancer chemotherapy, scientists have attempted to utilize many different strategies which are known to participate in any cellular process such as autophagy. Autophagy is a degradative process during stressful conditions intracellularly. Interestingly, autophagy has been linked to both tumor suppression and promotion (Yun and Lee, 2018; Chavez-Dominguez et al., 2020). Garcinol was demonstrated to inhibit autophagy in human prostate cancer cells through the activation of p-mTOR and p-PI3 Kinase/AKT (Wang et al., 2015). However, it was proven to induce autophagy in osteosarcoma cells by inhibiting acetyltransferase and increasing the expression of LC3-II (Pietrocola et al., 2015). Hence, future investigations should be focused to unravel the modulatory roles of garcinol in autophagy and cancer for anti-cancer drug discovery.

1.7 CONCLUSION AND FUTURE PERSPECTIVE

Although it grows extensively in India, GI has been recognized worldwide for its various benefits to humans such as culinary, medicinal, and industrial applications. GI has been shown to possess many pharmacological effects including antioxidant, anti-inflammatory, anti-neurodegenerative, and antitumor effects. Nevertheless, its secondary metabolite garcinol has been mostly researched so far in terms of anti-cancer activities. As shown in Figure 1.3, the anticancer mechanisms of garcinol are presented as promising anti-cancer drug candidates. Interestingly, the emerging roles of garcinol are extensively documented. However, elucidation of anti-cancer mechanisms exerted by garcinol is still needed for each type of cancer using different study designs and experimental models as well as various cutting-edge technologies. Recognized as a multifactorial disease, the mechanisms of cancer are not straightforward since the contribution of genetic factors and environmental factors toward cancer risk is distinct for each cancer type. More studies on garcinol can be achieved to fill gaps of knowledge, thus hoping to reach the phase of the clinical trials soon. It should be noted that researchers find difficulty in gaining a large quantity of garcinol from nature. So, improved extraction and purification are the key success factors in garcinol research since the insufficient quantity of garcinol hinders our progress nowadays.

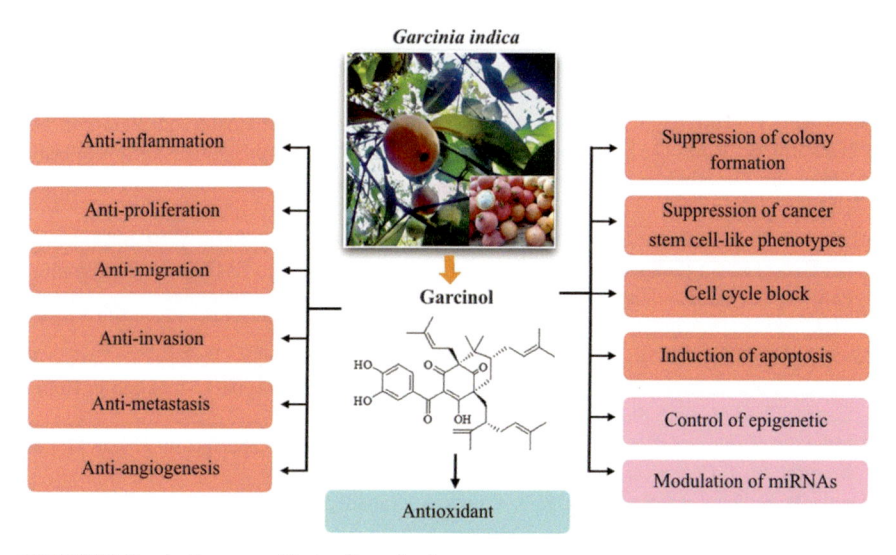

FIGURE 1.3 Anticancer effects of garcinol.

ACKNOWLEDGMENTS

The author wishes to thank the Rachadapisek Sompote Fund, Chulalongkorn University for supporting the work of Natural Products for Neuroprotection and Anti-Ageing Research Unit. Also, this chapter would have not be feasible without the images created by Dr. James M. Brimson and Ms. Chamaiphorn Wongwan. Finally, the author would like to thank Dr. Phaniendra Alugoju for his proofreading and English editing.

KEYWORDS

- *Garcinia indica*
- **plant extract**
- **garcinol**
- **anti-cancer**
- **molecular mechanisms**

REFERENCES

Aggarwal, S.; Das, S. N. Garcinol Inhibits Tumour Cell Proliferation, Angiogenesis, Cell Cycle Progression and Induces Apoptosis via NF-κB Inhibition in Oral Cancer. *Tumor Biol.* **2016,** *37* (6), 7175–7184.

Aggarwal, V.; Tuli, H. S.; Kaur, J.; Aggarwal, D.; Parashar, G.; Chaturvedi Parashar, N.; Kulkarni, S.; Kaur, G.; Sak, K.; Kumar, M.; Ahn, K. S. Garcinol Exhibits Anti-Neoplastic Effects by Targeting Diverse Oncogenic Factors in Tumor Cells. *Biomedicines* **2020,** *8* (5), 103.

Ahmad, A.; Sarkar, S. H.; Bitar, B.; Ali, S.; Aboukameel, A.; Sethi, S.; Li, Y.; Bao, B.; Kong, D.; Banerjee, S.; Padhye, S. B.; Sarkar, F. H. Garcinol regulates EMT and Wnt Signaling Pathways In Vitro and In Vivo, Leading to Anticancer Activity Against Breast Cancer Cells. *Mol. Cancer Therap.* **2012,** *11* (10), 2193–2201.

Ahmad, A.; Wang, Z.; Ali, R.; Maitah, M. Y.; Kong, D.; Banerjee, S.; Padhye, S.; Sarkar, F. H. Apoptosis-Inducing Effect of Garcinol Is Mediated by NF-kappaB Signaling in Breast Cancer Cells. *J. Cell. Biochem.* **2010,** *109* (6), 1134–1141.

Ahmed, W. Effect of *Garcinia Indica* Against Brain Ischemic Reperfusion in Rat Model. *Bangladesh J. Pharmacol.* **2020,** *15* (1), 39–40.

Antala, B. V.; Patel, M. S.; Bhuva, S. V.; Gupta, S.; Rabadiya, S.; Lahkar M. Protective Effect of Methanolic Extract of *Garcinia indica* Fruits in 6-OHDA Rat Model of Parkinson's Disease. *Indian J. Pharmacol.* **2012,** *44* (6), 683–687.

Balasubramanyam, K.; Altaf, M.; Varier, R. A.; Swaminathan, V.; Ravindran, A.; Sadhale, P. P.; Kundu, T. K. Polyisoprenylated Benzophenone, Garcinol, a Natural Histone Acetyltransferase Inhibitor, Represses Chromatin Transcription and Alters Global Gene Expression. *J. Biol. Chem.* **2004,** *279* (32), 33716–33726.

Baliga, M. S.; Bhat, H. P.; Pai, R. J.; Boloor, R.; Palatty, P. L. The Chemistry and Medicinal Uses of the Underutilized Indian Fruit Tree *Garcinia indica* Choisy (kokum), A Review. *Food Res. Int.* **2011,** *44* (7), 1790–1799.

Biersack B. *Critical Dietary Factors in Cancer Chemoprevention*, Ullah, M. F., Ahmad, A.; Springer International Publishing Switzerland: Cham, 2016; p 253.

Chate, M. R.; Kakade, S. B.; Neeha, V. S. Kokum (*Garcinia indica*) Fruit: A Review. *Asian J. Dairy Food Res.* **2019,** *38*, 329–332.

Cheng, A. C.; Tsai, M. L.; Liu, C. M.; Lee, M. F.; Nagabhushanam, K.; Ho, C. T.; Pan, M. H. Garcinol Inhibits Cell Growth in Hepatocellular Carcinoma Hep3B Cells Through Induction of ROS-Dependent Apoptosis. *Food Funct.* **2010,** *1* (3), 301–307.

Dhamija, I.; Parle, M.; Kumar, S. Antidepressant and Anxiolytic Effects of *Garcinia indica* Fruit Rind via Monoaminergic Pathway. *3 Biotech* **2017,** *7* (2), 131.

Duan, Y. T.; Yang, X. A.; Fang, L. Y.; Wang, J. H.; Liu, Q. Anti-Proliferative and Anti-Invasive Effects of Garcinol from *Garcinia indica* on Gallbladder Carcinoma Cells. *Pharmazie* **2018,** *73* (7), 413–417.

Farhan, M.; Malik, A.; Ullah, M. F. et al. Garcinol Sensitizes NSCLC Cells to Standard Therapies by Regulating EMT-Modulating miRNAs. *Int. J. Mol. Sci.* **2019,** *20* (4), 800.

Hemshekhar, M.; Sunitha, K.; Santhosh, M. S.; Devaraja, S.; Kemparaju, K.; Vishwanath, B. S.; Niranjana, S. R.; Girish, K. S. An Overview of Genus Garcinia: Phytochemical and Therapeutic Aspects. *Phytochem. Rev.* **2011,** *10*, 325–351.

Huang, C. C.; Lin, C. M.; Huang, Y. J.; Wei, L.; Ting, L. L.; Kuo, C. C.; Hsu, C.; Chiou, J. F.; Wu, A. T. H.; Lee, W. H. Garcinol Downregulates Notch1 Signaling via Modulating miR-200c and Suppresses Oncogenic Properties of PANC-1 Cancer Stem-Like Cells. *Biotechnol. Appl. Biochem.* **2017,** *64* (2), 165–173.

Huang, W. C.; Kuo, K. T.; Adebayo, B. O.; Wang, C. H.; Chen, Y. J.; Jin, K.; Tsai, T. H.; Yeh, C. T. Garcinol Inhibits Cancer Stem Cell-Like Phenotype via Suppression of the Wnt/β-catenin/STAT3 Axis Signalling Pathway in Human Non-Small Cell Lung Carcinomas. *J. Nutr. Biochem.* **2018,** *54*, 140–150.

Jayakar, V.; Lokapur, V.; Shantaram, M. Identification of the Volatile Bioactive Compounds by GC-MS Analysis from the Leaf Extracts of *Garcinia cambogia* and *Garcinia Indica*. *Medicinal Plants—Int. J. Phytomed. Related Ind.* **2020,** *12* (4), 580–590.

Jayakar, V.; Lokapur, V.; Shantaram, M. In Vitro Antioxidant and Selective Cytotoxicity of *Garciniacambogia* and *Garcinia indica* Leaf Extracts on Human Kidney Cancer Cell Line. *Int. J. Res. Pharma. Sci.*, **2021,** *12* (3), 1718–1728.

Kaur, R.; Chattopadhyay, S. K.; Tandon, S.; Sharma, S. Large Scale Extraction of the Fruits of *Garcinia indica* for the Isolation of New and Known Polyisoprenylated Benzophenone Derivatives. *Ind. Crops Products* **2012,** *37*, 420–426.

Khatib, N. A., Pawase, K.; Patil, P. A. Evaluation of Anti Inflammatory Activity of *Garcinia indica* Fruit Rind Extracts in Wistar Rats. *Int. J. Res. Ayurveda Pharm.* **2010,** *1* (2), 449–454.

Kim, S.; Seo, S. U.; Min, K. J. et al. Garcinol Enhances TRAIL-Induced Apoptotic Cell Death Through Up-Regulation of DR5 and Down-Regulation of c-FLIP Expression. *Molecules* **2018,** *23* (7), 1614.

Kopytko, P.; Piotrowska, K.; Janisiak, J.; Tarnowski M. Garcinol—A Natural Histone Acetyltransferase Inhibitor and New Anti-Cancer Epigenetic Drug. *Int. J. Mol. Sci.* **2021,** *22* (6), 2828.

Krishnamurthy, N.; Lewis, Y. S.; Ravindranath, B. On the Structures of Garcinol, Isogarcinol and Camboginol. *Tetrahedron Lett.* **1981,** *22,* 793–796.

Lakshmi, C.; Kumar, K. A.; Dennis, T. J.; Kumar, T. S. Antibacterial Activity of Polyphenols of *Garciniaindica. Indian J. Pharma. Sci.* **2011,** *73* (4), 470–473.

Li, F.; Shanmugam, M. K.; Siveen, K. S.; Wang, F.; Ong, T. H.; Loo, S. Y.; Swamy, M. M.; Mandal, S.; Kumar, A. P.; Goh, B. C.; Kundu, T.; Ahn, K. S.; Wang, L. Z.; Hui, K. M.; Sethi, G. Garcinol Sensitizes Human Head and Neck Carcinoma to Cisplatin in a Xenograft Mouse Model Despite Downregulation of Proliferative Biomarkers. *Oncotarget* **2015,** *6* (7), 5147–5163.

Liao, C. H.; Sang, S.; Ho, C. T.; Lin, J. K. Garcinol Modulates Tyrosine Phosphorylation of FAK and Subsequently Induces Apoptosis Through Down-Regulation of Src, ERK, and Akt Survival Signaling in Human Colon Cancer Cells. *J. Cell. Biochem.* **2005,** *96* (1), 155–169.

Liu, H. W.; Lee, P. M.; Bamodu, O. A.; Su, Y. K.; Fong, I. H.; Yeh, C. T.; Chien, M. H.; Kan, I. H.; Lin, C. M. Enhanced Hsa-miR-181d/p-STAT3 and Hsa-miR-181d/p-STAT5A Ratios Mediate the Anticancer Effect of Garcinol in STAT3/5A-Addicted Glioblastoma. *Cancers (Basel)* **2019,** *11* (12), 1888.

Maheshwari, B.; Reddy, S. Y. Application of Kokum (*Garcinia indica*) Fat as Cocoa Butter Improver in Chocolate. *J. Sci. Food Agric.* **2005,** *85* (1), 135–140.

Majeed, M.; Majeed, S.; Nagabhushanam, K.; Lawrence, L.; Mundkur, L. *Garcinia indica* Extract Standardized for 20% Garcinol Reduces Adipogenesis and High Fat Diet-Induced Obesity in Mice by Alleviating Endoplasmic Reticulum Stress. *J. Funct. Foods* **2020,** *67,* 103863.

Mishra, A.; Bapat, M.; Tilak, J.; Devasagayam, T. Antioxidant Activity of *Garcinia indica* (Kokam) and Its Syrup. *Curr. Sci.* **2006,** *91* (1), 90–93.

Oakes, S. A. Endoplasmic Reticulum Stress Signaling in Cancer Cells. *Am. J. Pathol.* **2020,** *190* (5), 934–946.

Padhye, S.; Ahmad, A.; Oswal, N.; Sarkar, F. H. Emerging Role of Garcinol, the Antioxidant Chalcone from *Garcinia indica* Choisy and Its Synthetic Analogs. *J. Hematol. Oncol.* **2009,** *2,* 38.

Panda, V.; Ashar, H.; Srinath, S. Antioxidant and Hepatoprotective Effect of *Garcinia indica* Fruit Rind in Ethanol-Induced Hepatic Damage in Rodents. *Interdisc. Toxicol.* **2012,** *5* (4), 207–213.

Panda, V.; Kundnani, K.; Islam, A. In Vivo Anti-Inflammatory Activity of *Garcinia indica* Fruit Rind (Kokum) in Rats. *J. Phytopharmacol.* **2013,** *2* (5), 8–14.

Parasramka, M. A.; Ali, S.; Banerjee, S.; Deryavoush, T.; Sarkar, F. H.; Gupta S. Garcinol Sensitizes Human Pancreatic Adenocarcinoma Cells to Gemcitabine in Association with microRNA Signatures. *Mol. Nutr. Food Res.* **2013,** *57* (2), 235–248.

Ranjbarnejad, T.; Saidijam, M.; Tafakh, M. S.; Pourjafar, M.; Talebzadeh, F.; Najafi, R. Garcinol Exhibits Anti-Proliferative Activities by Targeting Microsomal Prostaglandin E Synthase-1 in Human Colon Cancer Cells. *Human Exp. Toxicol.* **2017,** *36* (7), 692–700.

Reddy, S. Y.; Prabhakar, J. V. Cocoa Butter Extenders from Kokum (*Garcinia indica*) and Phulwara (*Madhucabutyracea*) Butter. *J. Am. Oil Chem. Soc.* **1994,** *71,* 217–219.

Schobert, R.; Biersack, B. Chemical and Biological Aspects of Garcinol and Isogarcinol: Recent Developments. *Chem. Biodiv.* **2019,** *16* (9), e1900366.

Sethi, G.; Chatterjee, S.; Rajendran, P.; Li, F.; Shanmugam, M. K.; Wong, K. F.; Kumar, A. P.; Senapati, P.; Behera, A. K.; Hui, K. M.; Basha, J.; Natesh, N.; Luk, J. M.; Kundu, T. K. Inhibition of STAT3 Dimerization and Acetylation by Garcinol Suppresses the Growth of Human Hepatocellular Carcinoma In Vitro and In Vivo. *Mol. Cancer* **2014**, *13*, 66.

Singh, N. P. Clusiaceae (Guttiferae nom. alt.) In *Flora of India*; Sharma, B. D., Balakrishnan, N. P., Eds., Vol. 3; Botanical Survey of India: Kolkatta, 1993; pp 86–151.

Sutar, R. L.; Mane, S. P.; Ghosh, J. S. Antimicrobial Activity of Extracts of Dried Kokum (*Garcinia indica* C). *Int. Food Res. J.* **2012**, *19* (3), 1207–1210.

Tharachand, C.; Selvaraj, C. I.; Abraham, Z. Comparative Evaluation of Anthelmintic and Antibacterial Activities in Leaves and Fruits of *Garcinia cambogia* (Gaertn.) Desr. and *Garcinia indica* (Dupetit-Thouars) Choisy. *Braz. Arch. Biol. Technol.* **2015**, *58* (3), 379–386.

Tu, S. H.; Chiou, Y. S.; Kalyanam, N.; Ho, C. T.; Chen, L. C.; Pan, M. H. Garcinol Sensitizes Breast Cancer Cells to Taxol Through the Suppression of Caspase-3/iPLA$_2$ and NF-κB/Twist1 Signaling Pathways in a Mouse 4T1 Breast Tumor Model. *Food Funct.* **2017**, *8* (3), 1067–1079.

Varalakshmi, K. N.; Sangeetha, C. G.; Shabeena, A. N.; Sunitha, S. R.; Vapika, J. Antimicrobial and Cytotoxic Effects of *Garcinia indica*. *World J. Agric. Sci.* **2011**, *7* (2), 193–196.

Wang, J.; Wang, L.; Ho, C. T.; Zhang, K.; Liu, Q.; Zhao H. Garcinol from *Garcinia indica* Downregulates Cancer Stem-like Cell Biomarker ALDH1A1 in Nonsmall Cell Lung Cancer A549 Cells Through DDIT3 Activation. *J. Agric. Food Chem.* **2017**, *65* (18), 3675–3683.

Wang, J.; Wu, M.; Zheng, D.; Zhang, H.; Lv, Y.; Zhang, L.; Tan, H. S.; Zhou, H.; Lao, Y. Z.; Xu, H. X. Garcinol Inhibits Esophageal Cancer Metastasis by Suppressing the p300 and TGF-β1 Signaling Pathways. *Acta PharmacologicaSinica* **2020**, *41* (1), 82–92.

Wang, Y.; Tsai, M. L.; Chiou, L. Y.; Ho, C. T.; Pan, M. H. Antitumor Activity of Garcinol in Human Prostate Cancer Cells and Xenograft Mice. *J. Agric. Food Chem.* **2015**, *63* (41), 9047–9052.

Ye, X.; Yuan, L.; Zhang, L.; Zhao, J.; Zhang, C. M.; Deng, H. Y. Garcinol, an Acetyltransferase Inhibitor, Suppresses Proliferation of Breast Cancer Cell Line MCF-7 Promoted by 17β-estradiol. *Asian Pac. J. Cancer Prev.* **2014**, *15* (12), 5001–5007.

Yu, S. Y.; Liao, C. H.; Chien, M. H.; Tsai, T. Y.; Lin, J. K.; Weng, M. S. Induction of p21 (Waf1/Cip1) by garcinol via downregulation of p38-MAPK signaling in p53-independent H1299 lung cancer. Journal of Agricultural and Food Chemistry **2014**, *62* (9), 2085–95.

Zhang, G.; Fu, J.; Su, Y.; Zhang X. Opposite Effects of Garcinol on Tumor Energy Metabolism in Oral Squamous Cell Carcinoma Cells. *Nutr. Cancer* **2019**, *71* (8), 1403–1411.

Zhang, J.; Fang, H.; Zhang, J.; Guan, W.; Xu G. Garcinol Alone and in Combination with Cisplatin Affect Cellular Behavior and PI3K/AKT Protein Phosphorylation in Human Ovarian Cancer Cells. *Dose Response* **2020**, *18* (2), 1559325820926732.

Zhang, M.; Lu, Q.; Hou, H.; Sun, D.; Chen, M.; Ning, F.; Wu, P.; Wei, D.; Duan, Y.; Pan, Y.; Lash, G. E. Garcinol Inhibits the Proliferation of Endometrial Cancer Cells by Inducing Cell Cycle Arrest. *Oncol. Rep* **2021**, *45* (2), 630–640.

Zhao, J.; Yang, T.; Ji, J.; Li, C.; Li, Z.; Li L. Garcinol Exerts Anti-Cancer Effect in Human Cervical Cancer Cells Through Upregulation of T-Cadherin. *Biomed. Pharmacother.* **2018**, *107*, 957–966.

Zheng, Y.; Guo, C.; Zhang, X.; Wang, X.; Ma, A. Garcinol Acts as an Antineoplastic Agent in Human Gastric Cancer by Inhibiting the PI3K/AKT Signaling Pathway. *Oncol. Lett.* **2020**, *20* (1), 667–676.

CHAPTER 2

Centella asiatica Linn

SAGAR BARGE[1], DHANANJAY JADE[2], and
NARAYAN CHANDRA TALUKDAR[1,3]

[1]Chemical Biology Lab I, Institute of Advanced Study in Science and Technology, Paschim Boragaon, Guwahati, Assam, India

[2]Biomedical Sciences, University of Leeds, Leeds, UK

[3]Assam Down Town University, Panikhaiti, Guwahati, Assam, India

ABSTRACT

Natural products are gaining a lot of attention in medicine due to their medicinal value. Natural products are mainly secondary metabolites and used for drug discovery for many years. *Centella asiatica* (CA) is a folk medicinal plant used for different phytotherapeutic treatments across southeast Asia. *Centella asiatica* L. is a herb-like plant traditionally assigned to the Apiaceae family. In India, species are locally known for their different names, such as Brahmi, mandukaparni, Indian pennywort or *Spadeleaf*, and *Gotu kola* in Southeast Asian countries, including China. Many tribal groups have been using traditional leaves, stem, and flower parts of the CA plant for hundreds of years. The records of the use of CA plants in both Indian Ayurveda and Chinese traditional medicine are available in the treatment of various medical emergencies. CA was further subjected to mechanistic and clinical investigation in modern medical research based on traditional knowledge. Several compounds have been isolated from CA, including polyphenolic, triterpenoids, saponins, and essential oils. Pharmacological reports reveal the efficacy of the CA plant against the antimicrobial, antiviral, antidepressive antidiabetic, wound-healing, memory-enhancing, and antioxidant activities.

Potent Anticancer Medicinal Plants: Secondary Metabolite Profiling, Active Ingredients, and Pharmacological Outcomes. Deepu Pandita and Anu Pandita (Eds.)

Compound, namely, asiatic acid (AA), asiaticoside has anticancer potential against Breast, lung, and ovarian cancer. These anticancer agents mainly affect the receptors, enzymes, transcription factors, cell signaling cascade, and apoptotic proteins. In the biosynthesis of compounds, plant hormones play an essential role in plant metabolism via the upregulation of genes related to secondary metabolites, and scientific efforts for manipulating these metabolic processes in order to increase the biosynthesis of these anticancer agents, have received much attention.

2.1 INTRODUCTION

The genus *Centella* consists of 53 species of flowering plants, including CA one of the popular medicinal herbs. The genus *Centella* initially belongs to the family of Araliaceae and Mackinlayoideae as a subfamily but the molecular studies assigned the genus to the Apiaceae family. CA is a green leafy vegetable medicinal plant known as Indian pennywort, with many nutritional values. Due to its high medicinal value, it is cultivated in many countries (Matsuda et al., 2001) for hundreds of years, including India, China, Sri Lanka, and Malaysia, for traditional purposes (Brinkhaus et al., 2000b). CA is reported as "*Sushruta Samhita*" in ayurvedic medicine, as ancient Indian text commonly called mandukparni (Diwan, 1991), which is commonly used for various diseases such as mental disorders, skin problems, liver ailments, epilepsy, asthma, hair loss, and tetanus (Meulenbeld and Wujastyk, 2001). In China and Indonesia, CA, also known as *Gotu kola,* and used as "miracle elixirs of life" (Diwan, 1991). The plant species' leaves are edible, grow in orbicular or oblong–elliptic shapes, and are yellowish–green (Chopra et al., 1956). The plant species are consumed as leaf juice, drink, and other food products. and it is most famous for its use as a "brain tonic." *Gotu kola* is also beneficial for the growth of connective tissues and wound healing (Somboonwong et al., 2012).

CA mainly consists of many pentacyclic triterpenoids saponins, also called centelloids. These include madecassic acid, asiatic acid (AA), brahmoside, brahminoside, madecassoside, sceffoleoside, and asiaticoside (Singh and Rastogi, 1968; Sun et al., 2020; Yu et al., 2006). Pharmacological evidence suggests that these compounds are used against many diseases (Sun et al., 2020), such as antidiabetic (Legiawati et al., 2020; Oyenihi et al., 2019), anticancer (Tariq et al., 2015), anti-inflammatory and antioxidant (Ariffin et al., 2011; Hafiz et al., 2020; Kumari et al., 2016; Ratz-Łyko et al., 2016), neuroprotective (Gray et al., 2018; Lokanathan et al., 2016; Yadav et al., 2019), etc. Cancer has caused many deaths globally, and it is estimated that by the end of the year 2025, there will be approximately 20 million new cancer cases globally (Sevgi and Nazım, 2019). However, natural product has played a vital role in cancer treatment and provided many natural drugs (Seca and Pinto, 2018). Therefore, many researchers mainly studied the anticancer effect of CA and its bioactive compound.

In many countries, including India, China, and South Africa, traditional healers' consultation and use of medicinal plants in disease is common health care practice. There are 80% of the population uses medicinal plant which is homegrown in their diet. Moreover, the treatment of various diseases with medicinal plants increases its market demand which caused harvesting and overexploitation of wild medicinal plants (Barbhuiya et al., 2009; Vasisht et al., 2016). To reduce the overexploitation of the CA plant, biotechnological manipulation of such plants using plant cell culture and organ culture is the best alternative source of natural products (Tan et al., 2010).

The manipulation of these metabolic pathways can be studied using plant metabolomics analytical techniques. These analytical techniques allow us to study and identify the changes in metabolic pathways upon treatment. This chapter has discussed the biosynthesis of secondary metabolites and metabolic profiling of the compounds present in the CA. We also highlight the omics approach, namely, metabolomics, by which almost the entire range of metabolites can be analyzed for bioactivity in an organism at a specific time and under particular conditions. Also, we will discuss the mechanism of action of anticancer agents present in the CA.

2.2 METABOLITE PROFILING OF *CENTELLA ASIATICA* PHYTOCONSTITUENTS

Metabolic profiling involves detecting the role of metabolites in a biochemical pathway, and metabolomics is the qualitative and quantitative analysis of metabolites in a plant crude extract (Bhalla et al., 2005). Metabolite profiling or metabolomics was used in various fields with various technical methods to identify and characterize secondary metabolites (Cerny et al., 2020; Sumner et al., 2007). The plant extracts consist of many complex secondary metabolites with different polarities in various concentrations; therefore, various techniques are employed to characterize and detect these compounds. This technique includes liquid chromatography–mass spectrometry (LC/

MS), ultrahigh-performance liquid chromatography (UHPLC), thin-layer chromatography (TLC), and nuclear magnetic resonance (NMR).

2.2.1 TLC AND HPLC-BASED METABOLITE PROFILING

The secondary metabolites from the CA plant have been identified and quantified by simple analytical techniques such as TLC and high-performance liquid chromatography (HPLC). In early 1991, high-speed counter-current chromatographic (HSCCC) techniques were introduced to separate madecassoside and asiaticoside from the CA plant (Diallo et al., 1991). Further, in 2004, the same technique separates pentacyclic triterpene aglycones and glycosides (Du et al., 2004). Gunther and Wagner, in 1996, quantified and characterized the AA, asiaticoside, madecassoside, and madecassic acid, the triterpene from CA plant using an acetonitrile/water as a mobile phase on RP18 column in a reverse phase separation system (Günther and Wagner, 1996). Later on, the separation and identification of triterpenes achieved by TLC were also reported (Brinkhaus et al., 2000b). Further, Bonfill et al. in 2006 identified four triterpenes using TLC and mass spectrometry matrix-assisted laser desorption/ionization-time of flight (MALDI-TOF) (Bonfill et al., 2006). Although during early studies of metabolite profiling, the HPLC was equipped with ultraviolet (UV) detection and used to characterize and quantify compounds but failed to detect some compounds due to a lack of solid UV-absorbing triterpenes. Therefore, some UHPLC systems are coupled with evaporative light scattering detection (ELSD) to better detect centellosides. Recently, metabolic profiling of ethanolic and aqueous extract of CA by HPLC analysis revealed that the presence of a flavonoid group of compounds such as rutin, gallic acid, quercetin, kaempferol, catechin, and luteolin (Mohammad Azmin and Mat Nor, 2020).

2.2.2 LC/MS-BASED METABOLITE PROFILING AND METABOLOMICS STUDY

The metabolomics approach allows for an analysis of the different and multiple metabolites present in the biological system. The LC-MS is the technique that tends to identify and analyze the multiple metabolites with different polarities. Metabolic profiling through LC-MS and gas chromatography–mass spectrometry (GC-MS) also addresses large group metabolites about the targeted metabolic pathway and its related compound (Eugster et al., 2011). High-resolution mass spectrometry (HRMS) is often used coupled with the LC to separate the compounds and the compounds and perform the analysis of compounds present in the crude extract. Metabolic profiling of CA plant extract has been studied extensively recently; screening of CA plant extract and fractions reveals the 22–33 flavonoid group of compounds through LS-MS and GCMS analysis (Ondeko et al., 2020). Similarly, 117 compounds are identified, and 24 are quantified through phytochemical analysis using a quadrupole time of flight analyzer coupled with an optimized HPLC separation (Alcazar Magana et al., 2020).

Furthermore, to analyze the changes during the biosynthesis of compounds by manipulating the pathways, the LCMS and GCMS techniques were employed. For example, changes in the metabolite profile were analyzed upon methyl jasmonate treatment to produce triterpenoids in the CA plant (James et al., 2013; Tugizimana et al., 2015). The CA plants' undifferentiated and differentiated leaf cells were also analyzed for their presence of 18 metabolites in methanolic extract using ultrahigh-performance liquid chromatography–quadrupole time-of-flight mass spectrometry (UHPLC/QTOF-MS) (Ncube et al., 2017).

2.2.3 NMR-BASED METABOLITE PROFILING AND METABOLOMICS STUDY

Secondary metabolites with structures are elucidated by NMR. NMR analysis for metabolite profiling is a simple and reproducible technique that provides direct detailed structural information, but this technique has the limitation that it has relatively low sensitivity, meaning that it generally enables profiling only of major constituents.

Different types of NMR are specifically used to identify the 2 dimensional or 1-dimensional structure of the compounds. For example, the two flavonoid molecules were isolated from the whole plant of CA and were identified as castilliferol 1 and castillicetin 2 using spectral analysis by 1D and 2D NMR (Subban et al., 2008). Similarly, metabolites from three different species of Centella, including CA, were analyzed using a ^{1}H NMR spectroscopy in various lighting conditions, which reveals the accumulation of certain secondary metabolites in some species making them more medicinal than other species (H et al., 2012). Further, in 2011, the researcher first time isolated the Arabic acid from the cell-cultured CA plant by MS/NMR spectroscopic analysis to confirm its structure (Antognoni et al., 2011).

2.3 SECONDARY METABOLITES OF *CENTELLA ASIATICA* AND ITS BIOSYNTHESIS

Centella asiatica consists of many significant phytoconstituents mainly found in the aerial and root parts of the plants (Brinkhaus et al., 2000a; James and Dubery, 2009). These secondary metabolites are involved in the defensive interaction between plants and the environment, including substances like attractants phytoalexins, antifeedants, and pheromones ("The biosynthesis of secondary metabolites," 2003; Yang et al., 2018). Therefore, studies on biosynthesis and the biological activity of these compounds are increased due to their importance in medicine and industry (Schäfer and Wink, 2009). Many medicinal plants consist of similar compounds, but their biochemical studies reveal different biosynthesis pathways based on their environmental conditions (Li et al., 2020).

2.3.1 BIOSYNTHESIS OF CENTELLA ASIATICA TRITERPENES AND TERPENOIDS

CA contains a high amount of triterpenoids; these are pentacyclic in nature and have also been called centelloids. The biosynthesis and level of saponins triterpenes change with the geographical origin and environmental conditions ("The biosynthesis of secondary metabolites," 2003). For the synthesis of saponins and terpins, they follow the isoprenoid pathway. This pathway mainly consists of three precursors: isoprenoid, isomeric 5-carbon precursors, isopentenyl diphosphate, and dimethylallyl diphosphate. In the first step, the prenyltransferase synthase enzyme converts these three C5 precursors into farnesyl diphosphate (C15) to create an assembly. In the next step, squalene synthase catalyzed the two molecules of farnesyl diphosphate to give squalene (C30).

At last, presence of O_2 and nicotinamide adenine dinucleotide phosphate (NADPH), squalene (C30) gets converted into the squalene 2,3-epoxide (2,3-oxidosqualene), which is the precursor for the synthesis of both sterol and triterpenoids. 2,3-oxidosqualene further cyclizes through cationic intermediates and converts into triterpenoids of CA via oxidosequalene cyclases enzyme, including α and β amyrin synthase. α-amyrin synthase is mainly involved in the synthesis of AA and the action of cycloartenol synthase engaged in the synthesis of sterol via protosteryl cation (Azerad, 2016; Gallego et al., 2014; Kim et al., 2005). Sterol and terpene biosynthesis general pathways are discussed by many researchers elsewhere (Collin, 2001; Gershenzon and Kreis, 1999; Kalinowska et al., 2005).

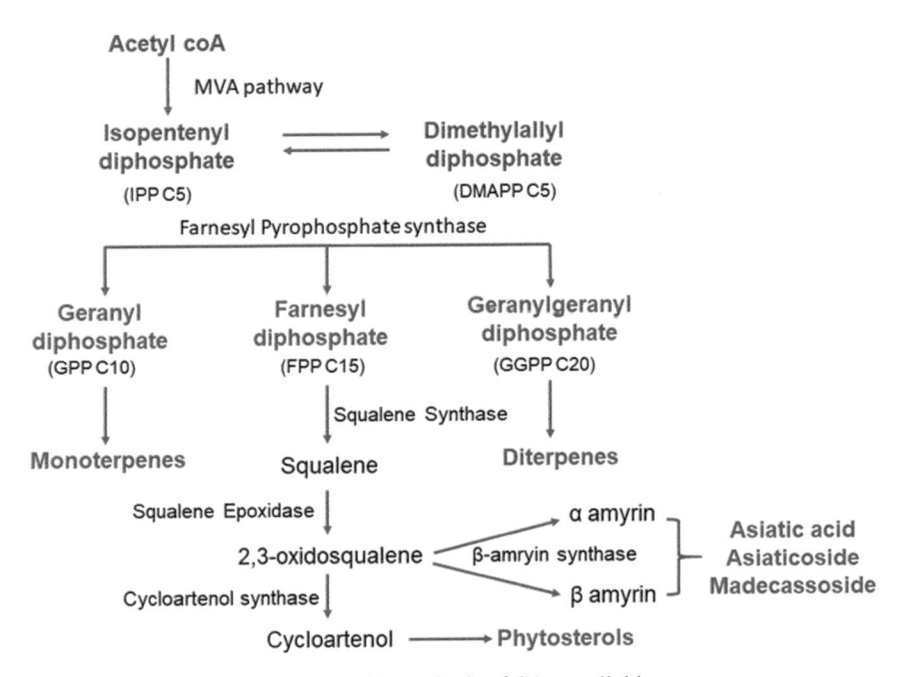

FIGURE 2.1   Pathways involved in biosynthesis of CA centelloids.

2.3.2 *BIOSYNTHESIS OF CENTELLA ASIATICA FLAVONOIDS AND PHENOLS*

CA contains many flavonoids, including catechin, epicatechin, kaempferol, quercetin (Idris and Mohd Nadzir, 2021). The biosynthesis of flavonoids occurs in CA plants via the shikimate pathway and acetate pathway: Both pathways use p-cinnamoyl-CoA and malonyl-CoA units as initial precursor molecules for biosynthesis. Further Claisen-type reaction forms a naringenin chalcone with the help of chalcone synthase, resulting in the formation of A ring of the naringenin and flavanone. Naringenin, via enzymatic reactions, gives flavons-like compounds called apigenin and flavonols (Dewick, 2002; Winkel-Shirley, 2001). From methanolic extract of the CA plant, Subban et al. have isolated a castilliferol, also known as kaempferol-3-p-coumarate, and castillicetin, also called quercetin-3-caffeate, esterified flavonoid and derivative of hydroxycinnamic (Subban et al., 2008).

Chlorogenic acids are another group of polyphenol-rich in CA plant formed by cinnamic acid derivative via esterification of quinic acid (Ncube et al., 2016). During the biosynthesis of chlorogenic acids, catalysis of L-phenylalanine occurs to eliminate the E2 ammonia by the enzyme called phenylalanine ammonia-lyase to produce trans-cinnamic acid further produces p-coumaric acid with the help of 4-cinnamic acid hydroxylase enzyme. Further hydroxycinnamic acid derivatives caffeic acid are formed via hydroxylation and methylation reactions. Additional hydroxyl groups on L-quinic acid get esterified, and caffeic acid carboxyl function yields caffeoylquinic acids. Finally, the ester of caffeoyl acid-containing 3-hydroxyl position of L-quinic acid produces chlorogenic acid (Xie et al., 2011).

2.4   MANIPULATION OF METABOLIC PATHWAYS FOR THE PRODUCTION OF PENTACYCLIC TRITERPENOIDS

Pentacyclic triterpenoids are pharmacologically active compounds with a complex structure, making the chemical synthesis of such compounds expensive and time-consuming. Therefore, increasing the production of such compounds in the plant via the manipulation of biosynthetic pathways is essential (Baltz, 2016; Birchfield and McIntosh, 2020). Secondary metabolite synthesis occurs during the specific time of plant development, and it is tissue-specific and also regulated by plant growth (Aziz et al., 2007; Khare et al., 2020). The metabolic interaction is essential during the Secondary metabolite production between the root and other plant parts (Pott et al., 2019). Plant cell culture is the approach used to manipulate biosynthetic pathways in CA by adding precursors in the cell culture medium (Bouhouche et al., 1998), but this approach has some limitations after prolonged culture, the cell's ability to produce compounds gets declined. Another approach is to increase the enzyme levels by using plant-specific molecules such as methyl jasmonate. A similar approach was applied to increase the asiaticoside levels in CA where the plant is treated with methyl jasmonate, and the sterols synthesizing enzymes are inhibited from increasing the synthesis of asiaticoside (Mangas et al., 2006). Another study reported that following treatment of methyl jasmonate in roots of CA inhibited the expression of cyclases responsible for synthesizing sterols and activated the expression of β-amyrin synthase responsible for AA production (Kim et al., 2005).

2.4.1 MANIPULATION OF BIOSYNTHESIS PATHWAY USING ELICITORS IN VITRO PLANT CULTURE

Elicitors are the external factors that stimulate physiological reactions and induce the stress response to protect the plants against disease, resulting in more production of secondary biological metabolites (Baenas et al., 2014). The plant tissue culture techniques use different types of elicitors such as CuCl2, MeJA, yeast extract, and fungal to increase the production of asiaticoside in CA.

The elicitors such as MeJA and yeast extract showed positive effects (Kim et al., 2004). In order to increase the yield of asiaticoside and madecassoside in CA, Kim et al. have performed various studies using cell culture techniques and treatment with methyl jasmonate. His studies demonstrated that the transformation of hairy root culture and methyl jasmonate treatment for 12h to 14 days increased gene expression of β-amyrin synthase, which significantly enhanced the production of asiaticoside. Further, in his other study, he raised the production of asiaticoside by using elicitors and yeast extract. (Kim et al., 2007, 2004). Methyl jasmonate (MeJA) about 100 μm along with cytokinin TDZ (0.025 mg/L) supplementation in plant cell culture increased 82% of the production of asiaticoside in leaves (Yoo et al., 2011).

The fungal elicitors have also been shown to positively affect biomass production and asiaticoside in CA plant shoot culture. The results revealed the use of fungal elicitors such as mycelial extract of *Colletotrichum lindemuthianum, Fusarium oxysporum, and Trichoderma harzianum* culture filtrate 3% v/v enhanced both asiaticose and biomass production (Prasad et al., 2013). Similarly, the fungus *Piriformospora indica* cocultivation with roots of CA plant shows an increase in the production of asiaticoside (2-fold) and biomass (Satheesan et al., 2012).

2.4.2 IN VITRO SECONDARY METABOLITE PRODUCTION IN CA

During CA tissue culture studies, manipulation of growth hormones such as auxin and cytokinin can increase centellosides production in CA via the influence of callus formation, somatic embryogenesis, and whole plantlet regeneration (Nirmal et al., 2007; Prasad et al., 2012). For example, increase in the concentration of 6-benzylaminopurine (BAP) and 1-naphthalene acetic acid (NAA) results in an enhanced axillary bud formation from nodal explant, which helped in the production of asiaticoside (Mukundan et al., 2004; Nath

Tiwari et al., 2000). Similarly, the callus and the cell suspension systems have increased the production of asiaticoside in vitro culture systems (Nath and Buragohain, 2004). Further different culture conditions are also necessary for the secondary metabolite production manipulation of these conditions, such as Ph and carbon source, resulting in metabolite production and increased expression of genes responsible for centelloside pathways. Leaves of the CA are an excellent source of asiaticoside; around 82.6% of the asiaticoside is produced from leaves compared to the whole plant culture (Kim et al., 2004).

Similarly, the triterpenoids (asiaticoside and madecassoside) localization was tissue-specific, and the highest amount of triterpenoids are present in leaves as compared to the transformed roots and undifferentiated callus; these observations are made comparing two Malaysian phenotypes of CA for root transformation and callus culture (Aziz et al., 2007). The study by James et al. in 2008 also shows that the triterpenoid content is higher in the leaves of callus, and cell suspensions in South African CA plants than the undifferentiated callus and cell suspensions (James et al., 2008). These reports suggest that the manipulation of growth hormones can increase the production of secondary metabolites.

2.5   ANTICANCER EFFECT OF CA INGREDIENTS

Cancer is commonly defined as the uncontrolled growth of abnormal and genetically unstable cells in the body and infecting nearby cells. CA poses many biological activities, including anticancer activity (Sun et al., 2020). The report suggests that triterpenoids in the ethanolic extract of the CA plant have an inhibitory action on the A549 cancer cells upon ionizing radiation IR induction in the lungs (Han et al., 2020). AA is a major compound found in the CA plant and is effective against various cancer. There are many reports on AA that were published showing its effect against lung cancer. AA is also effective against Nasopharyngeal carcinoma (NPC) by inducing apoptosis in the NCP cells via p38 phosphorylation (Liu et al., 2020). Similarly, AA has antiproliferative and apoptotic activity in KKU-156 and KKU-213 human cancer cells (Sakonsinsiri et al., 2018).

Furthermore, the AA treatment inhibits the growth of lung cancer tumor cells via mitochondrial-associated damage, which results in the formation of reactive oxygen species (ROS) in lung cancer cell lines such as A549 and H1299 cells also in C57BL/6J mice (Wu et al., 2017). AA also tends to in vitro downregulate the p-glycoprotein by reducing the expression of the MDR1 gene in human lung adenocarcinoma A546/DDP cells; the p-glycoprotein is responsible for multidrug resistance (MDR) during chemotherapy (Cheng et al., 2018). Kim et al., 2014, reports that the AA increases the expression of MicroRNA-1290 in A549 lung carcinoma cells resulting in apoptosis of cells (Kim et al., 2014).

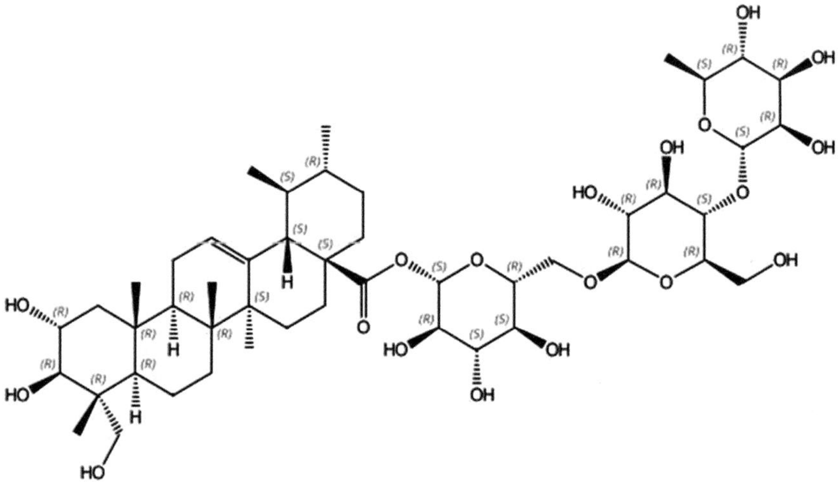

FIGURE 2.2   Asiaticoside ($C_{48}H_{78}O_{19}$) CAS RN 16830-15-2.

Source: https://scifinder-n.cas.org/

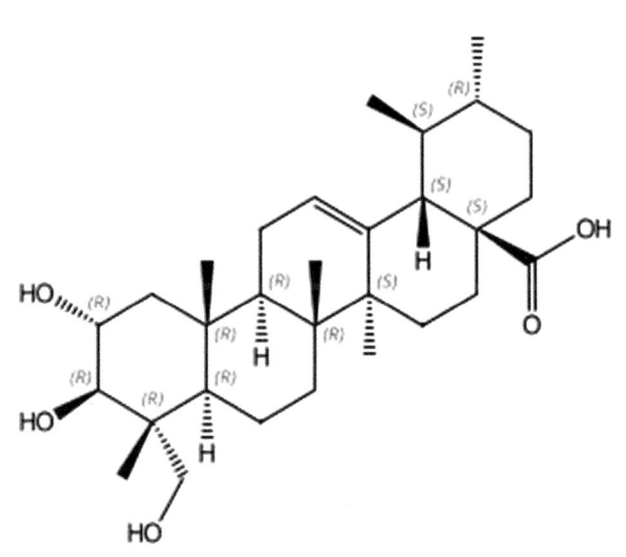

FIGURE 2.3   Asiatic acid ($C_{30}H_{48}O_5$) CAS RN 464-92-6.

Source: Ref. https://scifinder-n.cas.org/

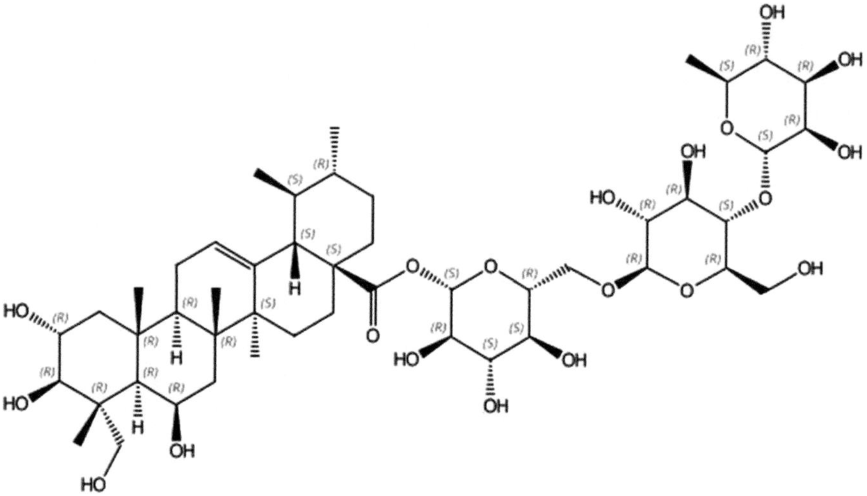

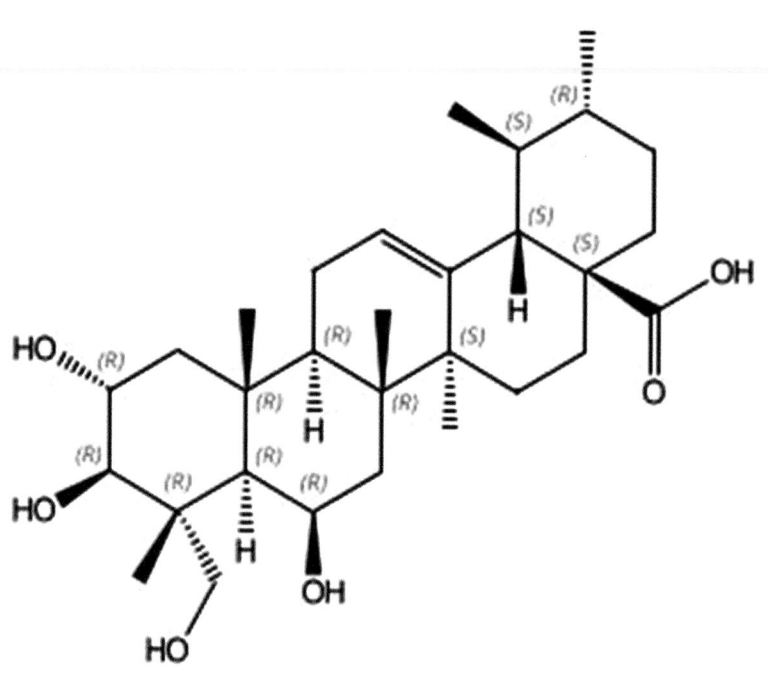

FIGURE 2.4   Madecassoside ($C_{48}H_{78}O_{20}$) CAS RN 34540-22-2.

Source: Ref. https://scifinder-n.cas.org/

FIGURE 2.5   Madecassic acid ($C_{30}H_{48}O_6$) CAS RN 18449-41-7.

Source: Ref. https://scifinder-n.cas.org/.

AA was also effective against colon cancer cells such as SW480 and HCT116 via the PI3K/Akt/mTOR/p70S6K signaling pathway by regulating Pdcd4 in cells. These pathways regulate cellular migration, proliferation, and apoptosis in cells; these cellular processes are found to be inhibited upon AA treatment in colon cancer cells (Hao et al., 2018). Similarly, AA induces apoptosis in the SW480 human colon cancer cells (Tang et al., 2009). AA was also influential on SK-MEL-2 skin cancer cells by inducing apoptosis and increasing ROS in cells (Park et al., 2005). AA exerts its effect on human ovarian cancer by (PItf3K)/Akt/mTOR cascade in vitro in SKOV3 and OVCAR-3 cells it causes 50% toxicity in the cells (Ren et al., 2016).

Another ingredient from the CA plant called asiaticoside has an anti-cancer effect on hepatocellular carcinoma (HCC); results have shown that the asiaticoside treatment has an antiproliferative effect on HCC cell lines and also induces apoptosis via G1 cell cycle growth arrest in the cells (Ma et al., 2020). Madecassoside major triterpene glycosides from CA plants have shown their in vitro bioactivity against HCC in HepG2 and SMMC-77 cancer cells. The evidence suggests that it works via the upregulation of the cMET-PKC-ERK1/2-COX-2-PGE2 pathway and inhibits the cells' hepatocyte growth factor (HGF) induced proliferation. These reports suggest that the major ingredients present in CA plants, such as centellocides are mainly involved in anticancer activity.

2.6 CONCLUSION

CA has a wide range of medicinal values in the market; due to that, the demand for the cultivation of CA plants increased in many countries. The excessive harvesting of CA plants from the wild makes them endangered species, and the conservation of such species is essential. Therefore, an alternative for producing secondary metabolites such as centellocides was achieved by genetic manipulation of metabolite pathways using in vitro plant tissue culture. The biosynthesis of metabolites includes using elicitation agents such as CuCl2, MeJA, yeast extract, and fungal elicitors to induce the production of secondary metabolites. Hairy root cultures are also effective for the biosynthesis of triterpenoids. The plant tissue culture techniques further manipulate the growth hormones such as auxin and cytokinin in different cultures like callus culture and cell suspension cultures to produce metabolites. These manipulations in metabolic pathways are mainly tracked using metabolomics studies that identify the changes in the manipulated pathways upon elicitation and also helped in metabolite profiling. The produced secondary metabolites were then characterized using different analytical techniques such as TLC, HPLC, LCMS, and NMR analysis. These techniques also help in the qualitative and quantitative identification of new compounds from the plant.

Centellosides are the enriched compounds in the CA plant that shows many biological activities, including anticancer activity. AA was majorly involved in vitro and in vivo regulation of pathways associated with anticancer activity.

KEYWORDS

- ***Centella asiatica* L.**
- **anticancer**
- **omics**
- **asiatic acid**
- **metabolomics**

REFERENCES

Alcazar Magana, A.; Wright, K.; Vaswani, A.; Caruso, M.; Reed, R. L.; Bailey, C. F.; Nguyen, T.; Gray, N. E.; Soumyanath, A.; Quinn, J.; Stevens, J. F.; Maier, C. S. Integration of Mass Spectral Fingerprinting Analysis with Precursor Ion (MS1) Quantification for the Characterisation of Botanical Extracts: Application to Extracts of *Centella asiatica* (L.) Urban. *Phytochem Anal.* **2020**, *31,* 722–738.

Anon. The Biosynthesis of Secondary Metabolites. In *Natural Products: The Secondary Metabolites*; Hanson, J. R., Ed.; The Royal Society of Chemistry, 2003; pp 105–130.

Antognoni, F.; Perellino, N. C.; Crippa, S.; Dal Toso, R.; Danieli, B.; Minghetti, A.; Poli, F.; Pressi, G. Irbic Acid, a Dicaffeoylquinic Acid Derivative from *Centella asiatica* Cell Cultures. *Fitoterapia* **2011**, *82,* 950–954.

Ariffin, F.; Heong Chew, S.; Bhupinder, K.; Karim, A. A.; Huda, N. Antioxidant Capacity and Phenolic Composition of Fermented *Centella asiatica* Herbal Teas. *J. Sci. Food Agric.* **2011**, *91,* 2731–2739.

Azerad, R. Chemical Structures, Production and Enzymatic Transformations of Sapogenins and Saponins from *Centella asiatica* (L.) Urban. *Fitoterapia* **2016**, *114,* 168–187.

Aziz, Z. A.; Davey, M. R.; Power, J. B.; Anthony, P.; Smith, R. M.; Lowe, K. C. Production of Asiaticoside and Madecassoside in *Centella asiatica* In Vitro and In Vivo. *Biologia Plantarum* **2007b**, *51,* 34–42.

Baenas, N.; García-Viguera, C.; Moreno, D. A. Elicitation: A Tool for Enriching the Bioactive Composition of Foods. *Molecules* **2014**, *19,* 13541–13563.

Baltz, R. H. Genetic Manipulation of Secondary Metabolite Biosynthesis for Improved Production in Streptomyces and Other Actinomycetes. *J. Ind. Microbiol. Biotechnol.* **2016,** *43,* 343–370.

Barbhuiya, A.; Sharma, G.; Arunachalam, A.; Deb, S. Diversity and Conservation of Medicinal Plants in Barak Valley, Northeast India, 2009.

Bhalla, R.; Narasimhan, K.; Swarup, S. Metabolomics and Its Role in Understanding Cellular Responses in Plants. *Plant Cell Rep.* **2005,** *24,* 562–571.

Birchfield, A. S.; Mcintosh, C. A.; Metabolic Engineering and Synthetic Biology of Plant Natural Products—A Minireview. *Curr. Plant Biol.* **2020,** *24,* 100163.

Bonfill, M.; Mangas, S.; Cusidó, R. M.; Osuna, L.; Piñol, M. T.; Palazón, J. Identification of Triterpenoid Compounds of *Centella asiatica* by Thin-Layer Chromatography and Mass Spectrometry. *Biomed. Chromatogr.* **2006,** *20,* 151–153.

Bouhouche, N.; Solet, J. M.; Simon-Ramiasa, A.; Bonaly, J.; Cosson, L. Conversion of 3-Demethylthiocolchicine Into Thiocolchicoside by *Centella asiatica* Suspension Cultures. *Phytochemistry* **1998,** *47,* 743–747.

Brinkhaus, B.; Lindner, M.; Schuppan, D.; Hahn, E. G. Chemical, Pharmacological and Clinical Profile of the East Asian Medical Plant *Centella asiatica. Phytomedicine* **2000a,** *7,* 427–448.

Brinkhaus, B.; Lindner, M.; Schuppan, D.; Hahn, E. G. Chemical, Pharmacological and Clinical Profile of the East Asian Medical Plant *Centella aslatica. Phytomedicine* **2000b,** *7,* 427–448.

Cerny, M. A.; Kalgutkar, A. S.; Obach, R. S.; Sharma, R.; Spracklin, D. K.; Walker, G. S. Effective Application of Metabolite Profiling in Drug Design and Discovery. *J. Med. Chem.* **2020,** *63,* 6387–6406.

Cheng, Q.; Liao, M.; Hu, H.; Li, H.; Wu, L.; Asiatic Acid (AA) Sensitizes Multidrug-Resistant Human Lung Adenocarcinoma A549/DDP Cells to Cisplatin (DDP) via Downregulation of P-Glycoprotein (MDR1) and Its Targets. *Cell Physiol. Biochem.* **2018,** *47,* 279–292.

Chopra, R. N.; Nayar, S. L.; Chopra, I. C.; Asolkar, L. V.; Kakkar, K. K.; Chakre, O. J.; Varma, B. S.; Council of, S.; Industrial, R. *Glossary of Indian Medicinal Plants; [with] Supplement.* Council of Scientific & Industrial Research: New Delhi, 1956.

Collin, H. A. Secondary Product Formation in Plant Tissue Cultures. *Plant Growth Regulation* **2001,** *34,* 119–134.

Dewick, P. M. *Medicinal Natural Products: A Biosynthetic Approach*; John Wiley & Sons, 2002.

Diallo, B.; Vanhaelen-Fastré, R.; Vanhaelen, M. Direct Coupling of High-Speed Counter-Current Chromatography to Thin-Layer Chromatography: Application to the Separation of Asiaticoside and Madecassoside from *Centella asiatica. J. Chromatogr. A* **1991,** *558,* 446–450.

Diwan, P. V. Anti-Anxiety Profile of Manduk Parni (*Centella asiatica*) in Animals. *Fitoterapia* **1991,** *62,* 253–257.

Du, Q.; Jerz, G.; Chen, P.; Winterhalter, P. Preparation of Ursane Triterpenoids from *Centella asiatica* Using High Speed Countercurrent Chromatography with Step-Gradient Elution. *J. Liquid Chromatogr. Related Technol.* **2004,** *27,* 2201–2215.

Eugster, P. J.; Guillarme, D.; Rudaz, S.; Veuthey, J. L.; Carrupt, P. A.; Wolfender, J. L. Ultra High Pressure Liquid Chromatography for Crude Plant Extract Profiling. *J AOAC Int* **2011,** *94,* 51–70.

Gallego, A.; Ramirez-Estrada, K.; Vidal-Limon, H. R.; Hidalgo, D.; Lalaleo, L.; Khan Kayani, W.; Cusido, R. M.; Palazon, J. Biotechnological Production of Centellosides in Cell Cultures of *Centella asiatica* (L) Urban. *Eng. Life Sci.,* **2014,** *14,* 633–642.

Gershenzon, J.; Kreis, W. Biochemistry of Terpenoids: Monoterpenes, Sesquiterpenes, Diterpenes, Sterols, Cardiac Glycosides and Steroid Saponins. *Biochem. Plant Secondary Metabol.* **1999**, *2,* 222–299.

Gray, N. E.; Alcazar Magana, A.; Lak, P.; Wright, K. M.; Quinn, J.; Stevens, J. F.; Maier, C. S.; Soumyanath, A. *Centella asiatica*—Phytochemistry and Mechanisms of Neuroprotection and Cognitive Enhancement. *Phytochem Rev,* **2018**, 17*,* 161–194.

Günther, B.; Wagner, H. Quantitative Determination of Triterpenes in Extracts and Phyto-preparations of *Centella asiatica* (L.) urban. *Phytomedicine* **1996**, *3,* 59–65.

H, M.; Khatib, A.; Shaari, K.; Abas, F.; Shitan, M.; Kneer, R.; Neto, V.; Lajis, N. H. Discrimination of Three Pegaga (Centella) Varieties and Determination of Growth-Lighting Effects on Metabolites Content Based on the Chemometry of 1H Nuclear Magnetic Resonance Spectroscopy. *J. Agric. Food Chem.* **2012**, *60,* 410–417.

Hafiz, Z. Z.; Amin, M. M.; Johari James, R. M.; Teh, L. K.; Salleh, M. Z.; Adenan, M. I. Inhibitory Effects of Raw-Extract *Centella asiatica* (RECA) on Acetylcholinesterase, Inflammations, and Oxidative Stress Activities via In Vitro and In Vivo. *Molecules* **2020**, 25.

Han, A. R.; Lee, S.; Han, S.; Lee, Y. J.; Kim, J. B.; Seo, E. K.; Jung, C. H. Triterpenoids from the Leaves of *Centella asiatica* Inhibit Ionizing Radiation-Induced Migration and Invasion of Human Lung Cancer Cells. *Evid Based Complement Altern. Med.* **2020**, 3683460.

Hao, Y.; Huang, J.; Ma, Y.; Chen, W.; Fan, Q.; Sun, X.; Shao, M.; Cai, H. Asiatic Acid Inhibits Proliferation, Migration and Induces Apoptosis by Regulating Pdcd4 via the PI3K/Akt/mTOR/p70S6K Signaling Pathway in Human Colon Carcinoma Cells. *Oncol. Lett.* **2018**, *15,* 8223–8230.

Idris, F. N.; Mohd Nadzir, M. Comparative Studies on Different Extraction Methods of *Centella asiatica* and Extracts Bioactive Compounds Effects on Antimicrobial Activities. *Antibiotics (Basel)* **2021**, 10.

James, J. T.; Dubery, I. A. Pentacyclic Triterpenoids from the Medicinal Herb, *Centella asiatica* (L.) Urban. *Molecules* **2009**, *14,* 3922–3941.

James, J. T.; Meyer, R.; Dubery, I. A. Characterisation of Two Phenotypes of *Centella asiatica* in Southern Africa Through the Composition of Four Triterpenoids in Callus, Cell Suspensions and Leaves. *Plant Cell, Tissue Organ Culture* **2008**, *94,* 91–99.

James, J. T.; Tugizimana, F.; Steenkamp, P. A.; Dubery, I. A.; Metabolomic Analysis of Methyl Jasmonate-Induced Triterpenoid Production in the Medicinal Herb *Centella asiatica* (L.) Urban. *Molecules* **2013**, *18,* 4267–4281.

Kalinowska, M.; Zimowski, J.; Pączkowski, C.; Wojciechowski, Z. A. The Formation of Sugar Chains in Triterpenoid Saponins and Glycoalkaloids. *Phytochem. Rev.* **2005**, *4,* 237–257.

Khare, S.; Singh, N. B.; Singh, A.; Hussain, I.; Niharika, K.; Yadav, V.; Bano, C.; Yadav, R. K.; Amist, N. Plant Secondary Metabolites Synthesis and Their Regulations Under Biotic and Abiotic Constraints. *J. Plant Biol.* **2020**, *63,* 203–216.

Kim, K. B.; Kim, K.; Bae, S.; Choi, Y.; Cha, H. J.; Kim, S. Y.; Lee, J. H.; Jeon, S. H.; Jung, H. J.; Ahn, K. J.; An, I. S.; An, S. MicroRNA-1290 Promotes Asiatic Acid-Induced Apoptosis by Decreasing BCL2 Protein Level in A549 Non-Small Cell Lung Carcinoma Cells. *Oncol. Rep.* **2014**, *32,* 1029–36.

Kim, O.-T.; Bang, K.-H.; Shin, Y.-S.; Lee, M.-J.; Jung, S.-J.; Hyun, D.-Y.; Kim, Y.-C.; Seong, N.-S.; Cha, S.-W.; Hwang, B. Enhanced Production of Asiaticoside from Hairy Root Cultures of *Centella asiatica* (L.) Urban Elicited by Methyl Jasmonate. *Plant Cell Rep.* **2007**, *26,* 1941–1949.

Kim, O. T.; Kim, M. Y.; Hong, M. H.; Ahn, J. C.; Hwang, B. Stimulation of Asiaticoside Accumulation in the Whole Plant Cultures of *Centella asiatica* (L.) Urban by Elicitors. *Plant Cell Rep.* **2004**, *23,* 339–344.

Kim, O. T.; Kim, M. Y.; Huh, S. M.; Bai, D. G.; Ahn, J. C.; Hwang, B.; Cloning of a cDNA Probably Encoding Oxidosqualene Cyclase Associated with Asiaticoside Biosynthesis from *Centella asiatica* (L.) Urban. *Plant Cell Rep.* **2005**, *24,* 304–311.

Kumari, S.; Deori, M.; Elancheran, R.; Kotoky, J.; Devi, R. In Vitro and In Vivo Antioxidant, Anti-Hyperlipidemic Properties and Chemical Characterization of *Centella asiatica* (L.) Extract. *Front Pharmacol.* **2016**, *7,* 400.

Legiawati, L.; Bramono, K.; Indriatmi, W.; Yunir, E.; Setiati, S.; Jusman, S. W. A.; Purwaningsih, E. H.; Wibowo, H.; Danarti, R. Oral and Topical *Centella asiatica* in Type 2 Diabetes Mellitus Patients with Dry Skin: A Three-Arm Prospective Randomized Double-Blind Controlled Trial. *Evid Based Complement Altern. Med,* **2020**, 7253560.

Li, Y.; Kong, D.; Fu, Y.; Sussman, M. R.; Wu, H. The Effect of Developmental and Environmental Factors on Secondary Metabolites in Medicinal Plants. *Plant Physiol. Biochem.* **2020**, *148,* 80–89.

Liu, Y.-T.; Chuang, Y.-C.; Lo, Y.-S.; Lin, C.-C.; Hsi, Y.-T.; Hsieh, M.-J.; Chen, M.-K.; Asiatic Acid, Extracted from *Centella asiatica* and Induces Apoptosis Pathway through the Phosphorylation p38 Mitogen-Activated Protein Kinase in Cisplatin-Resistant Nasopharyngeal Carcinoma Cells. *Biomolecules* **2020**, *10,* 184.

Lokanathan, Y.; Omar, N.; Ahmad Puzi, N. N.; Saim, A.; Hj Idrus, R. Recent Updates in Neuroprotective and Neuroregenerative Potential of Centella asiatica. *Malays. J. Med. Sci.* **2016**, *23,* 4–14.

Ma, Y.; Wen, J.; Wang, J.; Wang, C.; Zhang, Y.; Zhao, L.; Li, J.; Feng, X.; Asiaticoside Antagonizes Proliferation and Chemotherapeutic Drug Resistance in Hepatocellular Carcinoma (HCC) Cells. *Med. Sci. Monit.* **2020**, *26,* e924435.

Mangas, S.; Bonfill, M.; Osuna, L.; Moyano, E.; Tortoriello, J.; Cusido, R. M.; Teresa Piñol, M.; Palazón, J.; The Effect of Methyl Jasmonate on Triterpene and Sterol Metabolisms of *Centella asiatica*, Ruscus Aculeatus and Galphimia Glauca Cultured Plants. *Phytochemistry* **2006**, *67,* 2041–2049.

Matsuda, H.; Morikawa, T.; Ueda, H.; Yoshikawa, M. Medicinal Foodstuffs. XXVII. Saponin Constituents of *Gotu Kola* (2): Structures of New Ursane- and Oleanane-Type Triterpene Oligoglycosides, Centellasaponins B, C, and D, from *Centella asiatica* Cultivated in Sri Lanka. *Chem. Pharma. Bull.* **2001**, *49,* 1368–1371.

Meulenbeld, G. J.; Wujastyk, D. *Studies on Indian Medical History*; Motilal Banarsidass, 2001.

Mohammad Azmin, S. N. H.; Mat Nor, M. S. Chemical fingerprint of *Centella asiatica*'s bioactive Compounds in the Ethanolic and Aqueous Extracts. *Adv. Biomarker Sci. Technol.* **2020**, *2,* 35–44.

Mukundan, U.; Shrotri, M.; Gavhane, R. In Vitro Plant Regeneration of *Centella asiatica* and Determination of Asiaticoside Using HPTLC. *J. Trop. Med. Plants* **2004**, *5,* 1–89.

Nath, S.; Buragohain, A. K. In Vitro Method for Propagation of *Centella asiatica* (L) Urban by Shoot Tip Culture. *J. Plant Biochem. Biotechnol.* **2004**, *12,* 167–169.

Nath Tiwari, K.; Chandra Sharma, N.; Tiwari, V.; Deo Singh, B. Micropropagation of *Centella asiatica* (L.), a Valuable Medicinal Herb. *Plant Cell Tissue Organ Culture* **2000**, *63,* 179–185.

Ncube, E. N.; Steenkamp, P. A.; Madala, N. E.; Dubery, I. A. Chlorogenic Acids Biosynthesis in *Centella asiatica* Cells Is Not Stimulated by Salicylic Acid Manipulation. *Appl. Biochem. Biotechnol.* **2016**, *179,* 685–696.

Ncube, E. N.; Steenkamp, P. A.; Madala, N. E.; Dubery, I. A. Metabolite Profiling of the Undifferentiated Cultured Cells and Differentiated Leaf Tissues of *Centella asiatica*. *Plant Cell Tissue Organ Culture (PCTOC)* **2017**, *129,* 431–443.

Nirmal, J. Bipul, K. B.; Anand, K. Y. Somatic Embryogenesis and Plant Development in *Centella asiatica* L.; A Highly Prized Medicinal Plant of the Tropics. *HortScience Horts* **2007,** *42,* 633–637.

Ondeko, D. A.; Juma, B. F.; Baraza, L. D.; Nyongesa, P. K. LC-ESI/MS and GC-MS Methanol Extract Analysis, Phytochemical and Antimicrobial Activity Studies of *Centella asiatica*. *Asian J. Chem. Sci.* **2020,** *8,* 32–51.

Oyenihi, A. B.; Langa, S. O. P.; Mukaratirwa, S.; Masola, B. Effects of *Centella asiatica* on Skeletal Muscle Structure and Key Enzymes of Glucose and Glycogen Metabolism in Type 2 Diabetic Rats. *Biomed. Pharmacother.* **2019,** *112,* 108715.

Park, B. C.; Bosire, K. O.; Lee, E. S.; Lee, Y. S.; Kim, J. A. Asiatic Acid Induces Apoptosis in SK-MEL-2 Human Melanoma Cells. *Cancer Lett.* **2005,** *218,* 81–90.

Pott, D. M.; Osorio, S.; Vallarino, J. G. From Central to Specialized Metabolism: An Overview of Some Secondary Compounds Derived From the Primary Metabolism for Their Role in Conferring Nutritional and Organoleptic Characteristics to Fruit. *Front. Plant Sci.* **2019,** 10.

Prasad, A.; Mathur, A.; Kalra, A.; Gupta, M. M.; Lal, R. K.; Mathur, A. K. Fungal Elicitor-Mediated Enhancement in Growth and Asiaticoside Content of *Centella asiatica* L. Shoot Cultures. *Plant Growth Regulation* **2013,** *69,* 265–273.

Prasad, A.; Mathur, A.; Singh, M.; Gupta, M. M.; Uniyal, G. C.; Lal, R. K.; Mathur, A. K.; Growth and Asiaticoside Production in Multiple Shoot Cultures of a Medicinal Herb, *Centella asiatica* (L.) Urban, Under the Influence of Nutrient Manipulations. *J. Nat. Med.* **2012,** *66,* 383–387.

Ratz-Łyko, A.; Arct, J.; Pytkowska, K.; Moisturizing and Antiinflammatory Properties of Cosmetic Formulations Containing *Centella asiatica* Extract. *Indian J. Pharm. Sci.* **2016,** *78,* 27–33.

Ren, L.; Cao, Q. X.; Zhai, F. R.; Yang, S. Q.; Zhang, H. X. Asiatic Acid Exerts Anticancer Potential in Human Ovarian Cancer Cells via Suppression of PI3K/Akt/mTOR Signalling. *Pharm. Biol.* **2016,** *54,* 2377–2382.

Sakonsinsiri, C.; Kaewlert, W.; Armartmuntree, N.; Thanan, R.; Pakdeechote, P. Anti-Cancer Activity of Asiatic Acid Against Human Cholangiocarcinoma Cells Through Inhibition of Proliferation and Induction of Apoptosis. *Cell Mol Biol (Noisy-le-grand),* **2018,** *64,* 28–33.

Satheesan, J.; Narayanan, A. K.; Sakunthala, M. Induction of Root Colonization by *Piriformospora indica* Leads to Enhanced Asiaticoside Production in *Centella asiatica*. *Mycorrhiza* **2012,** *22,* 195–202.

Schäfer, H.; Wink, M. Medicinally Important Secondary Metabolites in Recombinant Micro-organisms or Plants: Progress in Alkaloid Biosynthesis. *Biotechnol. J.* **2009,** *4,* 1684–703.

Seca, A. M. L.; Pinto, D. C. G. A. Plant Secondary Metabolites as Anticancer Agents: Successes in Clinical Trials and Therapeutic Application. *Int. J. Mol. Sci.* **2018,** *19,* 263.

Sevgi, G.; Nazım, Ş.; Current Perspectives in the Application of Medicinal Plants Against Cancer: Novel Therapeutic Agents. *Anti-Cancer Agents Med. Chem.* **2019,** *19,* 101–111.

Singh, B.; Rastogi, R. P. Chemical Examination of *Centella asiatica* Linn—III: Constitution of Brahmic Acid. *Phytochemistry* **1968,** *7,* 1385–1393.

Somboonwong, J.; Kankaisre, M.; Tantisira, B.; Tantisira, M. H. Wound Healing Activities of Different Extracts of *Centella asiatica* in Incision and Burn Wound Models: An Experimental Animal Study. *BMC Complement Altern. Med.* **2012,** *12,* 103.

Subban, R.; Veerakumar, A.; Manimaran, R.; Hashim, K. M.; Balachandran, I. Two New Flavonoids from *Centella asiatica* (Linn.). *J. Nat. Med.* **2008,** *62,* 369–373.

Sumner, L. W.; Huhman, D. V.; Urbanczyk-Wochniak, E.; Lei, Z. Methods, Applications and Concepts of Metabolite Profiling: Secondary Metabolism. *Exs.* **2007,** *97,* 195–212.

Sun, B.; Wu, L.; Wu, Y.; Zhang, C.; Qin, L.; Hayashi, M.; Kudo, M.; Gao, M.; Liu, T. Therapeutic Potential of *Centella asiatica* and Its Triterpenes: A Review. *Front. Pharmacol.* **2020,** 11.

Tan, S.; Radzali, M.; Arbakariya, A.; Mahmood, M.; Effect of Plant Growth Regulators on Callus, Cell Suspension and Cell Line Selection for Flavonoid Production from Pegaga (*Centella asiatica* L. Urban). *Am. J. Biochem. Biotechnol.* **2010,** *6,* 284–299.

Tang, X. L.; Yang, X. Y.; Jung, H. J.; Kim, S. Y.; Jung, S. Y.; Choi, D. Y.; Park, W. C.; Park, H.; Asiatic Acid Induces Colon Cancer Cell Growth Inhibition and Apoptosis Through Mitochondrial Death Cascade. *Biol. Pharm. Bull.* **2009,** *32,* 1399–405.

Tariq, A.; Mussarat, S.; Adnan, M. Review on Ethnomedicinal, Phytochemical and Pharmacological Evidence of Himalayan Anticancer Plants. *J. Ethnopharmacol.* **2015,** *164,* 96–119.

Tugizimana, F.; Ncube, E. N.; Steenkamp, P. A.; Dubery, I. A. Metabolomics-Derived Insights Into the Manipulation of Terpenoid Synthesis in *Centella asiatica* Cells by Methyl Jasmonate. *Plant Biotechnol. Rep.* **2015,** *9,* 125–136.

Vasisht, K.; Sharma, N.; Karan, M.; Current perspective in the international trade of medicinal plants material: an update. *Current pharmaceutical design,* **2016,** *22,* 4288–4336.

Winkel-Shirley, B. Flavonoid Biosynthesis. A Colorful Model for Genetics, Biochemistry, Cell Biology, and Biotechnology. *Plant Physiol.* **2001,** *126,* 485–493.

Wu, T.; Geng, J.; Guo, W.; Gao, J.; Zhu, X. Asiatic Acid Inhibits Lung Cancer Cell Growth In Vitro and In Vivo by Destroying Mitochondria. *Acta Pharm. Sin. B* **2017,** *7,* 65–72.

Xie, C.; Yu, K.; Zhong, D.; Yuan, T.; Ye, F.; Jarrell, J. A.; Millar, A.; Chen, X. Investigation of Isomeric Transformations of Chlorogenic Acid in Buffers and Biological Matrixes by Ultraperformance Liquid Chromatography Coupled with Hybrid Quadrupole/Ion Mobility/Orthogonal Acceleration Time-of-Flight Mass Spectrometry. *J. Agric. Food Chem.* **2011,** *59,* 11078–11087.

Yadav, M. K.; Singh, S. K.; Singh, M.; Mishra, S. S.; Singh, A. K.; Tripathi, J. S.; Tripathi, Y. B.; Neuroprotective Activity of Evolvulus Alsinoides & *Centella asiatica* Ethanolic Extracts in Scopolamine-Induced Amnesia in Swiss Albino Mice. *Open Access Maced J. Med. Sci.* **2019,** *7,* 1059–1066.

Yang, L.; Wen, K. S.; Ruan, X.; Zhao, Y. X.; Wei, F.; Wang, Q. Response of Plant Secondary Metabolites to Environmental Factors. *Molecules* **2018,** 23.

Yoo, N. H.; Kim, O. T.; Kim, J. B.; Kim, S. H.; Kim, Y. C.; Bang, K. H.; Hyun, D. Y.; Cha, S. W.; Kim, M. Y.; Hwang, B. Enhancement of Centelloside Production from Cultured Plants of *Centella asiatica* by Combination of Thidiazuron and Methyl Jasmonate. *Plant Biotechnol. Rep.* **2011,** *5,* 283–287.

Yu, Q. L.; Duan, H. Q.; Takaishi, Y.; Gao, W. Y. A Novel Triterpene from *Centella asiatica*. *Molecules* **2006,** *11,* 661–665.

CHAPTER 3

The Anticancer Activity of *Nigella sativa* on Hepatocellular Carcinoma

IFFAT ZAREEN AHMAD[1] and HEENA TABASSUM[1,2]

[1]*Department of Bioengineering, Integral University, Dasauli, Lucknow, Uttar Pradesh, India*

[2]*Dr. D. Y. Patil Biotechnology and Bioinformatics Institute, Pune, Maharashtra, India*

ABSTRACT

Nigella sativa L. is a medicinal seed, commonly used as spice in many Indian cuisines including pickles, sauces, pulao, lentils, and vegetables. Its seed is a rich storehouse of several medicinal bioactive compounds, of which thymoquinone and thymol are the dominant ones. It has been shown to be highly effective in almost all the ailments and disorders. Since it is used in food, it is an important nutraceutical product and a source of dietary antioxidants. Hepatocellular carcinoma is among the leading cancers all around the globe and its mortality rate is very high. Besides, there is no treatment in advanced stages and whatever treatments are available, they show severe side effects. Several natural products have shown effects on hepatocellular carcinoma as both preventive and curative drugs. *N. sativa* has been explored extensively as an antihepatocellular cancer agent and it has a high potential to inhibit the growth. The studies have shown that *N. sativa* has a prospective effect against liver reperfusion and also on cholestatic liver damage; therefore, it could be a powerful cancer preventive agent. The researchers have given several mechanisms of its action on liver cancer cells. The receptors present on the liver cancer cells have been targeted to develop drugs. There is a huge scope

Potent Anticancer Medicinal Plants: Secondary Metabolite Profiling, Active Ingredients, and Pharmacological Outcomes. Deepu Pandita and Anu Pandita (Eds.)
© 2024 Apple Academic Press, Inc. Co-published with CRC Press (Taylor & Francis)

for the thymoquinone in drug designing and development so that it can be established into a commercialized drug that can specifically target the liver cancer cells.

3.1   INTRODUCTION

Nigella sativa is a herbal plant that produces flowers and grows up to 20–30 cm annually. It has finely divided green leaves with delicate pale blue and white blossoms. It is commonly grown in Middle Asia, Middle East, and North Africa for its aesthetic and ornamental value. The scientific term for the seeds is derived from Latin "Niger" meaning "black" (Aggarwal et al., 2009). Both the seed and its oil are exploited for the cure of various ailments including that of diuretic, lactagogue, vermifuge, and carminative.

3.2   *NIGELLA SATIVA* AND ITS BIOACTIVE MOLECULES

According to several studies, the seeds and oil obtained from *N. sativa* contain over 100 different bioactive compounds with healing potential. Out of these 100 compounds only 69 are well identified and characterized. It has been reported that the combined effect of these compounds is responsible for the promotion of good immune response. It is a noteworthy resource of carbohydrates, proteins, fatty acids, minerals, and vitamins. The *N. sativa* seeds comprise large quantities of sterols; particularly beta-sitosterol that is well recognized anticancer compound (Khan, 1999) as it is known for the presence of unsaturated esters of fatty acids along with alcohol of terpene. Moreover, alkaloids are reported which belong to two categories of alkaloids, that is, isochinoline alkaloids (nigellimin-N-oxide and nigellimin) and pyrazol alkaloids (nigellicin and nigellidin) (Yessuf, 2015). The compound, TQ, was recognized as the pivot component reported to be the constituent of the essential oil (about 0.5% of average with maximum 1.5%) besides pinene, p-cymene, diTQ, and thymohydroquinone. Similarly, some of the terpene derivatives such as limonene, carvacrol, citronellol, carvone, and 4-terpineol were found only in trace amounts. The seeds of *N. sativa* are the essential wellspring of unsaturated fatty acids chain in highest concentration, commonly including linoleic acid (50–60%), followed by about 20% of oleic acid, dihomolinoleic acid (10%), and eicodadienoic acid (3%). However, the amount of saturated fatty acid is to be present around 30%, with TQ as an ingredient of the essential oil which is responsible for a peculiar aroma and flavor. The seeds yield yellowish-brown volatile oil steam distillation with a characteristic smell and are reported to have the constituents of d-limonene, nigellone, and carvone (Yessuf, 2015).

Thymoquinone (TQ), thymohydroquinone (THQ), and thymol (THY) are three natural phytochemical compounds having tremendous medicin6+al significance as reported in previous research (Hussain and Hussain, 2016). Saponin, alkaloids, protein, and both essential and fixed oils are found in the blackseed (Ali and Blunden, 2003). Ghosheh et al. (1999) described the determination of certain medicinally significant constituents in blackseed oil, containing THY, diTQ, TQ, dihydrothymoquinone (DTQ), and THQ, using HPLC (High Performance Liquid Chromatography). Alkaloids, antioxidants, flavonoids, α-hederin, and fatty acids are among the other therapeutic compounds present in these seeds.

3.3 PHARMACOLOGICAL PROPERTIES OF THYMOQUINONE AND THYMOL FOR LIVER DISEASES

TQ is known to possess medicinal and toxic effects (Ali and Blunden, 2003). It shows exceptional anti-inflammatory and antioxidant properties. It also exhibits incredible anticancer potential. It has the capability to act together with a range of proteins and it can also inhibit protein–protein interaction (Reindl et al., 2008). It has the capacity to bind with the phosphoserine/phosphothreonine, α1-acid glycoprotein (orosomucoid), human bovine serum albumin (BSA), and recognition site of polo-box domain (PBD) (Yin et al., 2012). It inhibits the activity of polo-like kinase 1 (Plk1) and therefore prevents its localization by stopping the PDB-dependent association (Reindl et al., 2008). Some significant bioactivities of TQ and THY are highlighted in this chapter. Their mechanisms of action on hepatocellular carcinoma have also been discussed.

3.4 HEPATOCELLULAR CARCINOMA (HCC)

Hepatocellular carcinoma (HCC) is the most prevalent type primary cancer around the globe; however, high incidence of alcoholism, hepatitis infection, aflatoxin, and its side effects are the common reasons of liver cancer-related death. More than 100 out of 100,000 people have been affected by HCC in various regions of Asia and Africa. Secondary liver cancers are more common in Europe and the United States, whereas in Asia and Africa, primary liver cancer is more frequent. HCC is at number two and is most commonly associated with cancer-related death around the globe, with 746,000 deaths. It occurs more frequently in male rather than in females (with the ratio of about 2.4:1) (Torre et al., 2015). The diagnosis for liver malignancy is extremely poor with a total frequency of mortality of about 0.95, and in this way, the geological examples on the basis of occurrence and death are similar. Therefore, more effective approaches for the treatment of cancer are urgently required. The most promising approach is the application of the well-researched medicinally active compounds from plant sources (Cragg and Pezzuto 2016). Traditionally, medicinal plants have been the valuable source for many presently available cancer therapeutic agents such as vincristine, paclitaxel (Taxol®, Bristol-Myers Squibb Company, NY, USA), docetaxel, and topotecan (Newman and Cragg 2007). Of late, another natural compound that has been extensively studied and has been of interest to researchers is thymoquinone because of its broad-spectrum biological activities (Lutterodt et al., 2010), hepatoprotective, anticancer, antitumor, antimutagenic, and antibacterial (Woo et al., 2013).

Plant-derived compounds have potential anticancer activity, and can be developed into commercial drug (Banerjee et al., 2010). *N. sativa* is a flowering annual plant indigenous to the Mediterranean, India, and Pakistan (Poonia et al., 2016). In the Arabian region, its seed oil has been used conventionally as a natural medicine for the cure of lung-related illness, arthritis, and hypercholesterolemia (Pathak and Raghuvanshi, 2015). *N. sativa* has been reported to have antihypertensive, analgesic, diuretic, antibacterial, and liver-protective activities (Ahmad et al., 2003). However, the detailed scientific investigations are deficient. Previous research reported the inhibitory capacity of the extract of the *N. sativa* to inhibit the progression and spread of liver cancer cell (HepG$_2$). It is also demonstrated that *N. sativa* employs its anticancer properties through various mechanisms of action involving cell growth inhibition, apoptosis induction, arrest of cell cycle at different stages, ROS, as well as the fundamental molecular mechanism causing death of cells (Banerjee et al., 2010). Thymoquinone, the most active component of N. sativa, is consumed generously as a flavoring agent in many cuisines. In case of human breast cancer cell line, TQ has shown to suppress the growth and progression by inducing p38 and ROS signaling (Woo et al., 2013).

Liver carcinoma is one of the most familiar types of disease around the globe, and HCC comprises the bulk of them (70%– 90%) of them (Venook et al., 2010). HCC is a disease that is a challenging health problem due to the gradual rising rate of occurrence universally (Venook et al., 2010). The investigations by scientists have shown a significant reperfusion effect against hepatic cells, thereby making it notably a potential agent toward the cancer prevention (Yildiz et al., 2008). The researchers discovered in animal models that *N. sativa* had a possible effect of cholestatic liver damage (Coban et al., 2010). In CCl4-administered rats, *N. sativa* and *Urtica dioica* L show suppressed level of lipid peroxidase and liver enzymes and were effective cancer preventative agents (Kanter et al., 2005). A study has also reported the protective ability of Thymoquinone (TQ) on I/R in liver and decreased SSAT and CYP3A1 gene expression damage through a cancer preventive system (Abdel Wahab, 2013). It has been reported *N. sativa* is involved in stimulating the apoptosis cell signaling pathway by inducing the target p38 (Iyoda et al., 2003).

3.5   EFFECT OF BIOACTIVE MOLECULES ON HEPATOCELLULAR CARCINOMA

3.5.1   THYMOQUINONE

TQ contributes actively to the regulation of biochemical reactions and physiological processes that result in the ROS generation (Yu and Kim, 2015). It is an efficient scavenger of free radicals (Mansour et al., 2004). After interacting with amino acids and forming semiquinone, it undergoes numerous redox reactions that involve 1e-reduction and the other 2e-reduction with thymohydroquinone.

It undergoes several redox reactions after reacting with amino acids and with the synthesis of semiquinone: 1e⁻ reduction and thymo hydroquinone and 2e− reduction (Cremer et al., 1987). It also behaves as a potent anti-oxidant and effectively inhibits lipid peroxidation, producing radicals of superoxide (Woo et al., 2012; Banerjee et al., 2010). It effectively increases the activity of many antioxidant enzymes including superoxide dismutase (SOD), catalase (CAT), glutathione transferase, and glutathione (GSH). TQ produces two powerful antioxidants: glutathionyl-dihydro thymoquinone (DHTQ-GS) and DHTQ, upon reaction with antioxidant enzymes (GSH, NADH and NADPH). Taking into consideration the cellular microenvironment, TQ plays an essential function as both a pro-oxidant and an anti-oxidant. According to a study conducted by Badary et al. (2003), TQ was shown as a potent superoxide anion scavenger. TQ is capable of reversing the expression of mRNA and diethylnitrosamine-associated reduction in various antioxidant enzyme levels (Sayed-Ahmed et al., 2010). Extensive investigations have been aimed at determining the effect of TQ on tripeptide glutathione, which is widely used in the prevention of the damage caused by reactive oxygen species (ROS) and in the detoxification of medications (Pompella et al., 2003). TQ has also been shown to be helpful in the recovery of female Lewis rats after they had suffered from allergic encephalomyelitis (Mohamed et al., 2003).

TQ extracted from *N. sativa* incited caspase-3 via stimulation of cytochrome C release from mitochondria in squamous cell cancer (Das et al., 2012). Similar outcomes were observed in HL60 cells during apoptosis elicited by TQ (El-Mahdy et al., 2005). According to a study, TQ showed the decrement in the gene expression of SSAT and CYP3A1 while demonstrating a defensive activity against renal reperfusion-related injury or ischemia by an antioxidant defense system (Yildiz et al., 2008). The protective effect by a cancer preventive agent on renal ischemia/reperfusion-associated injury was shown (Yildiz et al., 2008). TQ also showed protection against tertbutyl hydroperoxide toxicity as reported in rodent hepatic cells. As a result of these observations, it was discovered that pretreatment of hepatic cells with TQ decreased the transit of cytoplasmic proteins, ALT, and AST (Daba and Abdel-Rahman, 1998). Owing to its antioxidant property, the in vivo administration of a single TQ dose to male albino rat initiated a protective effect toward CCl_4-induced hepatotoxicity (Sayed-Ahmed et al., 2010). In another study, the defensive role of TQ was evaluated on aflatoxin B1-associated hepatic toxicity in mice. This study showed that TQ totally diminished the level of ALP, ALT, MDA, and AST. This protective impact might be facilitated through a decrease in lipid peroxidation and also an increased tolerance to oxidative stress (Nili-Ahmadabadi et al., 2011). TQ demonstrated defensive activity against lipopolysaccharide-induced endotoxemia because of its anti-inflammatory, antioxidant, and anti-apoptotic activities (Helal, 2010). The protective activity of TQ on sodium fluoride-associated hepatic toxicity and oxidative stress in animals improved the antioxidant state and decreased the changes in biochemical factors. Oral administrations of TQ ameliorated the sodium fluoride-induced toxicity and stress in rats' liver probably by reducing the level of peroxidation and/or improving the activities of enzymatic and nonenzymatic antioxidants of the liver (Abdel-Wahab, 2013). Studies also showed that TQ markedly reduced the tamoxifen-induced depletion of hepatic glutathione and regulated SOD activity (Suddek, 2014). TQ worked against the hepatotoxicity induced by the acetaminophen and the studies reported reduced toxicity toward the liver by further decreasing the ALT activity in serum which was dependent on dose. Enhancement in tolerance to oxidative and nitrotative stress, as well as improved energy generation in mitochondria, was shown to assist the protective impact of TQ on hepatic cell viability (Nagi et al., 1999). The activity of ethanolic extracts of *Nigella sativa, Zingiber officinale*, and their combination in individuals infected with the hepatitis C virus were shown in clinical trials (HCV). TQ showed potential activity against I/R injury and neuronal damage. In addition to this, TQ also inhibited histological impairment, inflammation, and oxidative stress. Interestingly, it decreased the ER stress parameters expression that included caspase-12, CHOP, and GRP78. Along with this, it also increased mitochondrial functions and restricted the expression of apoptotic parameters. Additionally, TQ enhanced phosphorylation of ERK and P38 (Bouhlel et al., 2017). TQ reduced the combination of TNF-alpha, MCP-1, Cox-2, and interleukin (IL)-1b in pancreatic ductal adenocarcinoma (PDA) cells substantially, which is both dose and time dependent. TQ fully suppressed the production of the cytokines listed above, as well as the intrinsic activity of the MCP-1 promoter. It also inhibited NF-kB activation and decreased NF-kB transport to nucleus from the PDA cells (Chehl et al., 2009). TQ-induced apoptosis, as determined by flow cytometry and colorimetric estimate of Caspases 3 and 9, showed that an early G1/S arrest was typical for apoptosis (Hassan et al., 2010). The hepatoprotective effect of TQ on damaged liver through AMP-activated protein kinase (AMPK) signaling in HSCs was demonstrated. The data of in vitro study showed the time-dependent attenuation of Liver kinase B-1 (LKB1) by TGF-β and AMPK phosphorylation, which slowed down by pretreatment with an AMPK activator including 5-Aminoimidazole-4-carboxamide ribonucleotide (AICAR) and TQ. It can also considerably enhance the expression of MMP-13 while suppressing α-SMA, collagen-I, and tissue inhibitor metalloproteinase-1 (TIMP-1), to prevent the initiation of human HSCs induced via TGF-β. Furthermore, TQ increased the expression of the peroxisome proliferator activated receptor-γ (PPAR- γ), which was inhibited by a genetic deletion of the AMPK gene. In in vivo research on mice fed with ethanol, TQ significantly reduced elevated serum hepatic triglyceride and amino transferase levels, while also significantly increasing AMPK phosphorylation and LKB1 levels and further inducing an enhanced expression of the sirtuin 1 (SIRT1) gene (Yang et al., 2016). Different mechanisms have been shown in Figure 3.1.

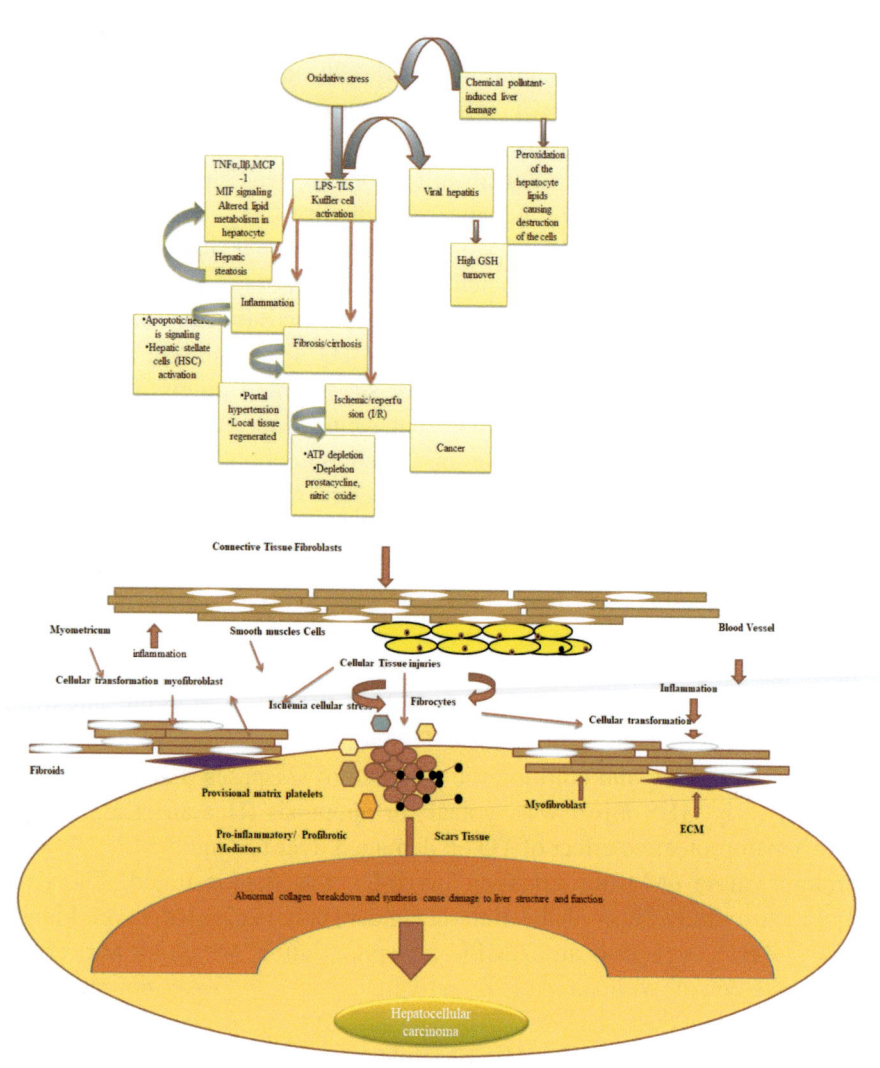

FIGURE 3.1   The molecular targets of thymoquinone on hepatocellular carcinoma cells.

3.5.2   THYMOL

Thymohydroquinone and TQ exhibited restricted inhibition of specifically COX-2 which pointed to the role of the *N. sativa* constituents (diTQ, THQ, THY, and TQas anti-inflammatory agents) (Marsik et al., 2005). With an IC_{50} value of 0.2 μM, THY shows the best inhibitory activity than TQ and thymohydroquinone against COX-1 with 0.1 and 0.3 μM of IC_{50}, respectively.

THQ showed cytotoxicity against human hepatic cells, that is, HepG$_2$ cells (Stammati et al, 1999), P815 mastocytoma cells (Jaafari et al., 2007), HepG$_2$, Caco-2, and V79 cells (Slamenova et al., 2007). THY showed antioxidative and cytotoxic potential against P388 cell line (Gedara, 2008; Hirobe et al., 1998). Jayakumar and collaborators (Jayakumar et al., 2012; Ozkan and Erdogan, 2011) demonstrated THY to be toxic to HepG$_2$, K562 cells, and colonic Caco-2 cells, believed to be associated with antioxidant action rather than the effect of DNA damage. HepG$_2$ cells were treated with both carvacrol and thymol, while the refence drug was taken at various doses (N-acetylcystene (NAC)). The data revealed that acetaminophen (APAP) effected the growth of HepG$_2$ cells by inflammation and oxidative stress. Carvacrol and THY treatment of HepG2 cells by exposing to acetaminophen for 24 hours reduced oxidative stress and inflammation. In addition, antioxidant enzymes such as ALT and LDH were studied in HepG2 cells and reported to provide protection toward the toxicity induced through APAP by improving antioxidant activity and decreasing pro-inflammatory cytokines, such as IL 1β, and TNF-α. In APAP toxicity, high-dosage THY and carvacrol were nearly identical to NAC therapy and shown increased effectiveness when compared to carvacrol (Palabiyik et al., 2016). A schematic representation of the effects of THY in different experimental models of cancer is shown in Figure 3.2.

3.6 MECHANISMS OF HEPATOPROTECTIVE ACTIVITY OF *N. SATIVA* AND ITS IMPORTANT BIOACTIVE COMPOUNDS

3.6.1 THE SIGNALING MECHANISMS UNDERLIE THE ANTICANCER PROPERTIES OF N. SATIVA AND ITS PRINCIPAL COMPONENT THYMOQUINONE

The disease of the liver encompassed several other cancerous cell signaling pathways including Notch (Strazzabosco and Fabris, 2012; Villanueva et al., 2012; Viatour, 2011), insulin-like growth factor 1/ IGF1 Receptor (IGF/ IGFR) (Enguita-Germán and Fortes, 2014; Sprinzl et al., 2015), c-Met/ HGF (hepatocyte growth factor) (You et al., 2011; Goyal et al., 2013), and epidermal growth factor receptor (EGFR) signaling pathway (Berasain and Avila, 2014). The biochemical mechanisms behind TQ's defence against HCC were unknown until it was discovered that the Notch signaling pathway is significantly suppressed by TQ (Ke et al., 2015). Overexpression of Notch 1 (NICD), the notch receptors resulted in the enhanced TQ's inhibitory effect on proliferation of the cells. Such evaluations were corroborated both in vivo

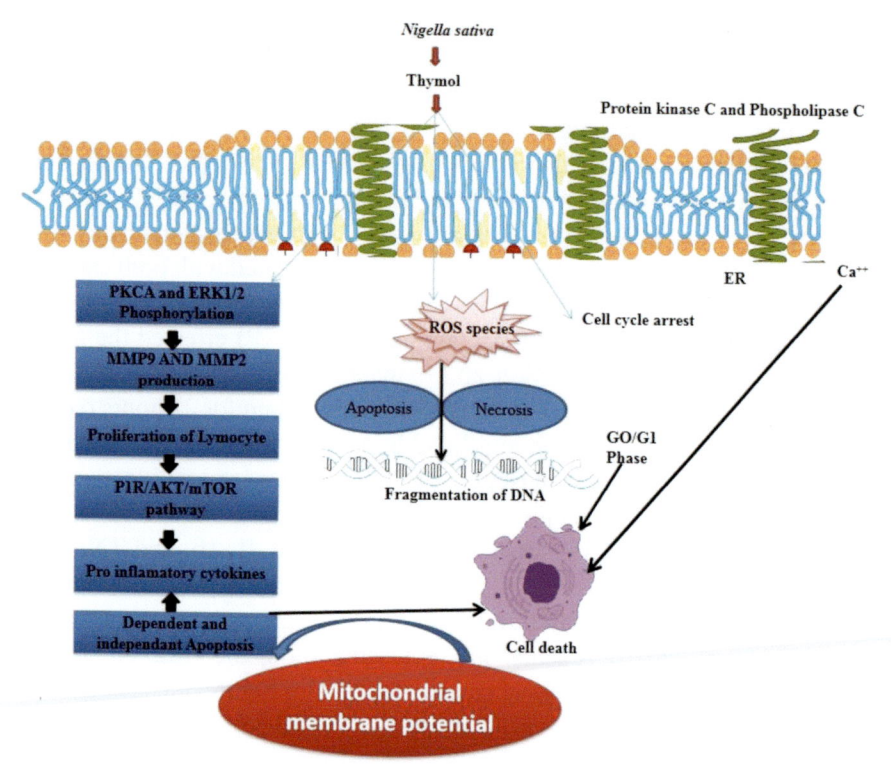

FIGURE 3.2 Schematic representations of the effects of thymol on different experimental models of hepatocellular carcinoma.

and in vitro studies in HCC patients where the notch pathway was targeted (Ke et al., 2015). Furthermore, TQ's efficacy was further investigated in murine leukemia, which further revealed the dose- and time-dependent manner reduction of the viability of WEHI-3 cells and also reported to induce apoptosis at an early stage accompanying arrest at the phase of G1/S of cell cycle, Bcl-2 downregulation, and Bax upregulation resulting in an enhanced ratio of the Bax/Bcl-2 (Salim et al., 2014). When administered orally to BALB/c mice, TQ substantially decreased neoplastic cells proliferation and elevated apoptosis in the liver and spleen, according to in vivo research. The activation of caspase 3 and the downregulation of XIAP showed that TQ had a strong pro-apoptotic activity. It also resulted in higher expression of p53 and p21, which caused arrest in G2/M and sub-G1 cell cycle phases. *N. sativa* oil along with TQ was also reported to significantly decrease 5-LO activity, which results in the powerful activity against cancer (El-Dakhakhny et al., 2002; Houghton et al., 1995; Mansour and Tornhamre, 2004; El-Mezayen et al., 2006). TQ has also shown to be an effective HDAC inhibitor, as evidenced by the activation of target genes of HDAC such as Maspin, p21, and Bax, and further reported to downregulate Bcl2 expression, as these are involved in inducing programmed cell death and arresting the cell cycle. The experiments were carried out, both in vivo and in vitro to evaluate the prospective of TQ along with Tamoxifen, to modulate the signaling of the control AKT in the cancer cells of breast (Rajput et al., 2013).

TQ-treated PEL cells were reported to associate with structural variation in Bax and further lead to MMP loss and excretion of Cyt C from the mitochondria to cytosol which results in elevated activity of Caspase-3/9, which further leads to the PARP cleavage, while the presence zVAD-FMK or NAC repels its effect (Hussain et al., 2011). A dose-dependent expression of prostaglandin E2 (PGE2) and COX-2 expression was reported to be induced in the MDA-MBA-231 cell line (Yu and Kim, 2013). SB203580 and LY294002 replicated the TQ's effect, shows a significant inhibition of p38 and P13K, respectively, which is related to the positive regulation of both PGE2 and COX-2 (Yu and Kim, 2013).

In another study a combination of paclitaxel and TQ demonstrated a significant triple activity in the cell lines of breast both in vitro in in vivo models (Sakalar et al., 2016). Another work suggested upregulation of genes involved in tumor suppressor (like BRCA1, hic 1 and p21) and increases caspases and PARP levels by TQ. TQ-induced autophagic cell death was halted by p38 and JNK inhibitors, indicating that TQ activated p38 and JNK signaling to cause the death of LoVo cells via autophagy (Chen et al., 2015). TQ-mediated autophagy was proven to be the result of caspase-independent activity, since co-treatment of the inhibitor of caspase-3 (z-DEVD-FMK) had no effect on cell death via autophagy induced by TQ (Chen et al., 2015). These results showed caspase-dependent apoptosis as the main type of cell death at low TQ treatment and it switches to caspase-independent autophagy at high concentrations of TQ treatment. Although TQ was shown to be the main anticancer compound it was established that some compounds other than TQ also significantly demonstrate the anticancer effects of *N. sativa* against HepG$_2$ cells (Samarakoon et al., 2010). Also, saponins, α-hederin, and a pentacyclic triterpene are all isolated from *N. sativa* seeds and reported to exhibit strong anticancer activities both in vitro and in vivo (Swamy and Tan, 2000, Cheng et al., 2014). Other bioactive compounds present in *N. sativa* which have demonstrated good cytotoxic and anticancer potential include thymol (THY), Thymohydroquinone (THQ), dithymoquinone (DTQ), carvacrol, nigellidine, nigellicine, and nigellimine-N-oxide (Randhawa and Alghamdi, 2011; Horvathova et al., 2014).

3.6.2   p53 SIGNALING PATHWAY

The anticancer and pro-apoptotic activities of TQ have been provided which involves its capability to alter the expression of many target proteins of cell cycle. The molecular mechanisms and signaling pathways involved in the TQ-associated anticancer activity have been revealed. It was shown that noncytotoxic concentrations of TQ markedly reduced the growth of papilloma and spindle carcinoma cells derived from mouse keratinocytes (Gali-Muhtasib et al., 2004). Treatment with TQ treatment was related to suppression of Bcl-2 expression and upregulation of p53 and its target p21, resulting in less CDK2 activity. Pifithrin-alpha (PFT-alpha) showed the role of p53 in TQ-induced cell cycle arrest and programmed cell death and the action of TQ on the expression level of Bcl2, p21, and p53 was studied on HCT-116 cells (Gali-Muhtasib et al., 2004). It was also shown that TQ was active on human breast cancer MCF-7/DOX cells by forcing G2/M phase cell cycle arrest resulting in damage of the DNA and thus triggering apoptosis (Arafa et al., 2011). The pro-proliferative and anti-apoptotic activity associated with TQ causes the enhanced p21 and p53 expression along with elevated PARP cleavage and caspase activation (Arafa et al., 2011). Furthermore, the activity also involves the downregulation of p53, BRCA1, and BRCA2 (Linjawi et al., 2015) along with the upregulated expression of Bax and a downregulated of Bcl-2, resulting in increase of Bax/Bcl-2 ratio (Arafa et al., 2011).

3.6.3   MAPK SIGNALING PATHWAY

TQ-induced apoptosis in five distinct human colon cancer cell lines was investigated for the direct connection between oxidants and MAPK signaling pathways (LoVo, HCT-116, Caco-2, HT-29, and DLD-1). TQ, interestingly, reduced the development of all cell lines of cancer examined with no impact on normal intestinal cell lines of human (El-Najjar et al., 2010). TQ-induced apoptosis was further shown by the substantial production of ROS, as treatment with N-acetyl cysteine (NAC), a powerful ROS scavenger, nearly entirely abolished TQ-induced apoptosis (El-Najjar et al., 2010). TQ increased activation of JNK and ERK but not p38 MAPK, and treatment with specific JNK and ERK inhibitors (SP600125 and PD98059) inhibited TQ-induced apoptosis (El-Najjar et al., 2010). Similarly, TQ treatment of HepG2 cells resulted in a substantial rise in ROS levels as well as the expression of NQO1 and HO-1 (oxidative stress-related genes), leading to apoptosis (Ashour et al., 2014). As a consequence of TQ use in HTLV-1 negative cells of Jurkat leukemia causes depletion of glutathione, cytochrome c release, loss of MMP, increase in ROS levels, activation of caspase-3/9, and PARP cleavage (Dergarabetian et al., 2013). TQ was found to potently promote apoptosis and ROS production in a time- and dose-dependent manner, with NAC co-treatment completely removing these effects (Yu and Kim, 2013). Furthermore, TQ administration increased the activity of JNK, p38, PI3K, and ERK1/2 (Yu and Kim, 2013). The findings indicated that TQ primarily targets MAPK and PI3K signaling pathways to mediate its pro-apoptotic actions.

3.6.4 HEPATOPROTECTIVE DISEASES AND REACTIVE OXYGEN SPECIES (ROS)

The liver is the body's central metabolic organ and is accountable for a variety of vital activities. The danger to the body's metabolic system may be life-threatening if the liver is damaged or injured. Antioxidants have been shown to delay hepatotoxicity by inhibiting lipid peroxidation, producing ROS, and lowering the activity of alkaline phosphatase (ALP), alanine aminotransferase (ALT), and aspartate aminotransferase (AST) (Yen et al., 2007; Park et al., 2012; El-Shafey et al., 2015). Antioxidant activities have been found in the bioactive components of *N. sativa* seed. The majority of liver damage is caused by the production of ROS, which depletes antioxidant enzyme sources and reduces the capacity of cells to resist harm (Muriel, 2009). The three enzymes that generate ROS in the body are myeloperoxidases, nitric oxide synthases (NOS), and NADPH oxidases. Different kinds of ROS generated as by-products of cellular metabolism include hydroxyl radical (OH-), superoxide anion radical (O2-), hypochlorous acid (HOCl-), peroxynitrite (ONOO-), and hydrogen peroxide (H2O2). Oxidative stress is a condition caused by an imbalance in the quantity of ROS and antioxidants in the body. Free radicals may interfere with cellular activities by disrupting the structural properties of polyunsaturated lipids and sulfhydryl groups (Muriel, 2009). Because the liver's metabolism is so fast, it is very vulnerable to oxidative stress and free radical damage. The negative consequences of ROS may be avoided by reducing the formation of free radicals, scavenging free radicals, or improving the antioxidant defense system (Tsukamoto and Lu 2001).

3.7 FUTURE PROSPECTS OF *NIGELLA SATIVA* SEEDS IN DRUG DISCOVERY

N. sativa seed is extensively studied for a broad range of medicinal properties (Darakhshan et al., 2015). It has shown high medicinal potential for the prevention and cure of many ailments (Ahmad et al., 2013). It has favorable characteristics of a potent drug candidate which is in the process of establishment as a new well-established drug (Dajani et al., 2016). Its bioactive compound, thymoquinone, is a widely studied compound and has crossed many critical steps involved in drug development like preclinical and clinical research and it has been reported by many scientists working on this compound (Goyal et al., 2017; Almajali et al., 2021).

Many analogs of thymoquinone have been developed in different studies that have shown incredible efficiency toward various classes of cancers. Recently, several cytotoxic terpenes, monoterpenes, and sesquiterpenes were combined with novel TQ analogues to see whether they might improve cancer-fighting efficacy (Banerjee et al., 2010).

Pharmacokinetics studies of thymoquinone have been done comprehensively on different animal models. Different doses of thymoquinone were administered to animals, by various routes including intraperitoneal, intravenous, and intragastric routes to study its efficacy in disease models (Alkharfy et al., 2015). The different factors of pharmacokinetics study in the blood samples were evaluated using high-performance chromatography extracted from rabbits (Alkharfy et al., 2015). Thymoquinone bound to proteins at a rate of 99%, with a sophisticated quick eradication and reduced absorption after oral administration. By considering the overall analysis, thymoquinone has demonstrated a substantial prospective for its clinical usage as an important drug for pharmaceutical development and commercialization. Moreover, thymoquinone has shown good solubility and bioavailability also which a common limitation in drug discovery and development is. So, it could be an effective drug if targeted drug delivery is done by adopting newer technologies, mainly nanotechnology (Elmowafy et al., 2016).

Concisely, it can be said that the wide range of medicinal properties, good pharmacokinetics, security and low toxic effects, the lipophilic nature, high therapeutic index, efficiency, and bioavailability make thymoquinone a favorable compound for the drug development. Apart from these, the exceptional probable benefit of this compound is its natural origin and its clinical trials will lead to the transformation of the experimental data into actuality in human being.

3.8 CONCLUSIONS

Autophagy is a significant mechanism of cell death that interacts closely with other prevalent programs of cell death that include apoptosis, to produce a cascade of events. It is a promising, but challenging, medication target for cancer, because autophagy plays a critical role in cell death. As a result, researchers tried to figure out whether autophagy and apoptosis were involved in the cancer cell inhibitory process. HepG2 cells were treated with a 5d sprout extract of N. sativa-loaded NLC, which caused cell cycle arrest, autophagy, mitochondrial membrane depolarization, and apoptosis. The data suggested that understanding of conventional medicinal systems can be used to investigate, recognize, develop, and take new medicinal compounds to commercial level that can be used safely and efficiently. Research into the potential benefits of nano-formulation in target-based curative uses such as anticancer drugs may further strengthen its use in these applications. As the mechanism of action, apoptosis or autophagy appears to be necessary machinery for cell death, and it is currently an objective in cancer treatment.

There has been considerable interest in the use of natural products as potential therapeutic agents for the treatment of human malignancies, owing to their low toxicity toward normal cells and the fact that they are nontoxic to cancer cells. In this respect, herbal extracts from various sections of plants have been thoroughly investigated for their possible anticancer properties (Zhou et al., 2009). TQ is the most important biological component extracted from *Nigella sativa*. The presence of active phytoconstituents such as TQ and THY in the methanolic extracts may explain the activity shown by the extracts. TQ and THY are found in the methanol soluble part of *N. sativa* oil, which is derived from the plant (Abou Basha et al., 1995). When TQ was shown to be the primary active component of black seed essential oil, researchers were encouraged since it has been shown to exhibit potential antineoplastic growth inhibition against a variety of tumor cell lines in vitro and in vivo, respectively (Salomi et al., 1992; Worthen et al., 1998; Shoieb et al., 2003; Gali-Muhtasib et al., 2004; Edris, 2004; El-Mahdy et al., 2005; Gali-Muhtasib et al., 2008). This action may be related to the drug's capacity to induce apoptosis as well as its inhibitory effects on the development of malignant cells (Gali-Muhtasib et al., 2004).

KEYWORDS

- *Nigella sativa*
- **thymoquinone**
- **thymol**
- **hepatocellular carcinoma**
- **drug discovery**

REFERENCES

Abdel-Wahab, W. M. Protective Effect of Thymoquinone on Sodium Fluoride-Induced Hepatotoxicity and Oxidative Stress in Rats. *J. Basic Appl. Zool.* **2013,** *66,* 263–270.

Aggarwal, B. B.; Kunnumakkara, A. B. *Molecular Targets and Therapeutic Uses of Spices: Modern Uses for Ancient Medicine*; World Scientific Publishing Co, 2009; 430 p.

Ahmad, A.; Husain, A.; Mujeeb, M. A Review on Therapeutic Potential of *Nigella sativa*: A Miracle Herb. *Asian Pac. J. Trop. Biomed.* **2013,** *3,* 337–352.

Ali, B. H.; Blunden, G. Pharmacological and Toxicological Properties of Nigella sativa. *Phytother. Res.* **2003,** *17,* 299–305.

Alkharfy, K. M.; Ahmad, A.; Khan, R. M.; Al-Shagha, W. M. Pharmacokinetic Plasma Behaviors of Intravenous and Oral Bioavailability of Thymoquinone in a Rabbit Model. *Eur. J. Drug Metab. Pharmacokinet* **2015,** *40,* 319–323.

Almajali, B.; Al-Jamal, H. A. N.; Taib, W. R. W.; Ismail, I.; Johan, M. F.; Doolaanea, A. A.; Ibrahim, W. N. Thymoquinone, as a Novel Therapeutic Candidate of Cancers. *Pharmaceuticals* **2021,** *14,* 369.

Arafa, E. S. A.; Zhu, Q.; Shah, Z. I.; Wani, G.; Barakat, B. M.; Racoma, I.; El-Mahdy, M. A.; Wani, A. A. Thymoquinone Up-Regulates PTEN Expression and Induces Apoptosis in Doxorubicin-Resistant Human Breast Cancer Cells. *Mutat. Res.* **2011,** *706,* 28–35.

Ashour, A. E.; Abd-Allah, A. R.; Korashy, H. M.; Attia, S. M.; Alzahrani, A. Z.; Saquib, Q.; Bakheet, S. A.; Abdel-Hamied, H. E.; Jamal, S.; Rishi, A. K. Thymoquinone Suppression of the Human Hepatocellular Carcinoma Cell Growth Involves Inhibition of IL-8 Expression, Elevated Levels of TRAIL Receptors, Oxidative Stress and Apoptosis. *Mol. Cell Biochem.* **2014,** *389,* 85–98.

Banerjee, S.; Azmi, A. S.; Padhye, S. Structure-Activity Studies on Therapeutic Potential of Thymoquinone Analogs in Pancreatic Cancer. *Pharm. Res.* **2010,** *27,* 1146–1158.

Banerjee, S.; Padhye, S.; Azmi, A.; Wang, Z.; Philip, P. A.; Kucuk, O.; Sarkar, F. H.; Mohammad, R. M. Review on Molecular and Therapeutic Potential of Thymoquinone in Cancer. *Nutr. Cancer* **2010,** *62,* 938–946.

Berasain, C.; Avila, M. A. The EGFR Signalling System in the Liver: from Hepatoprotection to Hepatocarcinogenesis. *J. Gastroenterol.* **2014,** *49,* 9–23.

Bouhlel, A.; Mosbah, I. B.; Abdallah, N. H.; Ribault, C.; Viel, R.; Mannaï, S.; Corlu, A.; Abdennebi, H. B. Thymoquinone Prevents Endoplasmic Reticulum Stress and

Mitochondria-Induced Apoptosis in a Rat Model of Partial Hepatic Warm Ischemia Reperfusion. *Biomed. Pharmacother.* **2017**, *94*, 964–973.

Chehl, N.; Chipitsyna, G.; Gong, Q.; Yeo, C. J.; Arafat, H. A. Anti-Inflammatory Effects of the Nigella Sativa Seed Extract, Thymoquinone, in Pancreatic Cancer Cells. *Int. Hepato-Pancreato-Biliary Assoc.* **2009**, *11*, 373–381.

Chen, M. C.; Lee, N. H.; Hsu, H. H.; Ho, T. J.; Tu, C. C.; Hsieh, D. J. Y.; Lin, Y. M.; Chen, L. M.; Kuo, W. W.; Huang, C. Y. Thymoquinone Induces Caspase-Independent, Autophagic Cell Death in CPT-11-Resistant Lovo Colon Cancer via Mitochondrial Dysfunction and Activation of JNK and p38. *J. Agric. Food Chem* **2015**, *63*, 1540–1546.

Cheng, L.; Xia, T. S.; Wang, Y. F.; Zhou, W.; Liang, X. Q.; Xue, J. Q.; Wang, M. The Anticancer Effect and Mechanism of α-Hederin on Breast Cancer Cells. *Int. J. Oncol.* **2014**, *45*, 757–763.

Coban, S.; Yildiz, F.; Terzi, A.; Al, B.; Aksoy, N.; Bitiren, M.; Celik, H. The Effects of *Nigella sativa* on Bile Duct Ligation Induced Liver Injury in Rats. *Cell Biochem. Funct.* **2010**, *28*, 83–88.

Cragg, G. M.; Pezzuto, J. M. Natural Products as a Vital Source for the Discovery of Cancer Chemotherapeuticand Chemopreventive Agents. *Med. Princ. Pract.* **2016**, *25*, 41–59.

Cremer, D.; Hausen, B. M.; Schmalle, H. W. Toward a Rationalization of the Sensitizing Potency of Substituted P-Benzoquinones: Reaction of Nucleophiles with P-Benzoquinones. *J. Med. Chem.* **1987**, *30*, 1678–1681.

Daba, M. H.; Abdel-Rahman, M. S. Hepatoprotective Activity of Thymoquinone in Isolated Rat Hepatocytes. *Toxic. Lett.* **1998**, *95*, 23–29.

Dajani, E. Z.; Shahwan, T. G.; Dajani, N. E. Overview of the Preclinical Pharmacological Properties of *Nigella sativa* (Black Seeds): A Complementary Drug with Historical and Clinical Significance. *J. Physiol. Pharmacol.* **2016**, *67*, 801–817.

Darakhshan, S.; Pour, A. B., Colagar, A. H.; Sisakhtnezhad, S. Thymoquinone and Its Therapeutic Potentials. *Pharmacol. Res.* **2015**, *95*, 138–158.

Das, S.; Dey, K. K.; Dey, G.; Pal, I.; Majumder, A.; Maiti Choudhury, S.; Mandal, M. Antineoplastic and Apoptotic Potential of Traditional Medicines Thymoquinone and Diosgenin in Squamous Cell Carcinoma. *PLoS One* **2012**, *7*, e46641.

Dergarabetian, E. M.; Ghattass, K. I.; El-Sitt, S. B.; Al-Mismar, R. M.; El-Baba, C. O.; Itani, W. S.; Melhem, N. M.; El-Hajj, H. A.; Bazarbachi, A. A.; Schneider-Stock, R.; Gali-Muhtasib, H.U. Thymoquinone Induces Apoptosis in Malignant T-Cells via Generation of ROS. *Front. Biosci. (Elite Ed.)* **2013**, *5*, 706–719.

El-Dakhakhny, M.; Madi, N. J.; Lembert, N.; Ammon, H. P. T. *Nigella sativa* Oil, Nigellone and Derived Thymoquinone Inhibit Synthesis of 5-Lipoxygenase Products in Polymorphonuclear Leukocytes from Rats. *J. Ethnopharmacol.* **2002**, *81*, 161–164.

El-Mahdy, M. A.; Zhu, Q.; Wang, Q. E.; Wani, G.; Wani, A. A. Thymoquinone Induces Apoptosis Through Activation of Caspase 8 and Mitochondrial Events in p53 Null Myeloblastic Leukemia HL 60 Cells. *Int. J. Cancer Res.* **2005**, *117*, 409–417.

El-Mezayen, R.; El Gazzar, M.; Nicolls, M. R.; Marecki, J. C.; Dreskin, S. C.; Nomiyama, H. Effect of Thymoquinone on Cyclooxygenase Expression and Prostaglandin Production in a Mouse Model of Allergic Airway Inflammation. *Immunol. Lett.* **2006**, *106*, 72–81.

Elmowafy, M.; Samy, A.; Raslan, M. A. Enhancement of Bioavailability and Pharmacodynamic Effects of Thymoquinone via Nanostructured Lipid Carrier (NLC) Formulation. *AAPS Pharm. Sci. Tech.* **2016**, *17*, 663–672.

El-Najjar, N.; Chatila, M.; Moukadem, H.; Vuorela, H., Ocker, M.; Gandesiri, M.; Schneider-Stock, R.; Gali-Muhtasib, H. Reactive Oxygen Species Mediate Thymoquinone-Induced Apoptosis and Activate ERK and JNK Signaling. *Apoptosis* **2010**, *15*, 183–195.

El-Shafey, M. M.; Abd-Allah, G. M.; Mohamadin, A. M.; Harisa, G. I.; Mariee, A. D. Quercetin Protects Against Acetaminophen-Induced Hepatorenal Toxicity by Reducing Reactive Oxygen and Nitrogen Species. *Pathophysiology* **2015**, *22*, 49–55.

Enguita-Germán, M.; Fortes, P. Targeting the Insulin-Like Growth Factor Pathway in Hepatocellular Carcinoma. *World J. Hepatol.* **2014**, *6*, 716.

Gali-Muhtasib, H.; Diab-Assaf, M.; Boltze, C. Thymoquinone Extracted from Black Seed Triggers Apoptotic Cell Death in Human Colorectal Cancer Cells via a p53-Dependent Mechanism. *Int. J. Oncol* **2004**, *25*, 857–866.

Gedara, S. R. Terpenoid Content of the Leaves of Thymus Algeriensis Boiss. Mans. *J. Pharm. Sci.* **2008**, *24*.

Goyal, L.; Muzumdar, M. D.; Zhu, A. X. Targeting the HGF/c-MET Pathway in Hepatocellular Carcinoma. *Clin. Cancer Res.* **2013**, *19*, 2310–2318.

Goyal, S. N.; Prajapati, C. P.; Gore, P. R.; Patil, C. R.; Mahajan, U. B.; Sharma, C.; Talla, S. P.; Ojha, S. K. Therapeutic Potential and Pharmaceutical Development of Thymoquinone: A Multitargeted Molecule of Natural Origin. *Front. Pharmacol.* **2017**, *8*, 656.

Hassan, S.; Ahmed, W.; M Galeb, F.; El-Taweel, M. In Vitro Challenge Using Thymoquinone on Hepatocellular Carcinoma (HepG2) Cell Line. *Iran. J. Pharm. Sci.* **2010**, *17*, 283–290.

Helal, G. K. Thymoquinone Supplementation Ameliorates Acute Endotoxemia-Induced Liver Dysfunction in Rats. *Pak. J. Pharm. Sci.* **2010**, *23*, 131–137.

Hirobe, C.; Qiao, Z. S.; Takeya, K.; Itokawa, H. Cytotoxic Principles from *Majorana syriaca*. *Nat. Med.* **1998**, *52*, 74–77.

Horvathova, E.; Navarova, J.; Galova, E.; Sevcovicova, A.; Chodakova, L.; Snahnicanova, Z.; Melusova, M.; Kozics, K.; Slamenova, D. Assessment of Antioxidative, Chelating, and DNA-Protective Effects of Selected Essential Oil Components (Eugenol, Carvacrol, Thymol, Borneol, Eucalyptol) of Plants and Intact *Rosmarinus officinalis* Oil. *J. Agric. Food Chem.* **2014**, *62*, 6632–6639.

Houghton, P. J.; Zarka, R.; de las Heras, B.; Hoult, J. R. S. Fixed Oil of *Nigella sativa* and Derived Thymoquinone Inhibit Eicosanoid Generation in Leukocytes and Membrane Lipid Peroxidation. *Planta Medica* **1995**, *61*, 33–36.

Hussain, A. R.; Uddin, S.; Bu, R.; Khan, O. S.; Ahmed, S. O.; Ahmed, M.; Al-Kuraya, K. S. Resveratrol Suppresses Constitutive Activation of AKT via Generation of ROS and Induces Apoptosis in Diffuse Large B Cell Lymphoma Cell Lines. *PLoS One* **2011**, *6*, e24703.

Hussain, D. A.; Hussain, M. M. *Nigella sativa* (Black Seed) Is an Effective Herbal Remedy for Every Disease Except Death—A Prophetic Statement Which Modern Scientists Confirm Unanimously: A Review. *AMPR* **2016**, *4*, 27–57.

Iyoda, K.; Sasaki, Y.; Horimoto, M.; Toyama, T.; Yakushijin, T.; Sakakibara, M.; Takehara T, Fujimoto, J.; Hori, M.; Wands, J. R.; Hayashi, N. Involvement of the p38 Mitogen Activated Protein Kinase Cascade in Hepatocellular Carcinoma. *Cancer* **2003**, *97*, 3017–3026.

Jaafari, A.; Mouse, H. A.; Rakib, E. M.; Tilaoui, M.; Benbakhta, C.; Boulli, A.; Abbad, A.; Zyad, A. Chemical Composition and Antitumor Activity of Different Wild Varieties of Moroccan Thyme. *Rev. Bras. Farmacogn.* **2007**, *17*, 477–491.

Jayakumar, S.; Madankumar, A.; Asokkumar, S.; Raghunandhakumar, S.; Kamaraj, S.; Divya, M. G. J.; Devaki, T. Potential Preventive Effect of Carvacrol Against Diethylnitrosamine-Induced Hepatocellular Carcinoma in Rats. *Mol. Cell. Biochem.* **2012**, *360*, 51–60.

Kanter, M.; Coskun, O.; Budancamanak, M. Hepatoprotective Effects of *Nigella sativa* L and *Urtica dioica* L on Lipid Peroxidation, Antioxidant Enzyme Systems and Liver Enzymes in Carbon Tetrachloride-Treated Rats. *World J. Gastroenterol.* **2005**, *11*, 6684.

Ke, X.; Zhao, Y.; Lu, X.; Wang, Z.; Liu, Y.; Ren, M.; Lu, G.; Zhang, D.; Sun, Z.; Xu, Z.; Song, J. H. TQ Inhibits Hepatocellular Carcinoma Growth In Vitro and In Vivo via Repression of Notch Signaling. *Oncotarget* **2015**, *6*, 32610.

Khan, M. A. Chemical Composition and Medicinal Properties of Nigella sativa Linn. Inflammopharmacology 1999, 7, 15–35.

Linjawi, S. A.; Khalil, W. K.; Hassanane, M. M.; Ahmed, E. S. Evaluation of the Protective Effect of *Nigella sativa* Extract and Its Primary Active Component Thymoquinone Against DMBA-Induced Breast Cancer in Female Rats. *AMS* **2015**, *11*, 220.

Lutterodt, H.; Luther, M.; Slavin, M.; Yin, J. J.; Parry, J.; Gao, J. M.; Yu, L. L. Fatty Acid Profile, Thymoquinone Content, Oxidative Stability, and Antioxidant Properties of Cold-Pressed Black Cumin Seed Oils. *LWT-Food Sci. Technol.* **2010**, *43*, 1409–1413.

Mansour, M.; Tornhamre, S. Inhibition of 5-Lipoxygenase and Leukotriene C4 Synthase in Human Blood Cells by Thymoquinone. *J. Enzyme Inhib. Med. Chem.* **2004**, *19*, 431–436.

Marsik, P.; Kokoska, L.; Landa, P.; Nepovim, A.; Soudek, P.; Vanek, T. In Vitro Inhibitory Effects of Thymol and Quinones of *Nigella sativa* Seeds on Cyclooxygenase-1-and-2-Catalyzed Prostaglandin E2 Biosyntheses. *Planta Medica* **2005**, *71*, 739–742.

Mohamed, A.; Shoker, A.; Bendjelloul, F.; Mare, A.; Alzrigh, M.; Benghuzzi, H.; Desin, T. Improvement of Experimental Allergic Encephalomyelitis (EAE) by Thymoquinone; An Oxidative Stress Inhibitor. *Biomed. Sci. Instrum.* **2003**, *39*, 440–445.

Muriel, P. Role of Free Radicals in Liver Diseases. *Hepatol. Int.* **2009**, *3*, 526–536.

Nagi, M. N.; Alam, K.; Badary, O. A.; AlShabanah, O. A.; Al Sawaf, H. A.; Al Bekairi, A. M. Thymoquinone Protects Against Carbon Tetrachloride Hetatotoxicity in Mice via an Antioxidant Mechanism. *IUBMB* **1999**, *47*, 153–159.

Newman, D. J.; Cragg, G. M. Natural Products as Sources of New Drugs Over the Last 25 Years. *J. Nat. Prod* **2007**, *70*, 461–477.

Nili-Ahmadabadi, A.; Tavakoli, F.; Hasanzadeh, G. R.; Rahimi, H. R.; Sabzevari, O. Protective Effect of Pretreatment with Thymoquinone Against Aflatoxin B1 Induced Liver Toxicity in Mice. *DARU J. Pharm. Sci.* **2011**, *19*.

Ozkan, A.; Erdoğan, A. A Comparative Evaluation of Antioxidant and Anticancer Activity of Essential Oil from *Origanum onites* (Lamiaceae) and Its Two Major Phenolic Components. *Turk. J. Biol.* **2011**, *35*, 735–742.

Palabiyik, S. S.; Karakus, E.; Halici, Z.; Cadirci, E.; Bayir, Y.; Ayaz, G.; Cinar, I. The Protective Effects of Carvacrol and Thymol Against Paracetamol–Induced Toxicity on Human Hepatocellular Carcinoma Cell Lines (HepG2). *HET* **2016**, *35*, 1252–1263.

Paramasivam, A.; Raghunandhakumar, S.; Priyadharsini, J. V.; Jayaraman, G. In Vitro Anti-Neuroblastoma Activity of Thymoquinone Against Neuro-2a Cells via Cell-Cycle Arrest. *Asian Pac. J. Cancer Prev.* **2015**, *16*, 8313–8319.

Park, H. M.; Kim, S. J.; Mun, A. R.; Go, H. K.; Kim, G. B.; Kim, S. Z.; Jang, S.; Lee, S. J.; Kang, H. S. Korean Red Ginseng and Its Primary Ginsenosides Inhibit Ethanol-Induced Oxidative Injury by Suppression of the MAPK Pathway in TIB-73 Cells. *J. Ethnopharmacol.* **2012**, *141*, 1071–1076.

Pathak, K.; Raghuvanshi, S. Oral Bioavailability: Issues and Solutions via Nanoformulations. *Clin. Pharma.* **2015**, *54*, 325–357.

Pompella, A.; Visvikis, A.; Paolicchi, A.; De Tata, V.; Casini, A. F. The Changing Faces of Glutathione, a Cellular Protagonist. *Biochem. Pharmacol.* **2003**, *66*, 1499–1503.

Poonia, N.; Kharb, R.; Lather, V.; Pandita, D. Nanostructured Lipid Carriers: Versatile Oral Delivery Vehicle. *Fut. Sci. OA* **2016**, *2*, FSO135.

Rajput, S.; Kumar, B. P.; Sarkar, S.; Das, S.; Azab, B.; Santhekadur, P. K.; Das, S. K.; Emdad, L.; Sarkar, D.; Fisher, P. B.; Mandal, M. Targeted Apoptotic Effects of Thymoquinone and Tamoxifen on XIAP Mediated AKT Regulation in Breast Cancer. *PloS One* **2013**, *8*, e61342.

Randhawa, M. A.; Alghamdi, M. S. Anticancer Activity of *Nigella sativa* (Black Seed)—A Review. *Am. J. Chin. Med.* **2011**, *39*, 1075–1091.

Reindl, W.; Strebhardt, K.; Berg, T. A High-Throughput Assay Based on Fluorescence Polarization for Inhibitors of the Polo-Box Domain of Polo-Like Kinase 1. *Anal. Biochem.* **2008**, *383*, 205–209.

Sakalar, C.; Yuruk, M.; Kaya, T.; Aytekin, M.; Kuk, S.; Canatan, H. Pronounced Transcriptional Regulation of Apoptotic and TNF–NF-Kappa-B Signaling Genes During the Course of Thymoquinone Mediated Apoptosis in HeLa Cells. *Mol. Cell Biochem.* **2013**, *383*, 243–251.

Salim, L. Z. A.; Othman, R.; Abdulla, M. A.; Al-Jashamy, K.; Ali, H. M.; Hassandarvish, P.; Hassandarvish, P.; Dehghan, F.; Ibrahim, M. Y.; Omer, F. A.; Mohan, S. Thymoquinone Inhibits Murine Leukemia WEHI-3 Cells In Vivo and In Vitro. *PLoS One* **2014**, *9*, e115340.

Samarakoon, S. R.; Thabrew, I.; Galhena, P. B.; De Silva, D.; Tennekoon, K. H. A Comparison of the Cytotoxic Potential of Standardized Aqueous and Ethanolic Extracts of a Polyherbal Mixture Comprised of *Nigella sativa* (Seeds), *Hemidesmus indicus* (Roots) and *Smilax glabra* (Rhizome). *Pharmacogn. Res.* **2010**, *2*, 335.

Sayed-Ahmed, M. M.; Aleisa, A. M.; Al-Rejaie, S. S.; Al-Yahya, A. A.; Al-Shabanah, O. A.; Hafez, M. M.; Nagi, M. N. Thymoquinone Attenuates Diethylnitrosamine Induction of Hepatic Carcinogenesis Through Antioxidant Signaling. *Oxid. Med. Cell. Longev.* **2010**, *3*, 254–261.

Singh, A. P.; Chaturvedi, P.; Batra, S. K. Emerging Roles of MUC4 in Cancer: A Novel Target for Diagnosis and Therapy. *Cancer Res.* **2007**, *67*, 433–436.

Slamenova, D.; Horvathova, E.; Sramkova, M.; Marsalkova, L. DNA-Protective Effects of Two Components of Essential Plant Oils Carvacrol and Thymol on Mammalian Cells Cultured In Vitro. *Neoplasma* **2007**, *54*, 108–112.

Sprinzl, M. F.; Puschnik, A.; Schlitter, A. M.; Schad, A.; Ackermann, K.; Esposito, I.; Lang, H.; Galle, P. R.; Weinmann, A.; Heikenwälder, M & Protzer, U. Sorafenib Inhibits Macrophage-Induced Growth of Hepatoma Cells by Interference with Insulin-Like Growth Factor-1 Secretion. *J. Hepatol.* **2015**, *62*, 863–870.

Stammati, A.; Bonsi, P.; Zucco, F.; Moezelaar, R.; Alakomi, H. L.; von Wright, A. Toxicity of Selected Plant Volatiles in Microbial and Mammalian Short-Term Assays. *Food Chem. Toxicol.* **1999**, *37*, 813–823.

Strazzabosco, M.; Fabris, L. Notch Signaling in Hepatocellular Carcinoma: Guilty in Association! *Gastroenterology* **2012**, *143*, 1430–1434.

Suddek, G. M. Protective Role of Thymoquinone Against Liver Damage Induced by Tamoxifen in Female Rats. *Can. J. Physiol. Pharmacol.* **2014**, *92*, 640–644.

Swamy, S. M. K.; Tan, B. K. H. Cytotoxic and Immunopotentiating Effects of Ethanolic Extract of *Nigella sativa* L. Seeds. *J. Ethnopharmacol.* **2000**, *70*, 1–7.

Torre, L. A.; Bray, F.; Siegel, R. L.; Ferlay, J.; Lortet Tieulent, J.; Jemal, A. Global Cancer Statistics, 2012. *CA: Cancer J. Clin.* **2015**, *65*, 87–108.

Torres, M. P.; Ponnusamy, M. P.; Chakraborty, S.; Smith, L. M.; Das, S.; Arafat, H. A.; Batra, S. K. Effects of Thymoquinone in the Expression of Mucin 4 in Pancreatic Cancer Cells: Implications for the Development of Novel Cancer Therapies. *Mol. Cancer Ther.* **2010**, *9*, 1419–1431.

Tsukamoto, H.; Lu, S. C. Current Concepts in the Pathogenesis of Alcoholic Liver Injury. *FASEB J.* **2001,** *15,* 1335–1349.

Venook, A. P.; Papandreou, C.; Furuse, J.; de Guevara, L. L. The Incidence and Epidemiology of Hepatocellular Carcinoma: A Global and Regional Perspective. *Oncologist* **2010,** *15,* 5–13.

Villanueva, A.; Alsinet, C.; Yanger, K.; Hoshida, Y.; Zong, Y.; Toffanin, S.; Rodriguez–Carunchio, L.; Solé.; M.; Thung, S.; Stanger, B. Z.; Llovet, J. M. Notch Signaling Is Activated in Human Hepatocellular Carcinoma and Induces Tumor Formation in Mice. *Gastroenterology* **2012,** *143,* 1660–1669.

Woo, C. C.; Kumar, A. P.; Sethi, G.; Tan, K. H. B. Thymoquinone: Potential Cure for Inflammatory Disorders and Cancer. *Biochem. Pharm.* **2012,** *83,* 443–451.

Yen, F. L.; Wu, T. H.; Lin, L. T.; Lin, C. C. Hepatoprotective and Antioxidant Effects of *Cuscuta chinensis* Against Acetaminophen-Induced Hepatotoxicity in Rats. *J. Ethnopharmacol.* **2007,** *111,* 123–128.

Yessuf, A. M. Phytochemical Extraction and Screening of Bio Active Compounds from Black Cumin (*Nigella Sativa*) Seeds Extract. *Am. J. Life Sci.* **2015,** *3,* 358–364.

Yildiz, F.; Coban, S.; Terzi, A.; Ates, M.; Aksoy, N.; Cakir, H.; Ocak, A. R.; Bitiren, M. *Nigella sativa* Relieves the Deleterious Effects of Ischemia Reperfusion Injury on Liver. *World J. Gastroenterol.* **2008,** *14,* 5204.

Yin, Z.; Song, Y.; Rehse, P. H. Thymoquinone Blocks pSer/pThr Recognition by Plk1 Polo-Box Domain as a Phosphate Mimic. *ACS Chem. Biol.* **2012,** *8,* 303–308.

You, H.; Ding, W.; Dang, H.; Jiang, Y.; Rountree, C. B. c-Met Represents a Potential Therapeutic Target for Personalized Treatment in Hepatocellular Carcinoma. *Hepatology* **2011,** *54,* 879–889.

Yu, S. M.; Kim, S. J. Thymoquinone-Induced Reactive Oxygen Species Causes Apoptosis of Chondrocytes via PI3K/Akt and p38 Kinase Pathway. *Exp. Biol. Med.* **2013,** *238,* 811–820.

Yu, S. M.; Kim, S. J. The Thymoquinone-Induced Production of Reactive Oxygen Species Promotes Dedifferentiation Through the ERK Pathway and Inflammation Through the p38 and PI3K Pathways in Rabbit Articular Chondrocytes. *Int. J. Mol. Med.* **2015,** *35,* 325–332.

Zhou, J.; Zhang, X. Q.; Ashoori, F.; McConkey, D. J.; Knowles, M. A.; Dong, L.; Benedict, W. F. Early RB94-Produced Cytotoxicity in Cancer Cells Is Independent of Caspase Activation or 50 kb DNA fragmentation. *Cancer Gene Ther.* **2009,** *16,* 13–19.

CHAPTER 4

Boswellia serrata Roxb

MANOHAR M. V.[1], ANU PANDITA[2], AMOGHA G. PALADHI[3],
BHOOMIKA INAMDAR[1], SUGUMARI VALLINAYAGAM[4], DEEPU PANDITA[5],
and K. M. SRINIVASA MURTHY[6]

[1]*JSS Medical College (Deemed to be University), Mysuru, Karnataka, India*

[2]*Vatsalya Clinic, Krishna Nagar, New Delhi, India*

[3]*Christ (Deemed to be University), Bengaluru, Karnataka, India*

[4]*Department of Biotechnology, Mepco Schlenk Engineering College, Sivakasi, Tamil Nadu, India*

[5]*Government Department of School Education, Jammu, Jammu and Kashmir, India*

[6]*Department of Microbiology and Biotechnology, Jnanabharathi Campus Bangalore University, Bengaluru, Karnataka, India*

ABSTRACT

Boswellia serrata Roxb belongs to the family Burseraceae. There are mentions of this plant in various ancient texts and its Sanskrit name "Gajabhakshya" and Salai guggal. In this chapter, anticancerous property of *Boswellia serrata* Roxb is discussed. The alkaloids *Boswellia serrata* Roxb has various components like oils, triterpenoids, and resins among which 25–35% is β-boswellic acid and also acetates of AKBA and ABA. Diene derivatives of boswellic acids like 11-dehydroxy β-boswellic acid, α-amyrin, 11-diene-24-oicacid are found. There are various tricyclic, tetracyclic components found in the extract of oleo gum resin that also shows a little anticancer activity. Understanding of chemistry of boswellic acid is the core knowledge to know the molecular

Potent Anticancer Medicinal Plants: Secondary Metabolite Profiling, Active Ingredients, and Pharmacological Outcomes. Deepu Pandita and Anu Pandita (Eds.)

activity of boswellic acid and its molecular interactions in various inhibition pathways. *Boswellia serrata* is known for multiple pharmacological effects. Experiments have shown that it can be helpful in treating various cancers like brain, CNS, breast, gastro-intestinal cancers like hepatic, colo-rectal cancer, prostate cancer, leukemia, and various other cancers. Molecular targets of boswellic acids are known to be topoisomerases-1 and -2, different apoptosis inducing cascades. Various clinical and preclinical analyses are also being mentioned, which show the rate of success of boswellic acids and its derivatives as treating agents of cancers and multiple myelomas. These experiments have reported inhibitions of certain pathways and also execution of apoptotic pathway that helps in curing of cancer and also in reducing the side effects of the tumor. *Boswellia serrata* extracts are known to show zero side effects in its natural form, thus making it a potential compound for therapeutic use of cancer treatment.

4.1 INTRODUCTION

Boswellia serrata Roxb is a tree that is usually found in the tropical parts of Asia and Africa. In India, it is found in dry hill forests like Madhya Pradesh, Rajasthan, Gujarat, Assam, Bihar, Orissa, and other regions. This tree secrets gum when incisions are made on tree trunks and the obtained gum are used in various fields like cosmetics, pharmaceuticals, etc. The obtained gum tastes bitter and is pleasant in flavor. The Frankinscence is used by different ancient civilizations of Greeks, Romans, and Egyptians as incense, various aromatic purposes, and fumigants. Currently, it is used in the production of incense sticks and incense powders. The preparations of oleo gum resin of *Boswellia serrata* have found a significant place in Unani and Ayurveda for its properties to treat various health conditions like asthma, cough, bronchitis, issues related to throat and digestive system. It is also used in treatment of cephalytes and pulmonary diseases. It acts as a stimulant of various processes to keep the system healthy. The gum is prescribed I cases of dysentery, dispenscia, diarrhea, jaundice, and hemorrhoids and can also be administered in weak and unhealthy subjects.

As cancer is an issue worldwide causing majority of morbidity and mortality, the interests in therapeutic using natural products contain bioactive constituents with anticancer and anti-inflammatory activity. Oleo gum resin of *Boswellia serrata* has found its importance. Common names of *Boswellia serrata* are Olebanum, Salai guggal, Kundur, Loban, and Gajabhakshya, oleo gum resin from *Boswellia serrata* consists of essential oil, resins, and gums that can cure various health conditions that have adversely affected the individual. The essential oil of Salai guggal has a mixture of mono, di, and triperpenes containing 33 essential components. The gum contains arabinose, galactose, and xylose containing some digestive enzymes. Resins are the most important essential compounds of *Boswellia serrata* Roxb extracts composed of pentacyclic triterpenic acids. The derivatives of this are found to be β-boswellic acids like 3-o-acetyl-β-boswellic acid, 3-o-acetyl-11-keto-β-boswellic acid, and 11-keto-β-boswellic acids. The residues of oleo resin present in Salai guggal possess anti-diarrhoeal, anti-hyperlipidemic, anti-asthematic, immunomodulatory, hepatoprotective, anti-inflammatory, antimicrobial, hypoglycemic, and anticancerous in nature. Various conducted experiments have shown that there are no side-effects found due to the administration of *Boswellia serrata* extracts and are also found to be administered in daily basis. Various groups of scientists have reported the use of AKBA isomer gives the more potent and efficient results (Nerkar, 2020). The boswellic acids are found to be actively involved in the inhibition of cell-proliferation by inducing various apoptotic pathways like caspase-3 activity, Bcl2, BAX. Despite cancer therapeutics, *Boswellia serrata* has also found its use in avoiding adverse effects of tumors like edema, necrosis, angiogenesis, etc. There are different nanoparticle-based studies that have evaluated boswellic acid for treatment of various purposes like cytotoxicity, anti-invasive activities, and cell-specific drug-resistant activities. Because boswellic acids are sourced from *Boswellia serrata* of various cultivators, the bioavailability of the extracts must be known to formulate and standardize to use them in clinical protocols.

Taxonomical hierarchy (Serrata et al., 2020)

Kingdom—*Plantae*

Subkingdom—*Tracheobionta*

Division—*Magnoliophyta*

Class—*Magnoliopsida*

Order—*Sapindales*

Family—*Burseraceae*

Genus—*Boswellia*

Species—*serrata*

Subspecies—Roxb

4.2   CHEMISTRY OF BOSWELLIC ACID

The alkaloids of *Boswellia serrata* Roxb mostly consist of essential oils, resins, and triterpenoids. The prime constituent of monoterpenoids is α-pinene (73.3%) and others include p-cymene (1.0%), limonene (1.42%), myrcene (1.71%), trans-pinocarveol (1.80%), β-pinene (2.05%), cis-verbenol (1.97%), verbenone (1.71%), borneol (1.78%), thuja-2,4 (10)-diene (1.18%), whereas α-copaene is the only sesquiterpene that has been recognized in the oil (Grant, 2009). Terpenoids are the major constituents of oleo gum resin and the fraction is found to be 25–35% including mainly β-boswellic acid [BA, 1] that also consists of 11-keto-b-boswellic acid along with its corresponding acetates AKBA and ABA. Tschirh et al. in 1898 first extracted terpenoids from boswellic acid (BA) (Sudharsan et al., 2005). After this first isolation of terpenoids from BA, many works have been done for structural elucidation of boswellic acid. Crystallographic X-ray studies were made to elucidate the structure of boswellic acid constituents (Syrovets et al., 2005a, 2005b). One of the boswellic acids, β-boswellic acid, belonging to the group of ursanes among triterpenic acids is lipophilic in nature that encompasses only one α- hydroxyl as the functional group. To vary the solubility and the acidic property of terpenes, it was subjected to modify the structural orientation by replacing a carboxyl functional group with an amino functional group, to make it a week basic compound (Garg and Deep, 2015). Boswellic acids are generally seen along with diene derivatives like with 3-O-acetyl-9 by the dehydration of 3-O-11-hydroxy-b-boswellic acid; a derivative is obtained namely 11-dehydro-beta-boswellic acid. NMR spectroscopy is known to elucidate the structures of all penta-cyclic triterpenes including boswellic acid and diene derivatives. X-ray crystallographic studies have also shown the structures of ABA and methyl esters of boswellic acid. 11-dien-24-oic acid, 3-hydroxy-urs-9, and α-amyrin were elucidated from the gum resins of *Boswellia serrata* Roxb. Various other tetracyclic terpenoids found in *Boswellia serrata* Roxb are found to be 3-keto-tirucall-8, 24-dien- 21-oic acid; 3a-acetoxy-tirucall-8, 24-dien-21-oic acid; 3a-hydroxy-tirucall-8, 24-dien-21-oic acid and 3b-hydroxytirucall-8, 24-dien-21-oic. Various other triterpenoids like urs-12-ene-3a, 24-diol, and 2, 3-dihydroxy-urs-12-ene-24-oic acid were found in *Boswellia serrata* Roxb that were structurally elucidated from the acidic and neutral fractions of the gum. urs-12-ene-3b,24-diol is present as an isomeric diol (Aktionen, 1967). *Boswellia serrata* extractions consist of gum resins containing around 50–60% of α- and β-boswellic acids. 1–3% of the total extracts are mostly bioactive AKBA fractions (Sterk et al., 2004).

The anticancerous activity of *Boswellia serrata* extractions is known to cure various cancers like prostrate, leukocytes, colon, liver, colo-rectal, brain, and CNS (central nervous system). The inhibition of tumor formation by AKBA acts on nuclear factors like NFκB kappa B, signal transducers and also induces activator of transcription-3 (STAT-3)-related pathways, and thus helps in apoptosis to avoid cell proliferation and inhibits angiogenesis to prevent neoplastic cell proliferation (Liu et al., 2006; Borrelli et al., 2006; Lu et al., 2008; Pang et al., 2009).

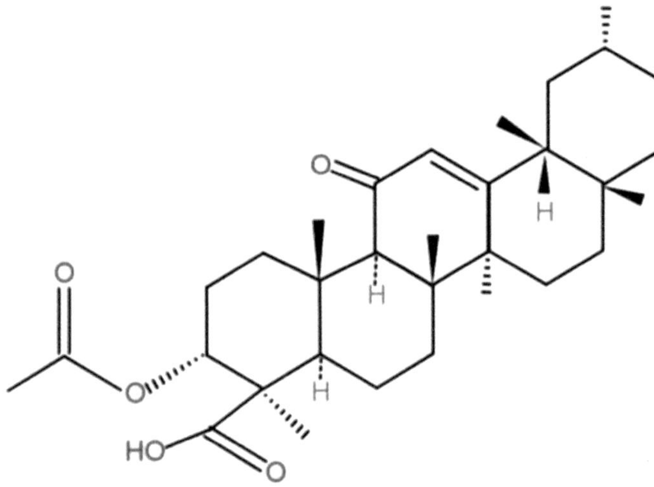

FIGURE 4.1   3-o-acetyl-11-keto-β-boswellic acid (AKBA).

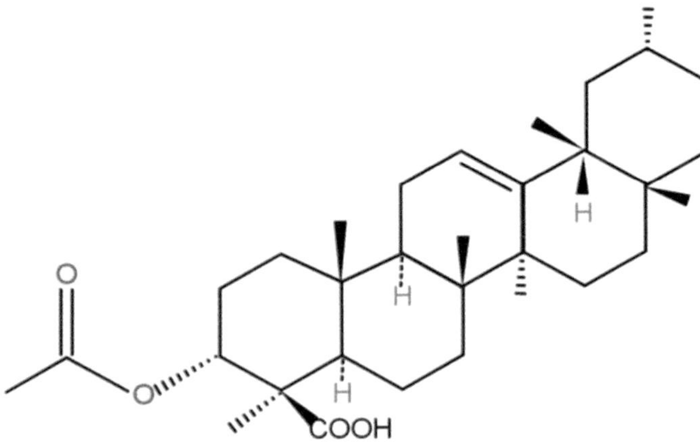

FIGURE 4.2   3-o-acetyl-β-boswellic acid (ABA).

4.3 APPLICATIONS

Indian frankincense or Salai guggal (*Boswellia serrata*) is known to be used in various medicinal practices like Ayurveda and Unani as a traditional medicine to treat inflammatory diseases since ages. References of use of *Boswellia serrata* can be seen in various ancient texts related to Ayurveda. "Gajabhakshya" is the Sanskrit name of the plant. The name is thus given because it is mostly consumed by elephants. In the original texts of Susruta Samhita, Astanga Hridaya, Astanga Sangraha, and Charaka Samhita the use of this plant is mentioned signifying its medicinal values. In Susruta Samhita and Charaka Samhita the reference of this plant is in the treatment of arthritis and other inflammatory diseases. In systems like Ayurveda and Unani, the internal and external administration of this plant is seen in the treatment of various inflammatory conditions (Moussaieff and Mechoulam, 2009). Later, based on these mentions, the resins and gums of these plants were evaluated for its medicinal value for conditions like arthritis, asthma, colitis, and other inflammatory issues (Moussaieff and Mechoulam, 2009). The gum resins usually contain triterpenoids, essential oils, and carbohydrates out of which fundamental constituent of oleo gum resin are boswellic acids like acetyl-11-keto-β-boswellic acid, 11-keto-β- boswellic acid, and β-boswellic acid (Kirste et al., 2011).

4.3.1 LEUKOTRIENE INHIBITION

Studies have shown that gum resins extracted from *Boswellia serrata* Roxb inhibit leukotriene B_4 formation of peritoneal neutrophils. Leukotrienes are forms when phagocytosis is stimulated particularly in neutrophils. Neutrophils are responsible for inflammation in body. Thus, BAs can also be used as nonsteroidal anti-inflammatory drug that is free from side effects (Upaganlawar and Ghule, 2009).

4.3.2 ANALGESIC AND PSYCHO-PHARMACOLOGICAL EFFECTS

Menon et al. have experimented the use of gum resin on animals. They found the extracts of *Boswellia serrata* had a sedative effect and analgesic activity on animals. The sedative effects of gum resins are due to its property of reducing spontaneous motor activity (Upaganlawar and Ghule, 2009).

4.3.3 ANTI-INFLAMMATORY AND ANTI-ARTHRITIC EFFECT

Extracts of *Boswellia serrata* are known to cure edema in mouse and also help in curing mycobacterium adjuvant-induced poly-arthritis in mice, thus decreasing paw swelling (Singh and Atal, 1986).

4.3.4 IMMUNO-MODULATORY ACTIVITY

Experiments have shown that gradual degranulation of mast cells when it is administered in different doses, thus resulting in stabilizing the activity of mast cells (Pungle et al., 2003).

4.3.5 HYPO-LIPIDEMIC AND HEPATO-PROTECTIVE ACTIVITY

The extracts of *Boswellia serrata* when used to study activities on lipidemia, the experiments resulted in decreased total cholesterol about 38–40% and also increased HDL, thus helping the movement of fats from the place of deposition to peripheral tissues. This causes decrease in liver damage and is known by reduced SGPT, SGOT anti-transferases and serum enzyme activities (Upaganlawar and Ghule, 2009).

4.3.6 HYPOGLYCEMIC ACTIVITY

Oleo gum resins extracted from *Boswellia serrata* are reported to show anti-diabetic activity in conditions like non-insulin-dependent diabetes mellitus. In rat models with induced diabetes, significant reduction of blood-glucose level is seen after the administration of this extract (Upaganlawar and Ghule, 2009).

4.3.7 ANTI-DIARRHOEAL ACTIVITY

Acetylcholine (ACh) and Barium chloride ($BaCl_2$) are toxic in nature due to which the peristalsis of the intestine increased causing inflammatory bowel syndrome which causes diarrhoea. *Boswellia serrata* extractions are found to be the inhibitors of ACh that inhibits rapid gastro intestinal movements. The same is also effective to inhibit castor oil-induced diarrhoea (Borrelli et al., 2006).

4.3.8   ANTIMICROBIAL ACTIVITY

Proteus mirabilis UCH 28, *E. coli* LASUTH 54, and *S. aureus* OGSUTH 108 were significantly inhibited by essential oils of *Boswellia serrata* and are also known to inhibit gram-positive and gram-negative bacteria (Upaganlawar and Ghule, 2009).

4.3.9   ANTI-ASTHMATIC ACTIVITY

A study was conducted by Gupta et al. in 1998 and the results were reported showing the patients with prolonged history of asthma were cured. Symptoms like dyspnea were inhibited due to increase in number of healthy bronchi and increased stimulation of MAPK (Mitogen Activated Protein Kinase) by the mobilization of intracellular calcium Ca^{2+} thus decreasing number of asthmatic attacks (Upaganlawar and Ghule, 2009).

4.4   ANTICANCER ACTIVITY

The extracts of *Boswellia serrata* induce anticarcinogenicity on various tumors. The mode of action by which it inhibits cell proliferation and cell growth is by inhibiting the biosynthesis of activities of RNA, DNA, and formed proteins (Tsukada et al., 1986). Boswellic acid when administered in particular doses is known to inhibit number of cancers by inhibiting tumor formation (Upaganlawar and Ghule, 2009). Boswellic acids are known to induce apoptosis by inhibiting protein synthesis and thus help to inhibit cell proliferation in patients of different cancers. The inhibitory on glioblastoma patients is primarily studied (Upaganlawar and Ghule, 2009).

4.4.1   ANTICANCER ACTIVITY IN VIVO AND IN VITRO

Various extracts and phytochemicals from *Boswellia* are being examined for its anticancerous and anti-inflammatory activities in comparison with other Frankinscence. Multiple experimental investigations were used in deter-mining the above properties. It is found that standardization of Frankinscence extracts has a very significant role in phytotherapy to cure cancer. Thus, it can serve as standardized drug. Synthetic drugs with synthetic compounds can also be used once these phytochemicals are completely analyzed. Preclinical experimental analysis of *Boswellia* extracts and isolated phytochemicals are found to actively inhibit various types of cancer malignancies like fibrosarcoma, cancers of liver, lung, bladder, breast, pancreas, prostate, glioblastoma, leukemia, multiple myeloma, and other meningioma. These experiments confirm that phytochemicals of *Boswellia serrata* possess the efficiency to cure cancer. The experimental analysis of boswellic acids actions clearly shows that the molecular inhibitions made by these phytochemicals have reduced oxidative stress and increase stress on proteins from endoplasmic reticulum (Hussain et al., 2009). *Boswellia* phytochemicals are known to cause de-methylation that leads to epigenetic changes by decreasing the action of DNMT1 (DNA methyl transferase-1) and increases histone protein activity (Thorsteinsdottir et al., 2011). Various signal transductor proteins and transcription factors were affected by inhibitory actions of *Boswellia serrata* extracts. This inhibition includes STAT-3, AMPK (AMP-activated protein kinase inhibition), protein kinase b (AKT) expression and phosphorylation, wingless int (WNT)/β-catenin, SRC (sarcoma tyrosine kinase), dephosphorylation of ERK1/2, ribosomal protein S6 kinase (RPS6), FAK (focal adhesion kinase) (Kunnumakkara et al., 2009) (B. Park et al., 2011). It also induces unfolding of proteins in endoplasmic reticulum by activating the signal transduction in unfolded protein response pathway (Saleh and Trinchieri, 2010). Boswellic acid causes modulations in NFκB and inhibitory factor κB-α (IκB-α) activity (Kunnumakkara et al., 2009; Zhao et al., 2003). It also inhibits the activity of transcription factors like mTOR, specificity protein 1 transcription factor (SP1) cellular myelocytomatosis protooncogene (c-MYC), HIF-1 (hypoxia-inducible factor 1) (Agrawal et al., 2011; Morad et al., 2013; Syrovets et al., 2000). Growth inhibition takes place due to induced apoptosis by p53 signaling pathway and DNA damage induced by proteins and p21 gene expression and also arrests the cell cycle in different phases thus inhibiting cell proliferation (Kunnumakkara et al., 2009; Syrovets et al., 2000). Cell proliferation is also inhibited by inhibition of few factors like proliferation cell nuclear antigen (PCNA) expression and mitosis, inhibition of Kiel proliferation antigen 67 (Ki-67), decreased fork-head box M1 protein (FOXM1) expression, inhibition of cyclin B/D, and cyclin-dependent kinases 2, 4, and 6 (CDK2/4/6), decreased DNA topoisomerases 1 and 2 (TOPO1/2A) expression, inhibition of Aurora A/B and Polo-like kinase 1, dephosphorylation of CDK1, decreased dual specificity phosphatase 25A (CDC25A) expression (Frank et al., 2009; Kunnumakkara et al., 2009; Park et al., 2011). Derivatives of *Boswellia serrata* inhibit inflation and also inflammatory markers like 5-LOX, TNF-α, leukotriene synthesis, COX-1/-2 IL6 signaling,

PGE2, and monocytic differentiation and increase the activity of IL12 expression (Conti et al., 2018; Weber et al., 2006). This elevates the expression of SMAD14, SRC homology region 2 domain-containing phosphatase 1 (SHP1), sphingomyelin phosphodiesterase 3 (SMPD3), glutathione-depleting ChaC glutathione-specific γ-glutamyl cyclotransferase 1 (CHAC1), homocysteine-inducible endoplasmic reticulum stress-inducible ubiquitin-like domain member 1 (HERPUD1), sestrin 2 (SESN2), cystathionine γ-lyase (CTH), tribbles homologues 3 (TRIB3) and decreases the activity of serum response factor (SRE), Praja ring finger ubiquitin ligase 2 (pJaC2), transglutaminase 2, endothelin 1 (EDN1), cell migration-inducing protein hyaluron-binding protein (CEMIP), inhibitor of DNA binding 1 (ID1), SRY box 9 protein (SOX9) (Lakka et al., 2011; Saleh and Trinchieri, 2010; Thorsteinsdottir et al., 2011).

Bosweelia serrata extracts and isolated phytochemicals induce apoptosis, metastasis, cell-proliferation, invasion, angiogenesis by increasing p53-upregulated modulator of apoptosis (PUMA) signaling and dynamin-related protein 1 (DRP1) translocation to mitochondria) and decreasing mitochondrial membrane potential, BCL-2 (B-cell Lymphoma protein), BH3 interacting domain death agonist (BID), BCL extra-large protein (Bcl-xL), inhibitor of apoptosis protein (IAP), survivin expression, activation of caspases 3, 8, and 9, increased BCL-2-associated X protein (BAX) expression, activation of caspases-3/-8/-9, cleavage of poly(ADP-ribose) polymerase (PARP), DNA fragmentation, increased expression of death receptor 4 and 5 (DR4/5), cytosolic release of pro-apoptotic factors (i.e., cytochrome c, Diablo IAP-binding mitochondrial protein (SMAC/DIABLO)), C/EBP homologous protein (CHOP), TNF receptor 1 (TNF-R1), inhibitor of caspase-activated DNAse (ICAD) expression (Kunnumakkara et al., 2009; Lakka et al., 2011; Xia et al., 2005).

4.4.2   CLINICAL INVESTIGATIONS

Preclinical analysis of *Boswellia serrata* has shown amusing results against tumor and cancer acting via inhibition of cell proliferation and induction of apoptosis. Thus, the compound was clinically investigated on human subjects for cancer therapy considering control and test. The way of approach in clinical investigations is different and varies when compared with laboratory analysis where the administration of compounds is either oral or intravenous. Thus, there is a bit of difficulty for the body to absorb these compounds in few patients. Boswellic acids when treated in various inflammatory conditions have reportedly cured the patients with various cancers or also found to inhibit rapid proliferation by increasing longevity of life. There is a long history of traditional use of *Boswellia serrata* for the conditions like cancers and tumors in Ayurveda and Unani as this also is considered as clinical trials when administered in particular dosages for certain time. Recent clinical investigations with various modes of administration have resulted in decreased response when compared to native medicines. Recently, analogues of boswellic acids are used in cancer therapy to cure the adverse effects of radiotherapy, chemotherapy, etc. where edematous and other inflammatory conditions are very well cured (Volm and Efferth, 2015). The effect of edema curing and tumor and antitumor response by *Boswellia serrata* extracts is confirmed when investigated on 29 glioma patients when administered with increasing doses consequently for a week (Kirste et al., 2011). There is another study of clinical treatment where 19 children were treated with 126mg/kg of boswellic acid for their condition of brain tumor showed very positive results and also reduced neurological symptoms with healthy weight gain, reduction in edema, improved muscle strength, and also general health improvements (Lalithakumari et al., 2006). In one of the studies, the patient suffered due to breast cancer with brain metastasis resistant to standard cancer treatment. Thus, boswellic acid administration was made which suppressed and controlled the re-occurrence of brain metastasis, for almost four years due to the activity of inhibition of lipoxygenase pathway (Singh et al., 2012). Thus, proper approach in the clinical administration in the treatment of cancer can remarkably cure the cancers and its adverse inflammatory effects and also symptoms. *Boswellia serrata* extracts thus possess properties of tumor regression and control the proliferation of cancer of almost any type of cancer (Efferth and Oesch, 2020).

4.4.3 BRAIN TUMOR

LN-18 and LN-229 were considered and treated with the ethanolic extract of *Boswellia serrata* gum resin. This experiment resulted in proving cytostatic and cytotoxic effect due to the induced apoptotic activity by the use of boswellic acid (Nand et al., 2016).

In brain tumor, a severe effect of edema is seen. This edematous condition when gets severe will be morbid to the life of patients. Various effects of *Boswellia serrata* include anti-inflammatory and antitumor activities and this principle can be used in clinical trials to treat edematous conditions in brain tumors. Studies have been conducted, where the patients were treated with *Boswellia serrata* extracts showing significantly low edematous condition than others with various comparative drugs (Kirste et al., 2011).

4.4.4 GLIOMA

Various boswellic acids like ABA (acetyl-β-boswellic acid), AKBA, and β-boswellic acid show anticancerous activity due to its property of cytotoxicity on malignant glioma cells. Even the low-concentration administrations of boswellic acids show anticancerous activity basically by induction of apoptosis and also apoptosis induced by BA is independent of formation of free radicals. Boswellic acids are known to induce the expressions of various tumor depressant genes like p21 that does not depend on p53 to cause apoptosis. BA-induced p21 expression does not alter the levels of BCL-2 and BAX proteins (Glaser et al., 1999). Further investigations were made on BCA; an *in vivo* study reported the efficiency of BA to reduce perifocal edema. The results of treatment were remarkable in varied concentrations of boswellic acid dosages (Winking et al., 2000). Another experiment was conducted to check the growth-inhibiting activity of extracts of *Boswellia serrata* was made in which immunocompromised mice contain C6 glioma tumor xenograft as targets. A derivative of boswellic acid containing a modified A ring (2 cyano, 3 enone), CEMB (cyanoenone of methyl boswellates) were administered that showed high-growth inhibition, pro-differentiative, anti-inflammatory, and antitumorous activity that cured the glioma of the subject (Ravanan et al., 2011).

4.4.5 BREAST

Suhail et al. (2011) showed that the essential oil extracts of *Boswellia serrata* reduce cell viability and increase cell death in tumor tissues, thus showing anticancerous activity in all human breast cancer cell-lines (Khan et al., 2016). Studies on the effect of AKBA to inhibit breast cancer were made and it is found that AKBA inhibits cell proliferation and invasion of tumors and confirmed by reduced CXCR4 protein levels (Park et al., 2011). A cream made of boswellic acid was used to study the prevention of breast cancer caused by radiations and it was found that it effectively reduced skin superficial symptoms and erythema (Khan et al., 2016).

4.4.6 GASTRO INTESTINAL CANCERS

Gastrointestinal system constitutes various glands and muscle systems which can be infected by cancer due to various reasons that are termed gastrointestinal cancers. The prevention and cure of such cancers using *Boswellia serrata* extracts are studied, such as hepatic, gastric, pancreatic, and colorectal cancers (Garcea et al., 2003).

4.4.7 HEPATIC CANCER

Hep G2 cells were subjected to administration of keto-β-boswellic acid, β-boswellic acid, and AKBA to study the pathways of inhibition of cell proliferation and apoptotic efficiency of *Boswellia serrata* extractions of these cell-lines. Studies showed that β-boswellic acids inhibit DNA histone complex formations at high concentration and keto-β-boswellic acid and AKBA increases the activity of caspase cascade (caspase-3, caspase-8, and caspase-9) which are associated with PARP cleavage (Khan et al., 2016).

4.4.8 COLO-RECTAL CANCER

Various *in vitro* and *in vivo* studies are conducted to prove the anticancerous properties of *Boswellia serrata* are also effective against colon cancer. HT-29 colo-rectal cancer cell-lines were subjected to the administration of the acetyl-keto-β-boswellic acid (AKBA), keto-β-boswellic acid (KBA), and β-boswellic acid (β-BA), to study apoptotic and antiproliferative efficiency where BA induces the activation of caspase-8-dependent apoptotic pathway and also Fas/Fas ligand-independent action in these cell-lines (Liu et al., 2002).

AKBA is investigated for anticancerous study using the xenograft model, where it was able to arrest cell cycles in certain stages thus preventing colo-rectal cancer (Toden et al., 2015).

4.4.9 PROSTATE

The chemiresistant androgen independent PC-3 cell-lines of prostate cancer were administered with 3-α-acetyl-11-keto-α-boswellic acid (AKBA) extracted from *Boswellia serrata in vitro* and *in vivo,* to study cytotoxic and anticancerous activity respectively. AKBA is found to cause toxicity in mitochondrial activity by releasing cytochrome c, causing DNA fragmentation, and also inhibiting NF-κB signaling (Syrovets et al., 2005b). The same is investigated using a chick chorioallantoic membrane with transplanted xenograft of prostate cancer cells (Büchele et al., 2006; Khan et al., 2016).

4.4.10   LEUKEMIA

Leukemia is one of the commonly seen cancers which is a global concern. Leukemia is known to be more incident in males than in females and also 10x higher in adults than in children. Various leukemia cell-lines of humans like (THP-1HL-60, K-562, MOLT-4, U-937) were administered using gum resin extract of *Boswellia serrata* to study its cytostatic, cytotoxic, pro-apoptotic, and anticancerous activity using flow cytometry (Khan et al., 2016).

The administration of AKBA is known to activate caspase-8 either by inducing activity of caspase-3 or Bid, thus causing the decrease in mitochondrial membrane potentials (Xia et al., 2005).

The triterpenediol induces nitric oxide formation that cleaves Bcl-2 and causes translocation of Bax in mitochondrial membrane leading to the loss of membrane potential thus releases cytochrome c. this action induces expression of death receptors like TNF-R1 and DR4 that activates caspase-8. Thus, the above study shows that triterpenediol induces apoptosis and also increases oxidative stress in both intrinsic and extrinsic apoptotic pathways (Bhushan et al., 2007).

4.5   MECHANISM OF ACTION AND MOLECULAR TARGETS OF BOSWELLIC ACID

Boswellic acids being anti-inflammatory and anticancerous have many molecular targets like enzymes, receptors, transcription factors, growth factors, and other biomolecules that are involved in cell growth and proliferation. Experiments have shown that boswellic acids and its analogues regulate apoptosis through inhibitory and supportive actions on the mentioned molecular targets, thus inhibiting tumor formation. Experiments show AKBA (acetyl-11-keto-β—boswellic acid) when administered to meningioma cells to study cytotoxic activity; it is found that the phosphorylation of extracellular signal induced kinase-1 and -2 prohibition and also it impairs the motility of meningioma cells that are induced by platelet-derived growth factors (Park et al., 2002). When HCT-116 was administered with boswellic acid as a treatment of colon cancer, it leads to the reduction in cyclin E, cyclin D, and Cyclin-dependent kinases like CDK2 and CDK4 and also reduces (pRb) phosphorylated Rb) (Liu et al., 2006). Syrovets and coworkers studied the effect of boswellic acid when administered to NF-κB which is a transcription factor and when inhibited it causes downregulation of activated human monocytes (Syrovets et al., 2005b). Boswellic acid also targets the activation of cell transducers, transcription factors STAT-3 (Signal Transducers and Activators of Transcription-3) and also regulates cell proliferation, survival, chemo resistance, and angiogenesis in tumor cells and MM (Multiple Myeloma). AKBA affects the genes, thus inhibiting cell proliferation and angiogenesis by targeting various factors (Kunnumakkara et al., 2009).

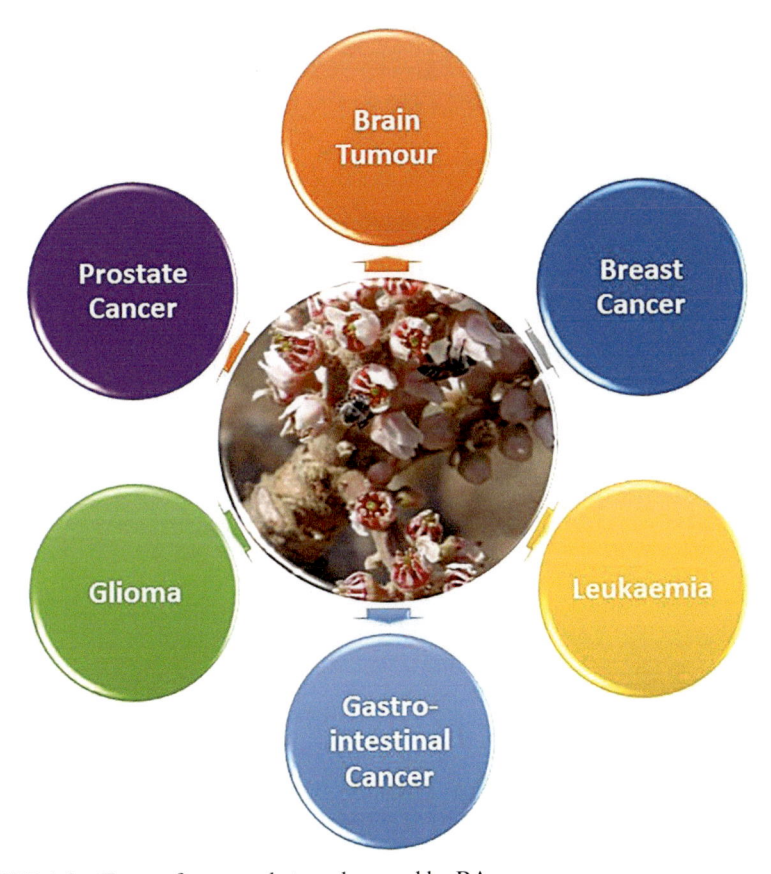

FIGURE 4.3 Types of cancers that can be cured by BA.

Molecular targets of AKBA in prostate cancer include DR5 (Death Receptor 5) androgen receptor, VEGFR2 (Vascular Endothelium Growth Factor Receptor 2) leading to the inhibition of cell proliferation by causing induced apoptosis and decreased angiogenesis (Lu et al., 2008; Yuan et al., 2008; Pang et al., 2009).

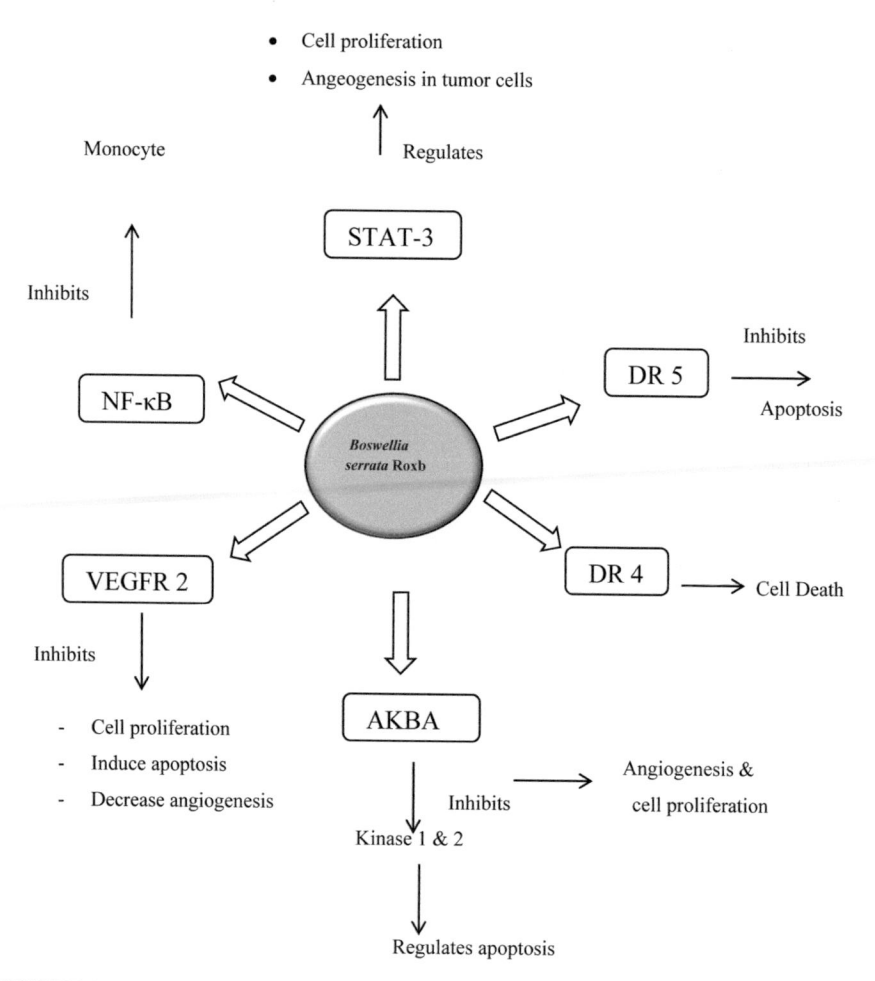

FIGURE 4.4   Molecular targets of Boswellic acid.

4.5.1 *INHIBITOR OF HUMAN TOPOISOMERASES I AND II*

Human topoisomerases are the enzymes that control the molecular structures and arrangements of DNA (Syrovets et al., 2000). Topoisomerases are the enzymes that act as ligases and lyases on DNA double strands during DNA recombination, replication fork formation, transcription, etc. DNA topoisomerases-1 causes DNA strands by opening the twists of DNA by changing the torsion. They also create torsions to pair up and wind the DNA. Rapidly dividing cells should generally have high DNA-replicating activities, where DNA topoisomerases are much needed. This condition is usually seen in tumor formation and cancerous condition. Boswellic acids are known to inhibit DNA topoisomerase activity. Thus, they can be pharmacologically used to inhibit DNA replication that helps in inhibiting self-proliferation (Nerkar, 2020).

4.5.2 TOPOISOMERASE II DNA CLEAVAGE COMPLEX

Topoisomerase II is the enzyme responsible for DNA supercoiling and entangling. Supercoiling and entangling inhibition is known to be target site to inhibit cancer cell-proliferation. These enzymes are ATP dependent. They pass through DNA acting on the strands using a set of internal interfaces or gates that are dissociable in nature. Topoisomerases have various analogues that can cleave single stranded DNA by unifying two essential enzyme families involved in cleavage mechanisms. This chromosomal detangling machine regulates breakage of DNA to avoid mutations and helps in cytotoxicity. These structures are analyzed by clearing the water molecules and co-catalyzed by a haem-ligand (Nerkar, 2020).

4.5.3 PRECLINICAL ACTIVITY

Hoernlein et al. (1999) discovered and reported their works on inhibition of topoisomerase enzyme by AKBA which is an extract of *Boswellia serrata* Roxb. AKBA in various laboratory experiments has reportedly exhibited the activity of constitutive signal transducer and transcription 3 activator inhibition (STAT-3) in human multiple myeloma cells. Thus it induces separation of STAT-3 by interleukins. AKBA inhibits phosphorylation of STAT-3 pathway. This leads to the separation of cyclin D1, Bcl-xL, VEGF separation (Kunnumakkara et al., 2009). Choi et al. have reported the inhibitory activity of extracts of *Boswellia serrata* that were examined to inhibit induced migration of PDGF (Platelet-Derived Growth Factor) and cell proliferation (Choi et al., 2009).

4.6   SAFETY AND BIOAVAILABILITY ENHANCEMENT OF *BOSWELLIA SERRATA*

Based on various clinical and preclinical activities of *Boswellia serrata*, activity on various animals and cell-lines of humans there are absolutely no mortality, side-effects, or adverse effects of this plant extract that was noticed (Singh et al., 2012). *Boswellia* is traditionally consumed as decoction in various doses, in later days; it can be considered taking in the form of tablets and capsules orally. The balance should be maintained in the doses for the medicine to be effective; the quantity and concentrations must be considered as suggested. Because *Boswellia serrata* is sourced from various places the quality may vary, considering this factor the effectiveness of the extracts from *Boswellia serrata* of different suppliers differs. Thus, the standardization becomes more complex. Clinical results cannot be compared when the extracts from different suppliers are used (Giles et al., 2005). Number of methods and way of approaches were used to discover the potential of pharmacological use of various analogues of boswellic acids to maximize their bioavailability (Du et al., 2015). Studies have also shown that the use of *Boswellia serrata* as a constituent of regular meal also helps in the effective mode of supplementation. The activity of anticancerous activity will be elevated when administered with anionic drugs (Hüsch et al., 2013; Skarke et al., 2012). There are various recent studies that are conducted which use nanoparticles, lipid carriers, micelles, poly (lactic-co-glycolic acid) nanoparticles, liposomes, and emulsions as vectors to help the effective utilization of the administered compound (Aqil et al., 2013).

4.7   STUDIES ON SEMISYNTHETIC BOSWELLIC ACID DERIVATIVES

Various studies have been conducted to assess semi-synthetic assay and alkyl derivatives of boswellic acid, BA145 (3-O-α-butyryl-11-keto- β-boswellic acid) as an efficient anticancerous agent that is toxic due to the inhibition of STAT proteins and NF-κB (Kumar et al., 2012).

Further, more studies conducted by Pathania et al. in 2013 showed that boswellic acids are highly potent in certain concentration ranges causing induced apoptosis via downregulation of PI3K/Akt and Erk. The experiment also proved that BA145 analogue induced autophagy which is a significant mechanism that can regulate angiogenesis and causes cell protection (Pathania et al., 2013).

Ravanan et al. (2011) used boswellic acid with a modified ring structure to develop cyanoenone modified methyl boswellates as an analogue to inhibit nitric oxide production caused by interferons in mice macrophage. It is found as a potent agent to induce apoptosis and to inhibit DNA synthesis when administered in doses to treat C6 rat glioma cells in mouse xenograft model (Ravanan et al., 2011).

Another experiment was conducted on a synthesized derivative of BA, butyl-2-cyano-3, 11-dioxours-1, 12-dien-24-oate and is witnessed to cause anti-proliferatory effect by upregulating apoptotic pathway by the activation of p53-p21-PUMA in HeLa cells. This is also known to inhibit cell signaling cascades like p-AKT and NF-κB (Khan et al., 2011). Khan et al., in 2012, also found that the cleaving activity of butyl-2-cyano-3,11-dioxours-1,12-dien-24-oate in PARP1 also induces caspase-3 cascade activity (Khan et al., 2012). The study of another derivative propionyloxy, that is, 11-keto-β-boswellic acid showed the inhibition of growth and proliferation of HL-60 pro-myelocytic leukemia cell-lines. It is known that the derivative inhibits the activity of topoisomerase-1 and -2 and also it is also found to induce apoptosis and is confirmed by the molecules that showed apoptotic activity when morphologically analyzed. The induced apoptosis is caused by the activation of caspase cascade and PARP cleavage by the derivative (Chashoo et al., 2011).

Boswellic acid semi-synthetic derivatives like 3-α-acet-oxy-4-β-amino-11-oxo-24-norurs-12-ene and 3-α-acetoxy-4-β-amino-24-norurs-12-ene are found to show efficient inhibition of histone deacetylases, thus arresting cell cycles at a certain concentration G1 phase and also causing loss of mitochondrial membrane potentials (Raina et al., 2014).

The semi-synthetic analogues of boswellic acids showed induced apoptosis and cytotoxicity in various cancer cell-lines. It is found that triterpenoid ring actively induces apoptosis by targeting mitochondria-dependent pathways in HL-60 cells and also induces DNA fragmentation, cell shrinkage, condensation of chromatin, fragmentation of nucleus, and membrane blebbing (Qurishi et al., 2010).

The experiments conducted showed that β-boswellic acids containing acyl substituents result in elevated property of chemotherapeutic actions. Another semi-synthetic triterpenoid derivative was synthesized, that is, 3-cinnamoyl-11- keto-β-boswellic acid which has its effect on rapamycin as a target and also induces apoptosis in both prostate and breast cancer. This compound induces antiproliferation and pro-apoptotic activities through inhibition of TOR signaling in a certain range of concentration (Morad et al.,

2013). 11-keto-β- boswellic acid induces the formation of endoperoxide that can be evaluated as an anticancer activity biomarker in many human cancer cell-lines by using sulforhodamine B assay (Csuk et al., 2010).

4.8 DISCUSSION

Boswellia serrata has found a significant importance in the field of medicine in various medicinal fields like Ayurveda, Unani, and other ethnopharmaceauticals. Various experiments provide evidence to the properties like antitumor, anticancer, anti-inflammatory, etc. *Boswellia serrata* is one of the indigenous species and also recently listed out as endangered species. Thus, the use of *Boswellia serrata* for the extractions of its components can be reduced by the use of artificial derivation and analogues of its components like oleo gum resin, boswellic acids. There are analogues of boswellic acid available in the market that can be manipulated accordingly for pharmacological uses. Various plant sources also contain boswellic acid and can be obtained from them. Increasing the number of dedicated plantation for extraction purposes can be done as an agricultural practice also increases socioeconomical per capita development of farmers as individuals. The current source of information on *Boswellia serrata* is obtained from a limited number of experiments conducted. Thus, scholars should continue more number of experiments regarding the inhibitory pathways of boswellic acids by which it inhibits cancers and other inflammations. Investigations on particular cancer during are more suggested by clinicians. Preclinical analysis has shown excellent results of anticancerous activity by the extracts of *Boswellia serrata* Roxb, but the clinically available data is very less known in current-day medicine practice. More investigations on cellular pathways in which boswellic acid inhibits cancer must be known. Molecular targets in induced apoptosis are clearer in the use of *Boswellia serrata* anticancer agents. Standardized extraction procedures should be improvised as various standards give different percentage of efficiency in the action of boswellic acids. Ethno medicines should be considered source where the information of properties of boswellic acids is already being mentioned and practiced since ancient times. Boswellic acids are found to have zero-toxic and no severe side effects except few mild symptoms like diarrhea. These can be used in the field of chemotherapeutic administration. *Boswellia serrata* is thus known as a potent, natural, and safe component to be administered.

4.9 CONCLUSIONS

Boswellia serrata Roxb is well known for gum resin secretions belonging to a certain class of plants known as Frankinscence. It is elucidated that pentacyclic triterpenes like boswellic acids and diene derivatives are responsible for its anticancerous activity. Other than triterpenoids there are also acidic and neutral fractions of the gum that assured 1–3% of boswellic acid constituents, out of which AKBA fractions are bioactive. They are known to cure cancers like Liver, Brain, Prostrate, Colon, and other Gliomas by inhibiting the formation of tumor by reducing the activities of DNA by inhibiting topoisomerases. They also act in apoptotic pathways to prevent neoplastic cell proliferation and also inhibit angiogenesis in which capillaries and blood vessel formation lead to the nourishment of these tumors that are inhibited. The mentions of this activity of Salai Guggal as Gajabhakshya in Sanskrit are found in the original texts of Susruta Samhita, and Charaka Samhita as references in systems like Ayurveda and Unani. Various effects of these gum extracts are known and can be listed as immunomodulatory anti-inflammatory and leukotriene inhibitors. Boswellic acid has also been reported as hypolepidemic and hepatoprotective in nature that helps in the reduced activity of tumor forming due to fat accumulation in hepatocellular cancers. These components also reduce Cyclin E, D, and cyclin-dependent kinases that help in apoptosis, and thus reduce the target tumors. Molecular targets of AKBA in apoptosis are death receptor-induced androgen receptors like VEGFR2; this also inhibits angiogenesis. The preclinical studies are conducted in various xenograft models in mice and rabbits and also in human cell-line studies, where the mode of action varies in every single experiment. The results show that almost every human-related cancer can be treated by using boswellic acids and its analogues. In *in vitro* conditions, boswellic acid analogues show its maximum potency in avoiding inflammation and cell-proliferation as per the known knowledge by ancient texts but when administered in clinical purposes, the activity of these acids is very low in nature. Thus, the way of approach in clinical treatments must be modulated so that the efficiency of the drug increases and also the utilization of the administered boswellic acid can be increased. The extractors should consider the bioavailability of the compound and the plant source and also should maintain balance. Various approaches of administration are seen in different systems of medication, considering the effectiveness the standardization should be varied accordingly. Due to its various properties, this is also suggested to be included in everyday meal or various lipid emulsions, lipid nanoparticles can also be used as carriers to reach the compounds to target organs. The use of *Boswellia serrata* extracts is also seen in various cosmetic products like perfumes, creams, lotions, soaps, and detergents (Serrata et al., 2020). In the above-mentioned studies, information about phytochemical analysis of gum portions of *Boswellia serrata* has reported very insufficient data. In later years, various pentacyclic triterpene compounds that are analogues of *Boswellia serrata* extracts are found, which needs more analysis in both structure and property-based data. Thus, special attention should be given on the clinical trials to know its therapeutic value that has tremendous potential to cure cancer.

KEYWORDS

- ***Boswellia serrate***
- **boswellic acid**
- **pharmacological effects**
- **cancer**
- **apoptotic pathway**

REFERENCES

Agrawal, S. S.; Saraswati, S.; Mathur, R.; Pandey, M. Antitumor Properties of Boswellic Acid Against Ehrlich Ascites Cells Bearing Mouse. *Food Chem. Toxicol.* **2011,** *49* (9), 1924–1934. https://doi.org/10.1016/j.fct.2011.04.007

Aktionen, S. *Untersuchungen fiber die Anteile.* **1967,** *298,* 263–298.

Aqil, F.; Munagala, R.; Jeyabalan, J.; Vadhanam, M. V. Bioavailability of Phytochemicals and Its Enhancement by Drug Delivery Systems. *Cancer Lett.* 2013. https://doi.org/10.1016/j.canlet.2013.02.032

Bhushan, S.; Kumar, A.; Malik, F.; Andotra, S. S.; Sethi, V. K.; Kaur, I. P.; Taneja, S. C.; Qazi, G. N.; Singh, J. A Triterpenediol from Boswellia Serrata Induces Apoptosis Through Both the Intrinsic and Extrinsic Apoptotic Pathways in Human Leukemia HL-60 Cells. *Apoptosis* **2007,** *12* (10), 1911–1926. https://doi.org/10.1007/s10495-007-0105-5

Borrelli, F.; Capasso, F.; Capasso, R.; Ascione, V.; Aviello, G.; Longo, R.; Izzo, A. A. Effect of Boswellia Serrata on Intestinal Motility in Rodents: Inhibition of Diarrhoea Without Constipation. *Br. J. Pharmacol.* **2006,** *148* (4), 553–560. https://doi.org/10.1038/sj.bjp.0706740

Büchele, B.; Zugmaier, W.; Estrada, A.; Genze, F.; Syrovets, T.; Paetz, C.; Schneider, B.; Simmet, T. Characterization of 3α-Acetyl-11-Keto-α-Boswellic Acid, a Pentacyclic Triterpenoid Inducing Apoptosis In Vitro and In Vivo. *Planta Med.* **2006,** *72* (14), 1285–1289.

Chashoo, G.; Singh, S. K.; Sharma, P. R.; Mondhe, D. M.; Hamid, A.; Saxena, A.; Andotra, S. S.; Shah, B. A.; Qazi, N. A.; Taneja, S. C.; Saxena, A. K. Chemico-Biological Interactions A Propionyloxy Derivative of 11-keto- % -Boswellic Acid Induces Apoptosis in HL-60 Cells Mediated Through Topoisomerase I & II Inhibition. *Chemico-Biol. Interact.* **2011,** *189* (1–2), 60–71. https://doi.org/10.1016/j.cbi.2010.10.017

Choi, O. B.; Park, J. H.; Lee, Y. J.; Lee, C. K.; Won, K. J.; Kim, J.; Lee, H. M.; Kim, B. Olibanum Extract Inhibits Vascular Smooth Muscle Cell Migration and Proliferation in Response to Platelet-Derived Growth Factor. *Korean J. Physiol. Pharmacol.* **2009,** *13* (2), 107–113. https://doi.org/10.4196/kjpp.2009.13.2.107

Conti, S.; Vexler, A.; Edry-Botzer, L.; Kalich-Philosoph, L.; Corn, B. W.; Shtraus, N.; Meir, Y.; Hagoel, I..; Shtabsky, A.; Marmor, S.; Earon, G.; Lev-Ari, S. Combined Acetyl-11-Keto-β-Boswellic Acid and Radiation Treatment Inhibited Glioblastoma Tumor Cells. *PLoS ONE* **2018,** *13*(7), 1–18. https://doi.org/10.1371/journal.pone.0198627

Csuk, R.; Niesen-barthel, A.; Barthel, A.; Kluge, R.; Ströhl, D. Synthesis of an Antitumor Active Endoperoxide from 11-keto- b -Boswellic Acid. *Eur. J. Med. Chem.* **2010,** *45* (9), 3840–3843. https://doi.org/10.1016/j.ejmech.2010.05.036

Du, Z.; Liu, Z.; Ning, Z.; Liu, Y.; Song, Z.; Wang, C.; Lu, A. Prospects of Boswellic Acids as Potential Pharmaceutics. *Planta Medica* **2015,** *81* (4), 259–271. https://doi.org/10.1055/s-0034-1396313

Efferth, T.; Oesch, F. Anti-Inflammatory and Anti-Cancer Activities of Frankincense: Targets, Treatments and Toxicities. *Seminars Cancer Biol.* **2020,** (January). https://doi.org/10.1016/j.semcancer.2020.01.015

Frank, M. B.; Yang, Q.; Osban, J.; Azzarello, J. T.; Saban, M. R.; Saban, R.; Ashley, R. A.; Welter, J. C.; Fung, K. M.; Lin, H. K. Frankincense Oil Derived from Boswellia Carteri Induces Tumor Cell Specific Cytotoxicity. *BMC Complement. Altern. Med.* **2009,** *9.* https://doi.org/10.1186/1472-6882-9-6

Garcea, G.; Dennison, A. R.; Steward, W. P.; Berry, D. P. Chemoprevention of Gastrointestinal Malignancies. *ANZ J. Surg.* **2003,** *73* (9), 680–686. https://doi.org/10.1046/j.1445-2197.2003.02739.x

Garg, P.; Deep, A. Anti-Cancer Potential of Boswellic Acid: A Mini Review. *Hygeia. J. D. Med.* **2015,** *7* (2), 3590. https://doi.org/10.15254/H.J.D.Med.7.2015.147

Giles, M.; Ulbricht, C.; Khalsa, K. P. S.; DeFranco Kirkwood, C.; Park, C.; Basch, E. Butterbur: An Evidence-Based Systematic Review by the Natural Standard Research Collaboration. *J. Herbal Pharmacother.* **2005,** *5* (3), 119–143. https://doi.org/10.1300/J157v05n03_12

Glaser, T.; Winter, S.; Groscurth, P.; Safayhi, H.; Sailer, E. R.; Ammon, H. P. T.; Schabet, M.; Weller, M. Boswellic Acids and Malignant Glioma: Induction of Apoptosis But No Modulation of Drug Sensitivity. *Br. J. Cancer* **1999,** *80* (5–6), 756–765. https://doi.org/10.1038/sj.bjc.6690419

Grant, S. K. Therapeutic Protein Kinase Inhibitors. *Cell. Mol. Life Sci.* **2009,** *66* (7), 1163–1177. https://doi.org/10.1007/s00018-008-8539-7

Hüsch, J.; Bohnet, J.; Fricker, G.; Skarke, C.; Artaria, C.; Appendino, G.; Schubert-zsilavecz, M.; Abdel-Tawab, M. Fitoterapia Enhanced Absorption of Boswellic Acids by a Lecithin Delivery form (Phytosome ®) of Boswellia Extract. *Fitoterapia* **2013,** *84*, 89–98. https://doi.org/10.1016/j.fitote.2012.10.002

Hussain, S.; Slevin, M.; Ahmed, N.; West, D.; Choudhary, M. I.; Naz, H.; Gaffney, J. Stilbene Glycosides Are Natural Product Inhibitors of FGF-2-Induced Angiogenesis. *BMC Cell Biol.* **2009,** *10*, 1–12. https://doi.org/10.1186/1471-2121-10-30

Khan, M. A.; Ali, R.; Parveen, R.; Najmi, A. K.; Ahmad, S. Pharmacological Evidences for Cytotoxic and Antitumor Properties of Boswellic Acids from *Boswellia serrata. J. Ethnopharmacol.* **2016,** *191*, 315–323. https://doi.org/10.1016/j.jep.2016.06.053

Khan, S.; Chib, R.; Shah, B. A.; Wani, Z. A.; Dhar, N.; Mondhe, D. M.; Lattoo, S.; Jain, S. K.; Taneja, S. C.; Singh, J. A Cyano Analogue of Boswellic Acid Induces Crosstalk Between p53/PUMA/Bax and Telomerase That Stages the Human Papillomavirus Type 18 Positive HeLa Cells to Apoptotic Death. *Eur. J. Pharmacol.* **2011,** *660* (2–3), 241–248. https://doi.org/10.1016/j.ejphar.2011.03.013

Khan, S.; Kaur, R.; Shah, B. A.; Malik, F.; Kumar, A.; Bhushan, S.; Jain, S. K.; Taneja, S. C.; Singh, J. A Novel Cyano Derivative of 11-Keto-β-Boswellic Acid Causes Apoptotic Death by Disrupting PI3K/AKT/Hsp-90 Cascade, Mitochondrial Integrity, and Other Cell Survival Signaling Events in HL-60 Cells. *Mol. Carcinogenesis* **2012,** *51* (9), 679–695. https://doi.org/10.1002/mc.20821

Kirste, S.; Treier, M.; Wehrle, S. J.; Becker, G.; Abdel-Tawab, M.; Gerbeth, K.; Hug, M. J.; Lubrich, B.; Grosu, A. L.; Momm, F. Boswellia Serrata Acts on Cerebral Edema in Patients Irradiated for Brain Tumors: A Prospective, Randomized, Placebo-Controlled, Double-Blind Pilot Trial. *Cancer* **2011,** *117* (16), 3788–3795. https://doi.org/10.1002/cncr.25945

Kumar, A.; Shah, B. A.; Singh, S.; Hamid, A.; Singh, S. K.; Sethi, V. K.; Saxena, A. K.; Singh, J.; Taneja, S. C. Bioorganic & Medicinal Chemistry Letters Acyl Derivatives of Boswellic Acids as Inhibitors of NF- J B and STATs. *Bioorg. Med. Chem. Lett.* **2012,** *22* (1), 431–435. https://doi.org/10.1016/j.bmcl.2011.10.112

Kunnumakkara, A. B.; Nair, A. S.; Sung, B.; Pandey, M. K.; Aggarwal, B. B. Boswellic Acid Blocks Signal Transducers and Activators of Transcription 3 Signaling, Proliferation, and Survival of Multiple Myeloma via the Protein Tyrosine Phosphatase SHP-1. *Mol. Cancer Res.* **2009,** *7* (1), 118–128. https://doi.org/10.1158/1541-7786.MCR-08-0154

Lakka, A.; Mylonis, I.; Bonanou, S.; Simos, G.; Tsakalof, A. Isolation of Hypoxia-Inducible Factor 1 (HIF-1) Inhibitors from Frankincense Using a Molecularly Imprinted Polymer. *Investig. New Drugs* **2011,** *29* (5), 1081–1089. https://doi.org/10.1007/s10637-010-9440-4

Lalithakumari, K.; Krishnaraju, A. V, Sengupta, K.; Subbaraju, G. V, & Chatterjee, A. Safety and Toxicological Evaluation of a Novel, Standardized 3-O-Acetyl-11-Keto-β-Boswellic Acid (AKBA)-Enriched Boswellia Serrata Extract (5-Loxin®). *Toxicol. Mech. Methods* **2006,** *16* (4), 199–226. https://doi.org/10.1080/15376520600620232

Liu, J. J.; Huang, B.; Hooi, S. C. Acetyl-Keto-β-Boswellic Acid Inhibits Cellular Proliferation Through a p21-Dependent Pathway in Colon Cancer Cells. *Br. J. Pharmacol.* **2006,** *148* (8), 1099–1107. https://doi.org/10.1038/sj.bjp.0706817

Liu, J. J.; Nilsson, Å.; Oredsson, S.; Badmaev, V.; Zhao, W. Z.; Duan, R. D. Boswellic Acids Trigger Apoptosis via a Pathway Dependent on Caspase-8 Activation But Independent on Fas/Fas Ligand Interaction in Colon Cancer HT-29 Cells. *Carcinogenesis* **2002,** *23* (12), 2087–2093. https://doi.org/10.1093/carcin/23.12.2087

Lu, M.; Xia, L.; Hua, H.; Jing, Y. Acetyl-keto-β-Boswellic Acid Induces Apoptosis Through a Death Receptor 5-Mediated Pathway in Prostate Cancer Cells. *Cancer Res.* **2008,** *68* (4), 1180–1186. https://doi.org/10.1158/0008-5472.CAN-07-2978

Morad, S. A. F.; Schmid, M.; Büchele, B.; Siehl, H. U.; Gafaary, M. El, Lunov, O.; Syrovets, T.; Simmet, T. A Novel Semisynthetic Inhibitor of the FRB Domain of Mammalian Target of

Rapamycin Blocks Proliferation and Triggers Apoptosis in Chemoresistant Prostate Cancer Cells. *Mol. Pharmacol.* **2013,** *83* (2), 531–541. https://doi.org/10.1124/mol.112.081349

Moussaieff, A.; Mechoulam, R. Boswellia Resin: From Religious Ceremonies to Medical Uses; A Review of In-Vitro, In-Vivo and Clinical Trials. *J. Pharm. Pharmacol.* **2009,** *61* (10), 1281–1293. https://doi.org/10.1211/jpp/61.10.0003

Nand, A.; Roy, K.; Deka, A.; Bordoloi, D.; Mishra, S.; Roy, N. K.; Deka, A.; Bordoloi, D.; Mishra, S.; Prem, A.; Sethi, G.; Kunnumakkara, A. B. The Potential Role of Boswellic Acids in Cancer Prevention and Treatment Cancer Biology Laboratory, Department of Biosciences and Bioengineering, IIT Guwahati, Assam. *Cancer Lett.* **2016.** https://doi.org/10.1016/j.canlet.2016.04.017

Nerkar, A. Docking of Boswellic Acids and their Derivatives on Anti-Inflammatory and Anti-Cancer Target Current Trends in Pharmacy and Pharmaceutical Chemistry Docking of Boswellic Acids and their Derivatives on Anti-Inflammatory and Anti-Cancer Target. *Curr. Trends Pharm. Pharma. Chem.* **2020,** *2* (1), 1–23.

Pang, X.; Yi, Z.; Zhang, X.; Sung, B.; Qu, W.; Lian, X.; Aggarwal, B. B.; Liu, M. Acetyl-11-Keto-β-Boswellic Acid Inhibits Prostate Tumor Growth by Suppressing Vascular Endothelial Growth Factor Receptor 2-Mediated Angiogenesis. *Cancer Res.* **2009,** *69* (14), 5893–5900. https://doi.org/10.1158/0008-5472.CAN-09-0755

Park, B.; Sung, B.; Yadav, V. R.; Cho, S. G.; Liu, M.; Aggarwal, B. B. Acetyl-11-keto-β-Boswellic Acid Suppresses Invasion of Pancreatic Cancer Cells Through the Downregulation of CXCR4 Chemokine Receptor Expression. *Int. J. Cancer* **2011,** *129* (1), 23–33. https://doi.org/10.1002/ijc.25966

Park, Y. S.; Lee, J. H.; Bondar, J.; Harwalkar, J. A.; Safayhi, H.; Golubic, M. Cytotoxic Action of Acetyl-11-keto-β-Boswellic Acid (AKBA) on Meningioma Cells. *Planta Med.* **2002,** *68* (05), 397–401.

Pathania, A. S.; Joshi, A.; Kumar, S.; Guru, S. K.; Bhushan, S.; Sharma, P. R.; Bhat, W. W.; Saxena, A. K.; Singh, J.; Shah, B. A.; Andotra, S. S.; Taneja, S. C.; Malik, F. A.; Kumar, A. Reversal of Boswellic Acid Analog BA145 Induced Caspase Dependent Apoptosis by PI3K Inhibitor LY294002 and MEK Inhibitor PD98059. *Apoptosis* **2013,** *18* (12), 1561–1573. https://doi.org/10.1007/s10495-013-0889-4

Pungle, P.; Banavalikar, M.; Suthar, A.; Biyani, M.; Mengi, S. Immunomodulatory Activity of Boswellic Acids of Boswellia Serrata Roxb. *Indian J. Exp. Biol.* **2003,** *41* (12), 1460–1462.

Qurishi, Y.; Hamid, A.; Zargar, M. A.; Singh, S. K.; Saxena, A. K. Potential Role of Natural Molecules in Health and Disease Importance of Boswellic Acid. *J. Med. Plants Res.* **2010,** *4* (25), 2778–2785.

Raina, H.; Soni, G.; Jauhari, N.; Sharma, N.; Bharadvaja, N. Phytochemical Importance of Medicinal Plants as Potential Sources of Anticancer Agents. *Turkish J. Bot.* **2014,** *38* (6), 1027–1035. https://doi.org/10.3906/bot-1405-93

Ravanan, P.; Singh, S. K.; Rao, G. S. R. S.; Kondaiah, P. Growth Inhibitory, Apoptotic and Anti-Inflammatory Activities Displayed by a Novel Modified Triterpenoid, Cyano Enone of Methyl Boswellates. *J. Biosci.* **2011,** *36* (2), 297–307. https://doi.org/10.1007/s12038-011-9056-7

Saleh, M.; Trinchieri, G. Innate Immune Mechanisms of Colitis and Colitis-Associated Colorectal Cancer. *Nat. Pub.Group* **2010,** *11* (1), 9–20. https://doi.org/10.1038/nri2891

Serrata, B.; Bioactive, R. A.; With, H.; Pharmacological, V.; Mishra, S.; Bishnoi, R. A. M. S.; Maurya, R.; Jain, D. *Activities* **2020,** *13* (11).

Singh, G. B.; Atal, C. K. Pharmacology of an Extract of Salai Guggal Ex-Boswellia Serrata, a New Non-Steroidal Anti-Inflammatory Agent. *Agents Actions* **1986**, *18* (3–4), 407–412. https://doi.org/10.1007/BF01965005

Singh, P.; Chacko, K. M.; Aggarwal, M. L.; Bhat, B.; Khandal, R. K.; Sultana, S.; Kuruvilla, B. T. A-90 Day Gavage Safety Assessment of Boswellia Serrata in Rats. *Toxicol. Int.* **2012**, *19* (3), 273–278. https://doi.org/10.4103/0971-6580.103668

Skarke, C.; Kuczka, K.; Tausch, L.; Werz, O.; Rossmanith, T.; Barrett, J. S.; Harder, S.; Holtmeier, W.; Schwarz, J. A. Increased Bioavailability of 11-Keto-β-Boswellic Acid Following Single Oral Dose Frankincense Extract Administration After a Standardized Meal in Healthy Male Volunteers: Modeling and Simulation Considerations for Evaluating Drug Exposures. *J. Clin. Pharmacol.* **2012**, *52* (10), 1592–1600. https://doi.org/10.1177/0091270011422811

Sterk, V.; Büchele, B.; Simmet, T. Effect of Food Intake on the Bioavailability of Boswellic Acids from a Herbal Preparation in Healthy Volunteers. *Planta Medica* **2004**, *70* (12), 1155–1160. https://doi.org/10.1055/s-2004-835844

Sudharsan, P. T.; Mythili, Y.; Selvakumar, E.; Varalakshmi, P. Cardioprotective Effect of Pentacyclic Triterpene, Lupeol and Its Ester on Cyclophosphamide-Induced Oxidative Stress. *Human Exp. Toxicol.* **2005**, *24* (6), 313–318. https://doi.org/10.1191/0960327105ht530oa

Syrovets, T.; Büchele, B.; Gedig, E.; Slupsky, J. R.; Simmet, T. Acetyl-Boswellic Acids Are Novel Catalytic Inhibitors of Human Topoisomerases I and IIα. *Mol. Pharmacol.* **2000**, *58* (1), 71–81. https://doi.org/10.1124/mol.58.1.71

Syrovets, T.; Gschwend, J. E.; Büchele, B.; Laumonnier, Y.; Zugmaier, W.; Genze, F.; Simmet, T. Inhibition of IκB Kinase Activity by Acetyl-Boswellic Acids Promotes Apoptosis in Androgen-Independent PC-3 Prostate Cancer Cells In Vitro and In Vivo. *J. Biol. Chem.* **2005a**, *280* (7), 6170–6180. https://doi.org/10.1074/jbc.M409477200

Syrovets, T.; Büchele, B.; Krauss, C.; Laumonnier, Y.; Simmet, T. Acetyl-Boswellic Acids Inhibit Lipopolysaccharide-Mediated TNF-α Induction in Monocytes by Direct Interaction with IκB Kinases. *J. Immunol.* **2005b**, *174* (1), 498–506. https://doi.org/10.4049/jimmunol.174.1.498

Thorsteinsdottir, S.; Gudjonsson, T.; Nielsen, O. H.; Vainer, B.; Seidelin, J. B. Pathogenesis and Biomarkers of Carcinogenesis in Ulcerative Colitis. *Nat. Pub. Group* **2011**, *8* (7), 395–404. https://doi.org/10.1038/nrgastro.2011.96

Toden, S.; Okugawa, Y.; Buhrmann, C.; Nattamai, D.; Anguiano, E.; Baldwin, N.; Shakibaei, M.; Boland, C. R.; Goel, A. Novel Evidence for Curcumin and Boswellic Acid-Induced Chemoprevention Through Regulation of miR-34a and miR-27a in Colorectal Cancer. *Cancer Prev. Res.* **2015**, *8* (5), 431–443. https://doi.org/10.1158/1940-6207.CAPR-14-0354

Tsukada, T.; Nakashima, K.; Shirakawa, S. Arachidonate 5-Lipoxygenase Inhibitors Show Potent Antiproliferative Effects on Human Leukemia Cell Lines. *Biochem. Biophys. Res. Commun.* **1986**, *140* (3), 832–836. https://doi.org/10.1016/0006-291X(86)90709-6

Upaganlawar, A.; Ghule, B. Pharmacological Activities of *Boswellia serrata* Roxb.-Mini Review. *Ethnobot. Leaflets* **2009**, *13*, 766–774.

Volm, M.; Efferth, T. Prediction of Cancer Drug Resistance and Implications for Personalized Medicine. *Front. Oncol.* **2015**, *5*, 1–14. https://doi.org/10.3389/fonc.2015.00282

Weber, C.-C.; Reising, K.; Müller, W. E.; Schubert-Zsilavecz, M.; Abdel-Tawab, M. Modulation of Pgp Function by Boswellic Acids. *Planta Med.* **2006**, *72* (06), 507–513.

Winking, M.; Sarikaya, S.; Rahmanian, A.; Jödicke, A.; Böker, D. K. Boswellic Acids Inhibit Glioma Growth: A New Treatment Option? *J. Neuro-Oncol.* **2000**, *46* (2), 97–103. https://doi.org/10.1023/A:1006387010528

Xia, L.; Chen, D.; Han, R.; Fang, Q.; Waxman, S.; Jing, Y. Boswellic Acid Acetate Induces Apoptosis Through Caspase-Mediated Pathways in Myeloid Leukemia Cells. *Mol. Cancer Therap.* **2005,** *4* (3), 381–388. https://doi.org/10.1158/1535-7163.mct-03-0266

Yuan, H. Q.; Kong, F.; Wang, X. L.; Young, C. Y. F.; Hu, X. Y.; Lou, H. X. Inhibitory Effect of Acetyl-11-Keto-β-Boswellic Acid on Androgen Receptor by Interference of Sp1 Binding Activity in Prostate Cancer Cells. *Biochem. Pharmacol.* **2008,** *75* (11), 2112–2121. https://doi.org/10.1016/j.bcp.2008.03.005

Zhao, W.; Entschladen, F.; Liu, H.; Niggemann, B.; Fang, Q.; Zaenker, K. S.; Han, R. Boswellic Acid Acetate Induces Differentiation and Apoptosis in Highly Metastatic Melanoma and Fibrosarcoma Cells. *Cancer Detection Prev.* **2003,** *27* (1), 67–75. https://doi.org/10.1016/S0361-090X(02)00170-8

CHAPTER 5

Catharanthus roseus

RAMACHANDRA REDDY PAMURU[1], RAJAGOPAL REDDY S.[2], AMBEDKAR[1,3], CHANDRASEKHAR T.[4], MADHUSUDHANA REDDY A.[2], and CHANDRAMATHI SHANKAR P.[3]

[1]*Department of Biochemistry, Yogi Vemana University, Kadapa, Andhra Pradesh, India*

[2]*Department of Botany, Yogi Vemana University, Kadapa, Andhra Pradesh, India*

[3]*Department of Biotechnology, Yogi Vemana University, Kadapa, Andhra Pradesh, India*

[4]*Department of Environmental Sciences, Yogi Vemana University, Kadapa, Andhra Pradesh, India*

ABSTRACT

Cancer, an alarming disease, occurs in humans without clear explanations causing deaths worldwide. Bioactive compounds are promising drugs to treat the cancer with less side effects compared with chemotherapy and radiotherapy in cancer patients. Several bioactive compounds are extracted from various plant sources with anticancerous activity. One of such plant holding a plethora of anticancer compounds is *Catharanthus roseus*. Over past decades the anticancer properties of this plant have been used for cancer treatment as traditional medicine and were later identified as the bioactive compounds. Moreover, producing bioactive anticancer pharmaceutical drugs is cost effective with no toxicity to the environment and less side effects in the patients treated with these drugs. An attempt is made to review the

Potent Anticancer Medicinal Plants: Secondary Metabolite Profiling, Active Ingredients, and Pharmacological Outcomes. Deepu Pandita and Anu Pandita (Eds.)

Catharanthus, a magic plant for its biology, distribution, phytochemicals, and anticancer properties along with the production of these compounds in submerged cultures in the present chapter.

5.1 INTRODUCTION

Cancer is the second highest mortality chronic disease worldwide. Uncontrolled growth takes up in cancerous tissue and later it spreads to other parts of the body. According to World Health Organization (WHO) about 10 million cancer deaths were recorded during the year 2020. Reasons for cancer are not clear and prevalence is high at Western countries. The available treatment methods of cancer are surgery, chemotherapy, and radiotherapy that cost more and cause severe side effects in patients. None of the scientific group invented/identified a correct drug for complete treatment of cancer without side effects. Still research groups are in search of a perfect drug to kill the tumors. However, out of all therapies natural plant-based medicine could be an alternative method of treating cancer. Several plant-based drugs have been extracted from various plant species and many were tested for their anticancer activity. Natural drugs that are used first time for cancer therapy are vincristine and vinblastine (Costa et al., 2008). Still the identified plant drugs are not so effective and were found with side effects.

Catharanthus roseus (*C. roseus*), a well-known plant available all over the world, belongs to family Apocynaceae. Purple flower is the meaning in Greek for the name of this plant *Catharanthus*. This plant is also commonly named as periwinkle and originated in Madagascar. This plant is a common medicinal and ornamental plant in many places and used in traditional medicine in India, China, and Africa. Common diseases like diabetes, Hodgkin's, and malaria are treated with extracts of *Catharanthus*. Several alkaloids extracted from this plant are in use as clinical medicine for its antispasmodic and antihypertensive properties (Nejat et al., 2015; Sain and Sharma, 2013).

The most important and potential anticancer plant identified is *C. roseus*. The most predominant alkaloid compounds that show antitumor function from this plant are vincristine and vinblastine (Das and Sharangi, 2017). Moreover, above 150 phytochemicals are identified and extracted from *Catharanthus* (Kabesh et al., 2015). There are studies finding the antitumor function of other compounds extracted from this plant. There are five drugs from this plant which are commercially available in the market for the treatment of cancer and other diseases. Combination of nanotechnology with bioactive compounds is another potential area to improve the efficacy of drugs against diseases. The importance *Catharanthus* bioactive compounds and their anticancer properties are presented in the present chapter.

5.2 CATHARANTHUS ROSEUS

Catharanthus roseus (*C. roseus*) common plant grows in Asia, Africa, Europe, and United States continents and is famous as *Vinca rosea, Ammocallis rosea, Lochnera rosea,* and *Pervinca rosea* (Plaizier, 1981). But it is endemic to Western Indian Ocean, that is, Madagascar. This plant has many common names and varies from one region/language to other. Some of the common names for this plant are cape periwinkle, Madagascar periwinkle, bright eyes, graveyard plant, pink periwinkle, old maid, rose periwinkle, everyday Jasmin, etc. (USDA-NGPS, 2021). In India this plant has many local names like sadabahar, sadaapushpa, nityapushpa, shavamnaari, sadaphuli, nayantara, kumtluang, nayantora, billaganneru, baahrama, asephool, etc. *Catharanthus*a flowering ornamental plant belongs to dicotyledon and grows elsewhere.

5.2.1 BOTANY

The evergreen plant *Catharanthus* is a small herb or shrub grows up to 60 cm. The leaves of this plant are long oval to oblong hairless shiny green with short petiole and pale midrib. Leaves are oppositely arranged and approximately 2.5–9.0 cm long and 1.0–3.5 cm broad. After germination *Catharanthus* grows very fast and reaches adult flowering stage in 15–25 days in normal conditions. Flowers hold basal tube (2.5–3.0 cm long) attached with five lobes like petalsin two colors (white or rosy-purple to dark pink) with centered dark red. Self- or insect-mediated (butterflies) pollination is common fertilization methods in this plant. Paired pod fruits of length of about 2.0–4.5 cm and width of 3.0 mm are seen in *Catharanthus*. The whole *Catharanthus* plant along with flowers and fruit is presented in Figure 5.1. Taxonomical classification of *Catharanthus* is as follows; Kingdom-Plantae, Division-Magnoliophyta, Class-Magnoliopsida, Order-Gentianales, Family-Apocynaceae, Genus-*Catharanthus*, and Species-*roseus*.

FIGURE 5.1 *Catharanthus roseus* complete plant, two types of flowers and fruit.

5.2.2 GEOGRAPHIC DISTRIBUTION OF PLANT

Catharanthus is a worldwide well-distributed ornamental tropical or subtropical plant and is a native of Madagascar, a large island located in Western Indian Ocean which is located near Africa. It mostly grows during rainy seasons and propagates widely through seeds at a soil pH 5.5–6.0. Leaves of this plant have thick vax coat which facilitates this plant to tolerate high abiotic stress (dry, frost-free, humid, and saline) and has no boundaries to grow even at different altitudes of land (0–900 m heights), during summers under sunlight, under shades and well-drained soils. It is a common plant to find everywhere in the Asian and African continents. Because of its easy growth and distribution in different habitats such as grass lands, dry waste lands, sandy soils, inland riverbanks, shrub lands, dunes in savannas, houses, beaches, roadsides, limestone rocks, deserts, and different other places, it was introduced into many countries and traveled across continents. Moreover, after the discovery of *Catharanthus* medicinal properties are cultivated widely as a commercial crop in countries like India,

Africa, United States, Spain, Australia, Southern Europe (PROTA, 2011; Łata, 2007; Joy et al., 2008). It survives in warm climates throughout the year, but it may not survive in cool springs, wet soils, and watery soils due to fungal and bacterial diseases (Thomas and Latimer, 1996; Whiting et al., 2011). Moreover, in many countries like India, South Africa, China, Mexico, and Malaysia *Catharanthus* is used as a medicinal and/or ornamental plant (Patel et al., 2012; Ong et al., 2011).

5.3 PHOTOCHEMISTRY AND MEDICINAL PROPERTIES

Phytochemicals are the secondary metabolites of plants. *C. roseus* is a treasure for phytochemicals. Since it is a common plant all over the world it is used as medicine for treating various ailments. All the parts of *Catharanthus* such as leaves, roots, stem, flowers, and fruits are used as medicine. Applications of modern biological methods are in use for developing traditional medicine industrially.

5.3.1 PHYTOCHEMICALS

Chemicals that are isolated from various parts of plants using modern biophysical techniques are collectively called phytochemicals. These chemicals are having high potentiality in treating various chronic diseases including cancer. Phytochemicals of *Catharanthus* are known and pharmaceutically important. The two major cytotoxic dimeric alkaloids isolated from this plant are vincristine and vinblastine. The amount of these two chemicals varies from different parts of the plant. Very low amounts were identified in the leaves of *Catharanthus* (Sevestre-Rigouzzo et al., 1992). There are different types of compounds such as steroids, polypehnols, flavonoids, anthocyanins, iridoid glucosides, and glycosides identified in different parts of *C. roseus* (Mustafa and Verpoorte, 2007). Natural indicator for acid–base testing has been isolated from the flowers of *Catharanthus* (Candido and Martinez, 2009). However, the compounds present in stems and leaves are similar, but they differ with petals and seeds (Ferreres et al., 2008). The two major compounds isolated from *Catharanthus* are alkaloids and phenolics. Figure 5.2 shows the structure of various phytochemicals isolated from *C. roseus*.

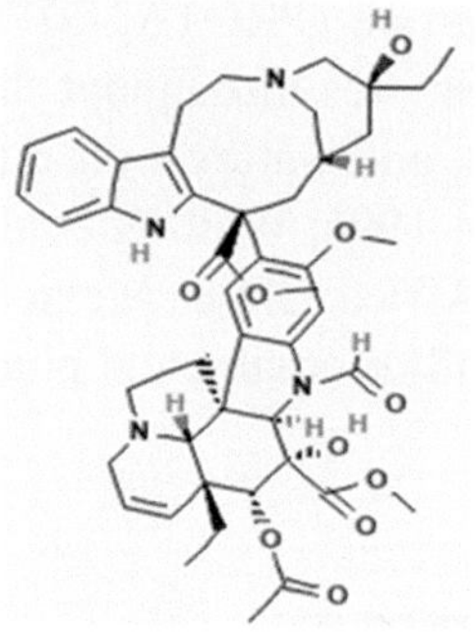

Vincristine

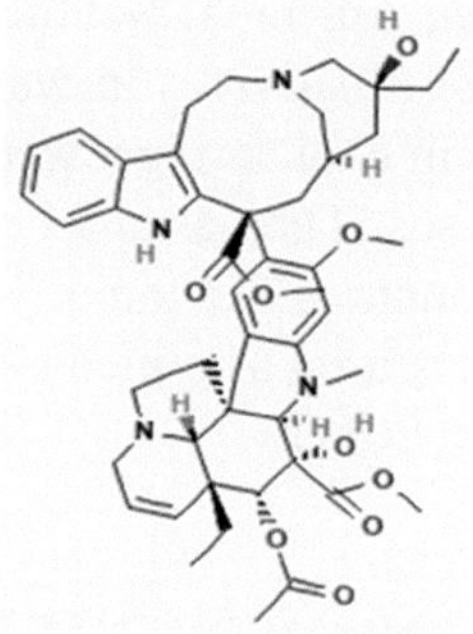

Vinblastine

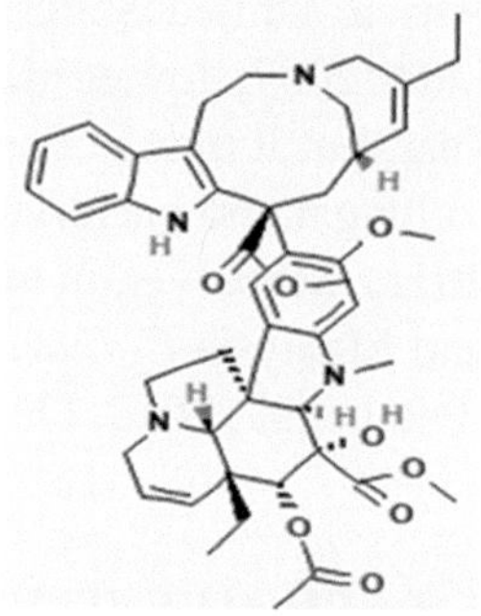

3,4–Anhydro vinblastine

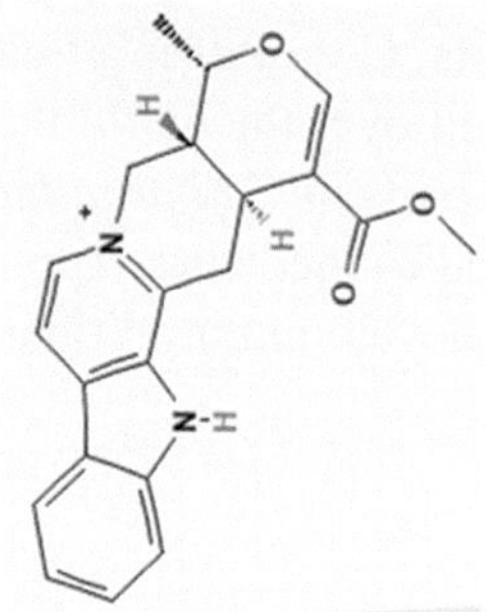

Serpentine

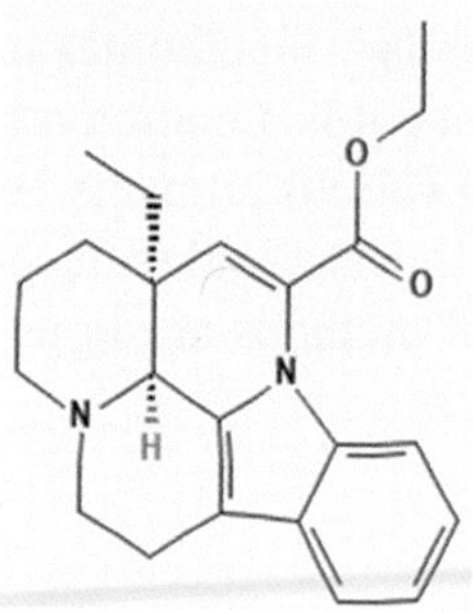

Vinopocetine

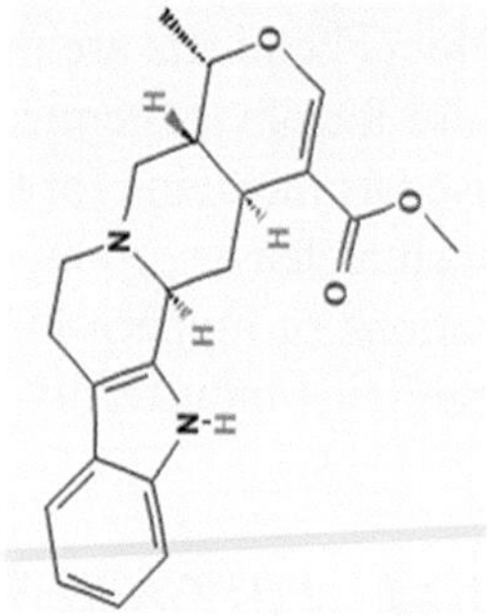

Ajmalicine

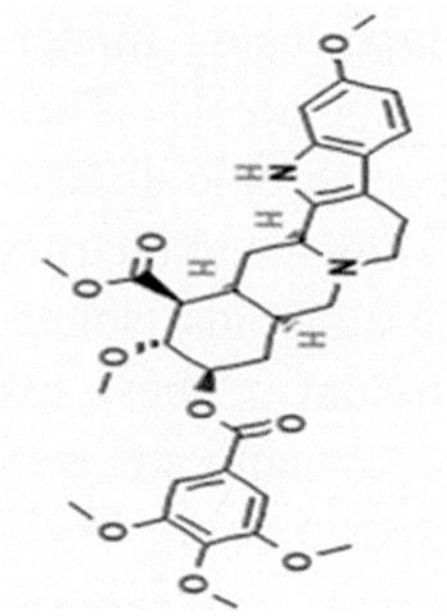

Reserpine

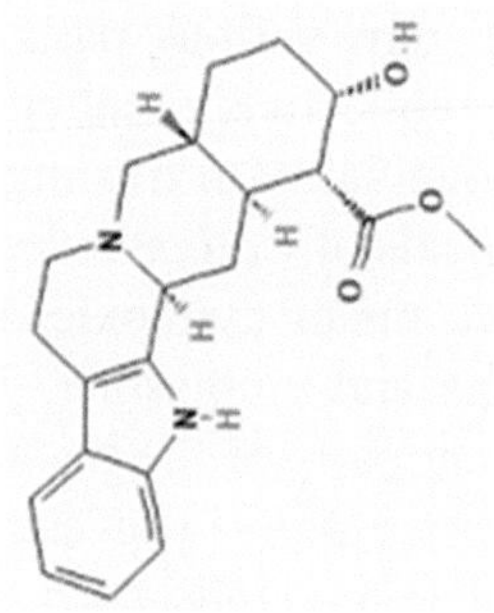

Yohimbine

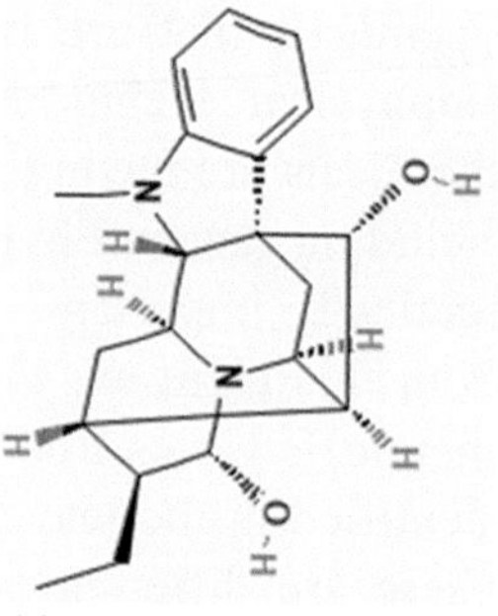

Ajamaline

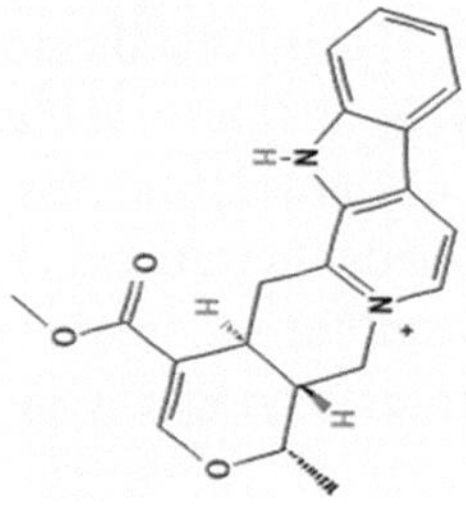

Serpentine

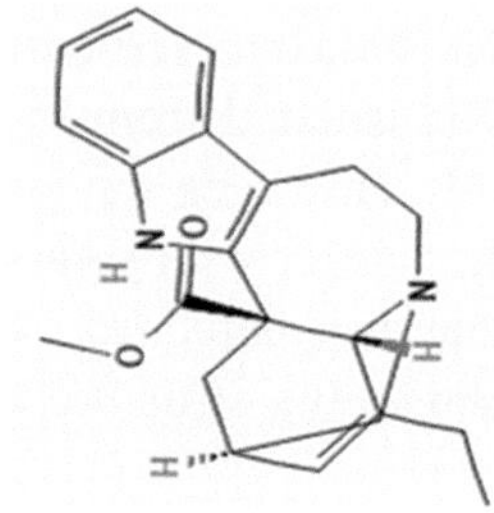

Catharanthine

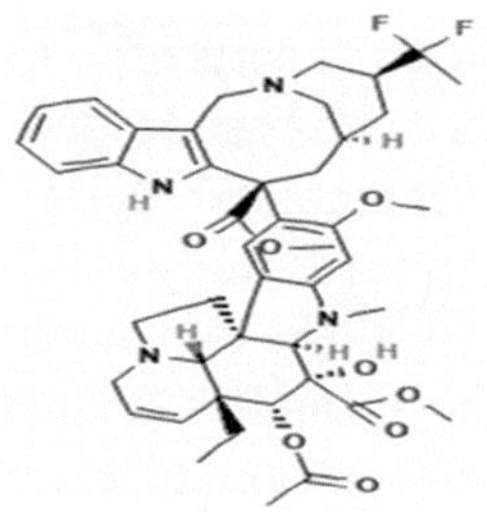

Vinflunine

FIGURE 5.2 *(Continued)*

Vindoline

Vindolicine

Vindolinine

Catharoseumine

Tabersonine

Tryptamine

Catharanthamine

17-Deacetoxycyclovinblastine

Vinorelbine

Cycloleurosine

17-Deaceto xyvinamidine

Vinposidin

FIGURE 5.2 *(Continued)*

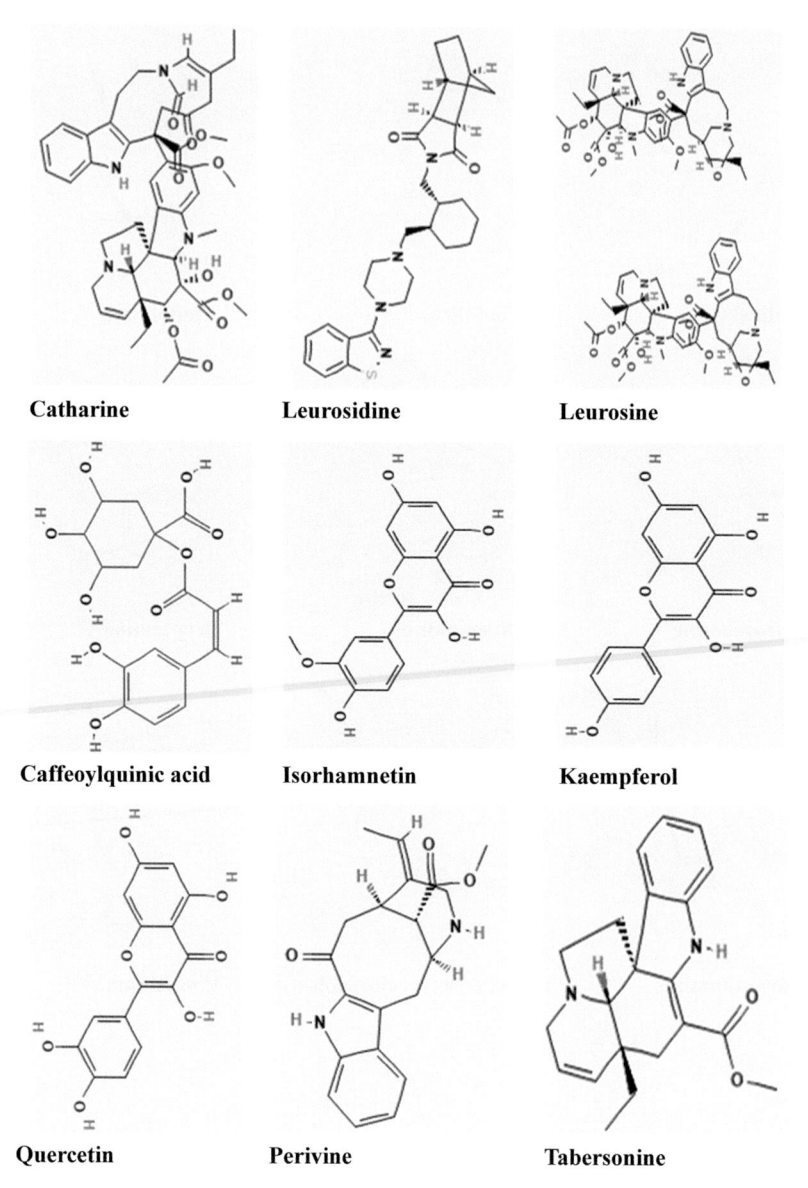

FIGURE 5.2 Identified *Catharanthus roseus* phytochemicals and their structure.

5.3.1.1 ALKALOIDS

An amount of 130 indole terpenoid alkaloids has been extracted from various parts of *Catharanthus* (Daniel, 2006; Hisiger and Jolicoeur, 2007; Renault et al., 1999; Wang et al., 2012). Majority of these alkaloids appear in high content during flowering stage which are holding distinct medicinal properties (Schmelzer and Gurib-Fakim, 2008). Many alkaloids that are extracted/isolated from this plant are presented in Table 5.1. Hisiger and Jolicoeur (2007) reported that out of all only five *Catharanthus* alkaloids isolated are commercialized such as vincristine, vinblastine, serpentine, 3,4–anhydro vinblastine, and ajmalicine. The two velbabanamine group alkaloids (vinorelbine and vinflunine) derived from precursor molecules vindoline and catharanthine are holding highest pharmacological role, compared to vinblastine that shows structurally with these molecules (Nirmala et al., 2011).

5.3.1.2 PHENOLICS

Another set of major phytochemicals along with alkaloids identified in *Catharanthus* are phenolic compounds. Phenolics are major secondary metabolites holding one to many phenolic groups and are found in all plant species. Flavonoids, cinnamic acid derivatives, and anthocyanins belong to phenyl-propanoids and 2,3-dihydroxy-benzoic acid is holding radical scavenging ability found in *Catharanthus* (Mustafa and Verpoorte, 2007). Table 5.1 shows various phenolic compounds isolated/extracted from *C. roseus*.

5.3.2 CATHARANTHUS IN TRADITIONAL MEDICINE

In the regions of India and Africa *Catharanthus* has been used since ancient days as medicinal plant for treating various ailments. This plant is well recognized in indigenous Indian Ayurveda for the treatment of cancer and diabetes due to its antioxidant, antimicrobial, antidiabetic, antitumor, and antimutagenic principles in leaves, roots, stem, and flowers (Chopra et al., 1956; Grover et al., 2002). Since the 1700s this plant is grown as an ornamental plant in Europe and later it became popular folk medicine with wide applications. Europeans also believe that it is a magic plant to ward off evil spirits. Violet of the sorcerers is referred by French people for this plant. A wide variety of diseases/disorders are treated using *Catharanthus* plant as traditional medicine and some are ocular inflammation, insect stings, tranquilizer, homeostasis (fever), diabetes, cancer, lowers hypertension, etc.; they are also used as disinfectant, to stop bleeding, comfort lung congestion, sore throats, eye infections and irritations, etc. The boiled leaves of this plant

TABLE 5.1 Alkaloids and Phenolic Compounds Isolated/Extracted from *Catharanthus roseus.*

S. no.	Part of the Plant	Method of extraction/ Identification	Phytochemicals	References
1	Protoplast-derived tissue	Yeast-elucidated extraction	Vinblastine and vincristine	Maqsood and Abdul (2017)
2	Flower petals	UPLC-Q-TOF	Vinblastine and vincristine	Schweizer et al. (2018)
3	Dried whole plant	Diode array detector HPLC	Vinblastine, vincristine, vindoline, yohimbine and catharanthine	Liu et al. (2016)
4	Roots, stem, and leaves	HPLC-Ultraquadrupole	Vindoline, vincristine, ajmalicine, serpentine, vinblastine, catharanthine	Jeong and Heung (2018)
5	Cambium culture	CMCs-ultraviolet extraction	Vincristine, vindoline, catharanthine, and vinblastine	Moon et al. (2018)
6	Leaves, roots and stem	Ultra HPLC-Tandem MS	Vincristine, vindoline, reserpine, vindesine, ajamaline, ajmalicine, and vinblastine	Kumar et al. (2018)
7	Leaves	IR, UV, NMR and MS	Vindolicine, perivine, vindolidine, vindoline, serpentine, and vindolinine	Tiong et al. (2015)
8	Dried whole plant	qRT-PCR, HPLCand UV-C	Vinblastine, vindoline, vincristine, catharanthineandvindogentianine	Moon et al. (2017)
9	Hairy root cultures	Reverse HPLC with UV detector	Vinblastine, vincristine, and catharanthine	Hanafy et al. (2016)
10	Crude extracts of aerial parts	Chromatographical centrifuge	Vindoline and catharanthine	Kotland et al. (2016)
11	Flowers, roots, stem and leaves	DAD-HPLC and analysis	Vindolinine, ajmalicine, vincristine, anhydrovinblastine, serpentine, catharanthine, and vinblastin	Pan et al. (2016)
12	Dried whole plant	HPLC with NF-κβ and JNK	Dimeric indole alkaloids and cathacunine	Wang et al. (2016)

TABLE 5.1 *(Continued)*

S. no.	Part of the Plant	Method of extraction/ Identification	Phytochemicals	References
13	Stem	Imaging MS and single-cell MS	Terpenoid Indole Alkaloids	Yamamoto et al. (2016)
14	Flowers, roots, stem, and leaves	Reviewed on methods of extraction	Terpenoid Indole Alkaloids	Almagro et al. (2015); Pham et al. (2020)

HPLC, High-Performance Liquid Chromatography; MS, Mass Spectrometry; NF-$\kappa\beta$, Nuclear Factor Kappa light chain enhancer of activated B lymphocytes; JNK, c-Jun N-terminal Kinase; DAD, Diode Array Detector; UV, Ultra-Violet; IR, Infra-Red; NMR, Nuclear Magnetic Resonance; qRT-PCR, quantitative Reverse Transcriptase-Polymerase Chain Reaction; UPLC-Q-TOF, Ultra HPLC-quadruple time of MS.

are believed to control diabetes effectively, which turns its role in modern medicine in the 1950s.

Besides its use in traditional medicine by various societies *Catharanthus* has been identified with a wide number of applications in pharmaceuticals. It is found with more than 120 medicinal/pharmaceutically important alkaloid compounds holding terpenoid and indole structures (Van der Heijden et al., 2004). Many phytochemicals of *Catharanthus* are isolated and characterized. Some are anticancer vincristine, vinblastine, antihypertensive ajmalicine, sedative serpentine, etc. (Sottomayor and Ros Barcelo, 2005). Though the vincristine and vinblastine are low in concentration in *Catharanthus*, they are widely used as drugs in combination with other chemicals used for the treatment of leukemia and lymphomas (Gidding et al., 1999).

5.4 ANTICANCEROUS ACTIVITY OF *CATHARANTHUS ROSEUS*

A significant improvement in medical oncology is the use of cytotoxic medications for cancer chemotherapy. Although these drugs are used to target tumor cells, the majority of them can also induce genotoxic, carcinogenic, and teratogenic outcomes in nontumor cells (Chung et al., 1998; Philip, 2005). These harmful consequences limit the application of chemotherapeutic agents despite their high effectiveness in the killing of target cancerous cells. Therefore, the search for alternative or corresponding drugs that are successful on cancer cells while showing insignificant toxicity to normal cells is an operational area of research (Tang et al., 2003). Many of these investigations are plant-based, folkloric medicine from various civilizations around the world. Moreover, a report from WHO (World Health Organisation, 1996) stated that about 80% of the world population is wholly or moderately dependent on plant-based drugs.

5.4.1 *RESEARCH ON CATHARANTHUS*

Anticancerous activity of biological compounds, especially alkaloid, mainly depends on the presence of antineoplatic elements like nitrogenous atoms. *Catharanthus*, a proven anticancerous plant, is rich with compounds holding antineoplastic atoms. Many studies have been focused on determining the anticancerous activity of natural *C. roseus* all over the world. Several indole alkaloids commonly named vinka alkaloids are widely used to stop the cell mitotic division which ultimately helps in controlling the growth of cancer cells (Almagro et al., 2015). Sottomayor et al. (2006) identified the two anticancer compounds vinblastine and vincristine from *Catharanthus*. But, Moudi et al. (2013) identified vindesine, vinflunine, and vionorelbine, the derivatives of the above two compounds. About six compounds isolated from *Catharanthus* showed potential anticancer activity and are released commercially for treating different cancer types (Table 5.2). The commercial names are oncovin (vincristine), velban (vinblastine), Navelbine (vinorelbine), vinflunine, cathachunine, and catharanthine. The first patented *Catharanthus* compound is vinposidin/leurosidin, an antimitotic molecule by Eli Lilly Company in 1974 (Keglevich et al., 2012).

5.4.2 DOSAGE AND ANTICANCER ACTIVITY

Studies confirmed that the anticancerous activity of *Catharanthus* alkaloids also emphasized on the dose and time of exposure-related function of these compounds. In general, an optimum dose of compounds can block the microtubule function effectively and causes apoptosis within a short span of time, whereas at a low concentration apoptosis may be induced after a long period of exposure (Attard et al., 2006). Almagro et al. (2015) and Van der Heijden et al. (2004) tested the anticancerous efficacy of vinblastine, vincristine, and vinflunine against human cancer xenografts and murine cancer tissue and found that vinflunine has great binding affinity toward calmodulin than vincristine and vinblastine which shows highest binding with tubulins. In vitro cytotoxic effects of catharoseumine holding unique peroxide bridge have been reported against human cancer cell lines (HL-60) (Wang et al., 2012).

5.4.3 MECHANISM OF ANTICANCER ACTIVITY

Microtubules of the cellular cytoskeletal components are responsible for mitotic/meiotic spindle formation during cell division. Spindle fibrils separate the chromosomes during anaphase. Besides this, microtubules are also held responsible for transportation, cell structure maintenance, and different other processes in the cell. Alkaloids of *Catharanthus* aggravate programmed cell death and lower the cell division through altering the microtubular dynamics (Wang et al., 2012). Microtubule formation and dissociation depends on polymerization and depolymerization of its basic heterodimeric units α- and β-tubulin. Potential binding of tubulins for the formation of microtubules is

TABLE 5.2 Anticancer Compounds of *Catharanthus roseus* and Their Testing.

S. no.	Compounds of *Catharanthus*	Method of testing	Type of cancer	References
1	Vinblastine	In vivo Chronic Lymphocytic Leukemia	Leukemia	Bates et al. (2013)
2	Vincristine	In vitro B cell lymphoma	Lymphomas	Qiu et al. (2018)
		In vitro EA hy926 h-umbilical vein cells; K562 h-chronic myelogenous leukemia cells; HL60 h-acute promyelocytic leukemia cells.	Leukemia	Wang et al. (2016)
3	Catharanthine	In vitro HCT116 h-colorectal cancer cells	Colorectal cancer	Siddiqui et al. (2010)
4	Vinflunine	In vivo human patients	Urethra cancer	Schinzari et al. (2018)
		C	Lung carcinoma	Krzakowski et al. (2010.)
5	Vinorelbine	In vivo clinical testing	Lung carcinoma	Nazir et al. (2016)
		In vivo clinical testing	Breast carcinoma	
		In vivo clinical testing	Solid carcinoma	Bahleda et al. (2018)
6	Cathachunine	NF-κβ and JNK pathways	Leukemia	Wang et al. (2016)

NF-κβ, Nuclear Factor Kappa light chain enhancer of activated B lymphocytes; JNK, c-Jun N-terminal Kinase.

maintained by guanosine triphosphate. Moreover, the stability of tubulins in the cellular system is maintained by electrostatic and van der Wall bonding and these can be easily destabilized by external molecules when they bind to it (Coderch et al., 2012). Jordan (2002) explained the destabilization of microtubule, which leads to programmed cell death in cancerous tissue. Moreover, the presence of depolymerizing agents causes no assembly of tubulins, thereby restricting microtubule formation (Perez, 2009). Cell apoptosis is induced in the cancerous tissue by *Catharanthus* alkaloids and their derivatives through depolymerization of microtubules when these bind with the surface of microtubular heterodimers near the GTP-binding site (Bolanos-Garcia, 2009). Gigant et al. (2005) reported the potential anticancerous functions of *Catharanthus* alkaloids through depolymerization of microtubules. Among tested alkaloids they ranked them as vincristine > vinblastine > vinorelbine > vinflunine for their anticancer activity. Schutz et al. (2011) and Barbier et al. (2014) reported a new mechanism for the action of alkaloids against cancer growth. They identified that calmodulin after binding with microtubule-associated protein can inhibit the protein synthesis through blocking internal amino acid biosynthetic pathways. However, the interactions of *Catharanthus* alkaloids with calmodulin are also reported. The mechanism of action of other *Catharanthus* alkaloids that are holding anticancerous activity is unknown.

5.4.4 *IN VITRO CULTURES AND ANTICANCEROUS COMPOUNDS*

Bioavailability of potent anticancerous compounds of *Catharanthus* is limited and immense interest has been created in producing these compounds using different biotechnological methods (Arora et al., 2010). Moreover, isolation and purification of these compounds from the plant is not cost effective. The effective method of producing high amount of quality product is found as plant cell cultivation in vitro. Due to its advantages in producing a large quantity of secondary metabolites into the culture medium, simple extraction process and well-controlled sterile systems plant tissue culture became the best for producing plant-based drugs within a short span of time. Initially, standard tissue culture methods were followed to produce secondary metabolites from the source materials. Later it is commercialized to produce large quantities of these products by cultivating plant cells in high-volume bioreactors. Submerged culture systems are favorable in producing high-value medicinal compounds from *Catharanthus* (Mujib et al., 2014). Verma et al. (2012) produced catharanthine, ajmalicine, and serpentine in a large scale using *Catharanthus* hairy root cultures at bioreactors. But for large production of vinblastine and vincristine stirred tank-type bioreactors were found suitable under controlled conditions. The only limitation in these processes is high amounts of excretory compounds may arrest the growth of cultured cells. This can be achieved by maintaining controlled conditions like aeration, pH, temperature, flow of nutrients in the case of continuous cultures, addition of antifoaming agents etc. during the fermentation process. Taha et al. (2014) observed the production of alkaloids vincristine (13.47 folds) and vinblastine (7.94 folds) is higher than the normal production of these alkaloids from intact plant. Chemical agents that are added to bioreactor can help to reduce the toxicity of secondary metabolites on the growth of culture cells and production of desired compounds may also increase. Sometimes, the added chemical may cause toxicity to the cultured cells if it is not appropriate for cultures. One of such chemical chromium (10–100 µg) was reported by Rai et al. (2014) for its cellular toxicity, while producing vincristine and vinblastine production. Furthermore, Fatima et al. (2015) noticed elevated enzyme activities related to stress with the addition of sodium chloride in the bioreactor cultivation of *Catharanthus* cells for its alkaloids. In contrast to chemicals, Maqsood and Abdul (2017) performed suspension cultures of *Catharanthus* with the addition of year extract (1.5 g/L) and found improved production of vincristine and vinblastine. Further, coculturing of *Aspergillus flavus* with *Catharanthus* cells elevated its growth and alkaloid production (Tonk et al., 2016).

5.5 CONCLUSIONS

Cancer treatments such as chemo and radiotherapies are costly and showing many side effects on patient's health. Alternative found best is bioactive compounds of holding antitumor functions. *Catharanthus*, a wonderful plant, is having more than 150 bioactive compounds and holding anticancerous activity. Major anticancer compounds found in this plant are vincristine and vinblastine which are available in low content in the plant. Simple extraction and enhanced quantity need to be improved to meet the demands of bioactive anticancerous drugs with an affordable price to the ever-growing human cancer patients worldwide. In vitro culture methods were found promising to produce high amounts of *Catharanthus* alkaloids and were upgraded to commercial bioreactor level. By-product inhibition of cell growth is one of the greatest obstacles to producing high amounts of *Catharanthus* alkaloids in large-scale fermenters. Addition of chemicals may improve this but can also create cellular toxicity. One of the promising methods to improve *Catharanthus* anticancerous alkaloids production is coculturing of other suitable organisms in the culture media or addition of biological-based compounds/extracts. Further, studies in this direction are essential to find promising organisms and biological extracts/compounds.

ACKNOWLEDGMENTS

The authors are highly thankful to the academic and administrative support provided by Yogi Vemana University, Kadapa, Andhra Pradesh, India, in completing this book chapter. The authors are also grateful to the departmental facility and support of staff for successful completion of this project. The authors are grateful to PubChem for the structure of *Catharanthus* alkaloids and other compounds.

CONFLICT OF INTEREST

The authors declared to have no conflict of interest and have contributed their part in this book chapter.

KEYWORDS

- **cancer**
- **Catharanthus**
- **biology**
- **phytochemicals**
- **anticancer compounds**
- **tissue culture**

REFERENCES

Almagro, L.; Fernández-Pérez, F.; Pedreño, M. A. Indole Alkaloids from *Catharanthus roseus*: Bioproduction and Their Effect on Human Health. *Molecules* **2015**, *20* (2), 2973–3000.

Arora, R. A. J. E. S. H.; Malhotra, P.; Mathur, A. K.; Mathur, A.; Govil, C. M.; Ahuja, P. S. Anticancer Alkaloids of *Catharanthus roseus*: Transition from Traditional to Modern

Medicine. *Herbal Medicine: A Cancer Chemopreventive and Therapeutic Perspective*; Jaypee Brothers Medical Publishers Pvt. Ltd: New Delhi, India, 2010; pp 292–310.

Attard, G.; Greystoke, A.; Kaye, S.; De Bono, J. Update on Tubulin-Binding Agents. *Pathologie Biologie.* **2006**, *54* (2), 72–84.

Bahleda, R.; Varga, A.; Bergé, Y.; Soria, J. C.; Schnell, D.; Tschoepe, I.; Delord, J. P. Phase I Open-Label Study of Afatinib Plus Vinorelbine in Patients with Solid Tumours Overexpressing EGFR and/or HER2. *Br. J. Cancer* **2018**, *118* (3), 344–352.

Barbier, P.; Tsvetkov, P. O.; Breuzard, G.; Devred, F. Deciphering the Molecular Mechanisms of Anti-Tubulin Plant Derived Drugs. *Phytochem. Rev.* **2014**, *13* (1), 157–169.

Bates, D. J.; Danilov, A. V.; Lowrey, C. H.; Eastman, A. Vinblastine Rapidly Induces NOXA and Acutely Sensitizes Primary Chronic Lymphocytic Leukemia Cells to ABT-737. *Mol. Cancer Therap.* **2013**, *12* (8), 1504–1514.

Bolanos-Garcia, V. M. Assessment of the Mitotic Spindle Assembly Checkpoint (SAC) as the Target of Anticancer Therapies. *Curr. Cancer Drug Targets* **2009**, *9* (2), 131–141.

Candido, R. S.; Martinez, D. M. The Stability of an Acid-Base Indicator Paper from *Catharanthus roseus* (Periwinkle) Flower Extract. *WMSU Res. J.* **2009**, *28* (1), 1–1.

Chopra, R. N.; Nayar, S. L.; Chopra, I. C.; *Glossary of Indian Medicinal Plants*; CSIR: New Delhi, 1956.

Coderch, C.; Morreale, A.; Gago, F. Tubulin-Based Structure-Affinity Relationships for Antimitotic Vinca Alkaloids. *Anti-Cancer Agents Med. Chem. (Formerly Current Medicinal Chemistry-Anti-Cancer Agents).* **2012**, *12* (3), 219–225.

Costa, M. M. R.; Hilliou, F.; Duarte, P.; Pereira, L. G.; Almeida, I.; Leech, M.; Sottomayor, M. Molecular Cloning and Characterization of a Vacuolar Class III Peroxidase Involved in the Metabolism of Anticancer Alkaloids in *Catharanthus roseus*. *Plant Physiol.* **2008**, *146* (2), 403–417.

Daniel, M. *Medicinal Plants: Chemistry and Properties*; Science Publishers, 2006.

Das, S.; Sharangi, A. B. Madagascar Periwinkle (*Catharanthus roseus L.*). Diverse Medicinal and Therapeutic Benefits to Humankind. *J. Pharmacogn. Phytochem.* **2017**, *6* (5): 1695–1701.

Fatima, S.; Mujib, A.; Tonk, D. NaCl Amendment Improves Vinblastine and Vincristine Synthesis in *Catharanthus roseus*: A Case of Stress Signalling as Evidence by Antioxidant Enzymes Activities. *Plant Cell Tissue Organ Culture (PCTOC).* **2015**, *121* (2), 445–458.

Ferreres, F.; Pereira, D. M.; Valentão, P.; Andrade, P. B.; Seabra, R. B.; Sottomayor, M. New Phenolic Compounds and Antioxidant Potential of *Catharanthus roseus*. *J. Agric. Food Chem.* **2008**, *56* (21), 9967–9974.

Gidding, C. E. M.; Kellie, S. J.; Kamps, W. A.; de Graaf, S. S. N. Vincristine Revisited. *Crit. Rev. Oncol./Hematol.* **1999**, *29* (3), 267–287.

Gigant, B.; Wang, C.; Ravelli, R. B.; Roussi, F.; Steinmetz, M. O.; Curmi, P. A. Structural Basis for the Regulation of Tubulin by Vinblastine. *Nature* **2005**, *435* (7041), 519.

Grover, J. K.; Yadav, S.; Vats. V. Medicinal Plants of India with Anti-Diabetic Potentia. *J. Ethnopharmacol.* **2002**, 81–100.

Hanafy, M. S.; Matter, M. A.; Asker, M. S.; Rady, M. R. Production of Indole Alkaloids in Hairy Root Cultures of *Catharanthus roseus* L. and Their Antimicrobial Activity. *SA J. Bot.* **2016**, *105*, 9–18.

Hisiger, S.; Jolicoeur, M. Analysis of *Catharanthus roseus* Alkaloids by HPLC. *Phytochem. Rev.* **2007**, *6* (2–3), 207–234.

Jacobs, D. I.; Snoeijer, W.; Hallard, D.; Verpoorte, R. The Catharanthus Alkaloids: Pharmacognosy and Biotechnology. *Curr. Med. Chem.* **2004**, *11* (5), 607–628.

Jeong, W. T.; Heung, B. L. A UPLC-ESI-Q-TOF Method for Rapid and Reliable Identification and Quantification of Major Indole Alkaloids in *Catharanthus roseus*. *J. Chromatogr. B.* **2018**, *1080*, 27–36.

Jordan, M. A. Mechanism of Action of Antitumor Drugs That Interact with Microtubules and Tubulin. *Curr. Med. Chem.-Anti-Cancer Agents* **2002**, *2* (1), 1–17.

Joy, P. P.; Mathew, S.; Skaria, B. P. *Kerala Agricultural University Aromatic and Medicinal Plants Research Station Odakkali*. Asamannoor PO Ernakulam District, Kerala, India, 2008.

Kabesh, K.; Senthilkumar, P.; Ragunathan, R.; Kumar, R. R. Phytochemical Analysis of *Catharanthus roseus* Plant Extract and Its Antimicrobial Activity. *Int. J. Pure Appl. Biosci.* **2015**, *3* (2), 162–172.

Kotland, A.; Chollet, S.; Diard, C.; Autret, J. M.; Meucci, J.; Renault, J. H.; Marchal, L. Industrial Case Study on Alkaloids Purification By Ph-Zone Refining Centrifugal Partition Chromatography. *J. Chromatogr. A.* **2016**, *1474*, 59–70.

Krzakowski, M.; Ramlau, R.; Jassem, J.; Szczesna, A.; Zatloukal, P.; Von Pawel, J. et al. Phase III Trial Comparing Vinflunine with Docetaxel in Second-Line Advanced Non–Small-Cell Lung Cancer Previously Treated with Platinum-Containing Chemotherapy. *J. Clin. Oncol.* **2010**, *28* (13), 2167–2173.

Kumar, S.; Singh, A.; Kumar, B.; Singh, B.; Bahadur, L.; Lal, M. Simultaneous Quantitative Determination Of Bioactive Terpene Indole Alkaloids in Ethanolic Extracts Of *Catharanthus roseus (L.)* G. Don by Ultrahigh Performance Liquid Chromatography-Tandem Mass Spectrometry. *J. Pharm. Biomed. Analys.* **2018**, *151*, 32–41.

Łata, B. Cultivation, Mineral Nutrition and Seed Production of *Catharanthus roseus* (L.) G. Don in the Temperate Climate Zone. *Phytochem. Rev.* **2007**, *6* (2), 403–411.

Liu, Z.; Wu, H. L.; Li, Y.; Gu, H. W.; Yin, X. L.; Xie, L. X.; Yu, R. Q. Rapid and Simultaneous Determination of Five Vinca Alkaloids in *Catharanthus roseus* and Human Serum Using Trilinear Component Modeling of Liquid Chromatography–Diode Array Detection Data. *J. Chromatogr. B.* **2016**, *1026*, 114–123.

Maqsood, M.; Abdul, M. Yeast Extract Elicitation Increases Vinblastine and Vincristine Yield in Protoplast Derived Tissues and Plantlets in *Catharanthus roseus*. *Revista Brasileira de Farmacognosia* **2017**, *27* (5), 549–556.

Moon, S. H.; Pandurangan, M.; Kim, D. H.; Venkatesh, J.; Patel, R, V.; Mistry, B. M. A Rich Source of Potential Bioactive Compounds with Anticancer Activities by *Catharanthus roseus* Cambium Meristematic Stem Cell Cultures. *J. Ethnopharmacol.* **2018**, *217*, 107–117.

Moudi, M.; Go, R.; Yien, C. Y. S.; Nazre, M. Vinca Alkaloids. *Int. J. Prev. Med.*. **2013**, *4* (11), 1231.

Mujib, A.; Ali, M.; Isah, T.; Dipti. Somatic Embryo Mediated Mass Production of *Catharanthus roseus* in Culture Vessel (Bioreactor)—A Comparative Study. *Saudi J. Biol. Sci.* **2014**, *21*, 442–449.

Mustafa; Rianika, N.; Verpoorte, R. Phenolic Compounds in *Catharanthus roseus*. *Phytochem. Rev.* **2007**, *6* (2–3), 243–258.

Nazir, T.; Taha, N.; Islam, A.; Abraham, S.; Mahmood, A.; Mustafa, M. Monocytopenia; Induction by Vinorelbine, Cisplatin and Doxorubicin in Breast, Non-Small Cell Lung and Cervix Cancer Patients. *Int. J. Health Sci.* **2016**, *10* (4), 542–547.

Nejat, N.; Valdiani, A.; Cahill, D.; Tan, Y-H.; Maziah, M.; Abiri, R. Ornamental Exterior Versus Therapeutic Interior of Madagascar Periwinkle (*Catharanthus roseus*): The Two Faces of a Versatile Herb. *Sci. World J.* **2015**, Article ID 982412.

Nirmala, M. J.; Samundeeswari, A.; Sankar, P. D. Natural Plant Resources in Anti-Cancer Therapy-A Review. *Res. Plant Biol.***2011**, *1* (3).1–14.

Ong; Chooi, H; Ahmad, N.; Milow, P. Traditional Medicinal Plants Used by the Temuan Villagers in Kampung Tering, Negeri Sembilan, Malaysia. *Stud. Ethno-Med.* **2011**, *5* (3), 169–173.

Pan, Q.; Saiman, M. Z.; Mustafa, N. R.; Verpoorte, R.; Tang, K. A Simple and Rapid HPLC-DAD Method for Simultaneously Monitoring the Accumulation of Alkaloids and Precursors in Different Parts and Different Developmental Stages of *Catharanthus roseus* Plants. *J. Chromatogr. B.* **2016**, *1014*, 10–16.

Patel, D. K.; Kumar, R.; Laloo, D.; Hemalatha, S. Natural Medicines from Plant Source Used for Therapy of Diabetes Mellitus: An Overview of Its Pharmacological Aspects. *Asian Pac. J. Trop. Dis.* **2012**, *2* (3), 239–250.

Perez, E. A. Microtubule Inhibitors: Differentiating Tubulin-Inhibiting Agents Based on Mechanisms of Action, Clinical Activity, and Resistance. *Mol. Cancer Therap.* **2009**, *8* (8), 2086–2095.

Pham, H. N. T.; Vuong, Q. V.; Bowyer, M. C.; Scarlett, C. J. Phytochemicals Derived from *Catharanthus roseus* and Their Health Benefits. *Technologies* **2020**, *8* (4), 80.

Plaizier, A. C. *A Revision of Catharanthus roseus (L.) G. Don (Apocynaceae)*; Landbouwhoge School 1981; pp 81–89.

PROTA, Plant Resources of Tropical Africa. African Ornamentals. *Proposals and Examples*; PROTA Foundation: Wageningen, The Netherlands, 2011.

Qiu, L.; Dong, C.; Kan, X. Lymphoma-Targeted Treatment Using a Folic Acid-Decorated Vincristine-Loaded Drug Delivery System. *Drug Design Dev. Therap.* **2018**, *12*, 863–872.

Rai, V.; Tandon, P. K.; Khaltoon, S. Effect of Chromium on Antioxidant Potential of Catharanthus Roseus Varieties and Production of Their Anticancer Alkaloids: Vincristine and Vinblastine. *BioMed. Res. Int.*. **2014**, Article ID 934182, 10p.

Renault, J. H.; Nuzillard, J. M.; Le Crouérour, G.; Thépenier, P.; Zéches-Hanrot, M.; Le Men-Olivier, L. Isolation of Indole Alkaloids from *Catharanthus roseus* By Centrifugal Partition Chromatography in the Ph-Zone Refining Mode. *J. Chromatogr. A* **1999**, *849* (2), 421–431.

Sain, M.; Sharma, V. *Catharanthus roseus* (An Anti-Cancerous Drug Yielding Plant)—A Review of Potential Therapeutic Properties. *Int. J. Pure Appl. Biosci.* **2013**, *1*, 139–142.

Schinzari, G., Rossi, E, Pierconti, F., Garufi, G., Monterisi, S. et al. Monoinstitutional Real World Experience in Management of Vinflunine as Second Line Therapy for Transitional Cell Carcinoma of the Urothelium. *Oncotarget* **2018**, *9* (9), 8765.

Schmelzer, G. H.; Gurib-Fakim, A. *Plant Resources of Tropical Africa*; Medicinal Plants PROTA Foundation, 2008; pp 205–206.

Schutz, F. A.; Bellmunt, J.; Rosenberg, J, E.; Choueiri, T. K. Vinflunine: Drug Safety Evaluation of This Novel Synthetic Vinca Alkaloid. *Expert Opin. Drug Safety* **2011**, *10* (4), 645–653.

Schweizer; Fabian; Colinas, M.; Pollier, J.; Van Moerkercke, A.; Vanden Bossche, R.; De Clercq, R.; Goossens, A. An Engineered Combinatorial Module of Transcription Factors Boosts Production of Monoterpenoid Indole Alkaloids in *Catharanthus roseus*. *Metab. Eng.* **2018**, *48*, 150–162.

Sevestre-Rigouzzo, M.; Nef-Campa, C.; Ghesquière, A.; Chrestin, H. Genetic Diversity and Alkaloid Production in *Catharanthus roseus*, C. *Trichophyllus* and Their Hybrids. *Euphytica.* **1992**, *66* (1), 151–159.

Siddiqui, M. J.; Ismail, Z.; Aisha, A. F. A.; Majid, A. M. Cytotoxic Activity of *Catharanthus roseus* (Apocynaceae) Crude Extracts and Pure Compounds Against Human Colorectal Carcinoma Cell Line. *Int. J. Pharmacol.* **2010**, *6* (1), 43–47.

Sottomayor, M.; Ros Barcelo, A. The Vinca Alkaloids: From Biosynthesis and Accumulation— in Plant Cells, To Uptake, Activity and Metabolism in Animal Cells. In *Studies in Natural Products Chemistry (Bioactive Natural Products)*; Attaur, R., Ed.; Elsevier Science Publisher: Amsterdam, 2005; 813–857.

Sottomayor, M.; Ros Barceló, A. The Vinca Alkaloids: From Biosynthesis and Accumulation in Plant Cells, to Uptake, Activity and Metabolism in Animal Cells. *Stud. Nat. Products Chem.* **2006,** *33*, 813–857.

Taha, H. S.; Shamas, K. A.; Nazif, N. M.; Seif El-Nasr, M. M. In Vitro Studies on Egyptian *C. Roseus* (L.) G.Don V: Impact Of Stirred Reactor Physical Factors on Achievement of Cells Proliferation and Vincristine and Vinblastine Accumulation. *Res. J. Pharma. Biol. Chem. Sci.* **2014,** *5* (2), 330–340.

Thomas, P. A.; Latimer, J. G. Growth of Vinca as Affected by Form of Nitrogen, Presence of Bark, and Type of Micronutrients. *J. Plant Nutr.* **1996,** *18*, 2127–2134.

Tiong, S. H.; Looi, C. Y.; Arya, A.; Wong, W. F.; Hazni, H.; Mustafa, M. R.; Awang, K. Vindogentianine, a Hypoglycemic Alkaloid from *C. roseus* (L.) G. Don (Apocynaceae). *Fitoterapia.* **2015,** *102*, 182–188.

Tonk, D.; Mujin, A.; Maqsood, M.; Ali, M.; Zafar, N.s Aspergillus Flavus Fungus Elicitation Improves Vincristine and Vinblastine Yield by Augmenting Callus Biomass Growth in *Catharanthus roseus. Plant Cell, Tissue Organ Culture* (PCTOC) **2016,** *126* (2), 291–303.

USDA, Agricultural Research Service, National Plant Germplasm System. *Germplasm Resources Information Network (GRIN Taxonomy)*; National Germplasm Resources Laboratory, Beltsville, MD, October 2021. https://npgsweb.arsgrin.gov/gringlobal/taxon/taxonomydetail?id=70159

Van der Heijden, R; Jacobs, D. I.; Snoeijer, W; Hallard, D; Verpoorte, R. The Catharanthus Alkaloids: Pharmacognosy and Biotechnology. *Curr. Med. Chem.* **2004,** *11* (5), 607–628.

Verma, P; Mathur, A. K.; Shanker, K. Growth, Alkaloid Production, Rol Genes Integration, Bioreactor Up-Scaling and Plant Regeneration Studies in Hairy Root Lines of *Cathranthus roseus. Plant Biosyst.* **2012,** *146*, 27–40.

Wang, L.; He, H. P.; Di, Y. T.; Zhang, Y.; Hao, X. J. Catharoseumine, a New Monoterpenoid Indole Alkaloid Possessing a Peroxy Bridge from *Catharanthus roseus. Tetrahedron Lett.* **2012,** *53* (13), 1576–1578.

Wang, X. D.; Li C. Y.; Jiang, M. M.; Li, D.; Wen, P.; Song, X. et al., Induction of Apoptosis in Human Leukemia Cells Through an Intrinsic Pathway by Cathachunine, A Unique Alkaloid Isolated from *Catharanthus roseus. Phytomedicine* **2016,** *23* (6), 641–653.

Whiting, D., O'Connor, A.; Jones, J.; McMulkin, L.; Potts, L. Taxonomic Classification. *CMG Garden Notes* **2011,** *122*, 1–8.

Yamamoto, K.; Takahashi, K.; Mizuno, H.; Anegawa, A.; Ishizaki, K.; Fukaki, H.; Ohnishi, M.; Yamazaki, M.; Masujima, T.; Mimura, T. Cell-Specific Localization of Alkaloids in *Catharanthus roseus* Stem Tissue Measured with Imaging MS and Single-Cell MS. *Proc. Natl. Acad. Sci.* **2016.**

CHAPTER 6

Withania somnifera (L.) Dunal

PULALA RAGHUVEER YADAV[1], LEPAKSHI M. D. BHAKSHU[2],
K. VENKATA RATNAM[3], ANU PANDITA[4], DEEPU PANDITA[5], and
K. M. SRINIVASA MURTHY[6]

[1]Department of Biotechnology, Indian Institute of Technology Hyderabad, Kandi, Telangana State, India

[2]Department of Botany, PVKN. Government College (Autonomous), Chittoor, Andhra Pradesh, India

[3]Department of Botany, Rayalaseema University, Kurnool, Andhra Pradesh, India

[4]Vatsalya Clinic, Krishna Nagar, New Delhi, India

[5]Government Department of School Education, Jammu, Jammu and Kashmir, India

[6]Department of Microbiology and Biotechnology, Jnanabharathi Campus Bangalore University, Bengaluru, India

ABSTRACT

Among noncommunicable diseases in humans, the second most cause of death is cancer. Vital organs are affected by cancer and various cancers are increasing worldwide, accompanied by the death rate. So there is a need for effective drugs for cancer treatment. Chemoprotective drugs are routinely used but have side effects, whereas plant-based medications have minimal side effects. In Ayurveda, plants with medicinal properties have been utilized to cure ailments for centuries. Therefore, plant-based products may reduce adverse side effects in cancer treatment. *Withania somnifera* is a shrub, erect

Potent Anticancer Medicinal Plants: Secondary Metabolite Profiling, Active Ingredients, and Pharmacological Outcomes. Deepu Pandita and Anu Pandita (Eds.)

with lengthy tuberous roots, from the Solanaceae family commonly called Indian winter cherry/Indian ginseng/Ashwagandha, a familiar Ayurvedic medicinal plant. In Ayurveda, extracts of root/leaf from *Withania somnifera* are included in formulations in India. *Withania somnifera* is used as antioxidant, anti-inflammatory, antistress, antiepileptic, antiarthritic, antidepressant, anticoagulant, antidiabetic, antipyretic, rejuvenative, immunomodulatory, and antiangiogenesis. The existing research shows that the plant extracts and their active components have demonstrated anticarcinogenic activity. The anticarcinogenic activities of *Withania somnifera* inhibit cancer cell survival, proliferation, motility, angiogenesis, metastasis, cell cycle arrest, apoptosis and autophagy induction, etc., on cancers in the last 20 years. Withaferin A is the most studied drug for anticancer effects in various cancers due to its therapeutic effect. Herein, we review the anticancer properties of *Withania somnifera* extracts and their active components, which have shown inhibitory properties against several cancers and their mechanism of action *in vitro and in vivo*.

6.1   INTRODUCTION

Cancer is a non-communicable disease in humans and is the second reason of death. Cancer affects various human organs in the body. There is a development in the field of drugs toward cancer treatment, but various cancers are increasing worldwide along with the death rate. Chemoprotective drugs have side effects, whereas drugs from plants show fewer side effects. In Ayurveda the *Withania somnifera* extracts of root/leaf are included in formulations in India. *Withania somnifera* (L.) Dunal is a shrub, which is erect with lengthy tuberous roots and comes from the Solanaceae family commonly called Indian winter cherry/Ashwagandha/Indian ginseng known for its several beneficial health activities in India from olden days which is shown in Figure 6.1. *Withania somnifera* is used as antioxidant, anti-inflammatory, antistress, antiepileptic, antiarthritic, antidepressant, anticoagulant, antidiabetic, antipyretic, rejuvenative, immunomodulatory, and antiangiogenesis (Subbaraju et al., 2006). *Withania somnifera* and its components have an anticancer effect.

Withania somnifera is still under investigation for its anticancer effects inhibiting cell survival, proliferation, motility, angiogenesis, metastasis, cell cycle arrest, apoptosis, autophagy induction, etc. *Withania somnifera* bioactive components are shown in Figure 6.2.

FIGURE 6.1   *Withania somnifera* (L.) Dunal.

Withania somnifera bioactive compounds

Alkaloids: Withanine Anahygrine Cuscohygrine Isopelletierine Tropine	Steroidal lactones: Withaferin A Withanolides	Steroids: Sitoinosides VII and VIII Glycol-withanolides/ Sitoinosides IX/X	Flavonoids: Kaempferol Quercetin	Nitrogen containing compounds	Salts, coagulins and Trace elements

FIGURE 6.2   *Withania somnifera* bioactive compounds.

The withanolides are the bioactive compounds, C28-steroidal lactones and Withaferin A is the most studied drug for the anticancer effects in various cancers due to its therapeutic effect. The active component structures (Tao et al., 2015; Motiwala et al., 2013) involved in the anticarcinogenic effect are shown in Figure 6.3.

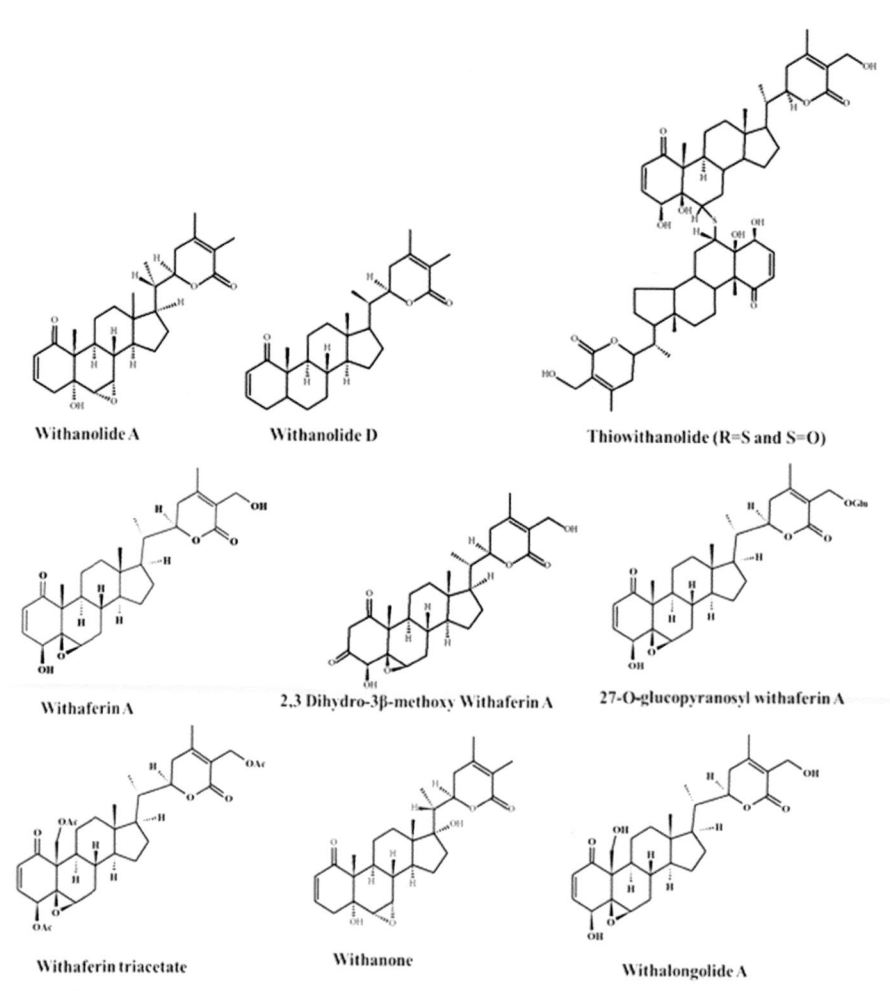

Withanolide A   Withanolide D   Thiowithanolide (R=S and S=O)

Withaferin A   2,3 Dihydro-3β-methoxy Withaferin A   27-O-glucopyranosyl withaferin A

Withaferin triacetate   Withanone   Withalongolide A

FIGURE 6.3   Structures of active constituents of *Withania somnifera* against cancer.

The chemistry of the therapeutic effect of Withaferin A is yet to be understood. Withaferin A chemical structural analysis suggests alkylation reactions at nucleophilic sites, probably at three positions. Structure–activity relationship experiments propose Ring A α, β-unsaturated ketone moiety, OH group at C27 and C5(6) epoxide involvement in the biological activity of Withaferin A (Berghe et al., 2012). This chapter reviews the *Withania somnifera* extracts and their components anticancer effects on various cancers and their mechanisms of actions.

6.2 BREAST CANCER

Breast cancers were diagnosed in 2.3 million women in 2020 and there were 685,000 deaths worldwide. In the last 5 years 7.8 million were diagnosed with breast cancer. These data reveal breast cancer as the most prevalent cancer in the world. Worldwide breast cancer is a severe ailment in women affecting lakhs of women. Though there are medications, there is still a need to explore new therapeutic and preventive ways required to reduce the deaths and suffering from this disease. The medicinal plants and their constituents are researched for therapy and/or chemoprevention of breast cancer. Here, we discuss the *Withania somnifera* and its component's anticancer effects on cellular processes and molecular targets.

Kuttan (1996) reported *Withania somnifera* methanol root extract (Withanolide sulfoxide) treatment on breast cancer cell lines suppressed NF-κB activation. Dar et al. (2019) investigated *Withania somnifera* aqueous root extract administration which led to induction of apoptosis in MDA-MB-231 breast cancer cells. Alfaifi et al. (2016) explored *Withania somnifera* methanol leaf extract administration which led to cell cycle arrest and apoptosis in breast cancer cell lines. Srivastava et al. (2016) investigated the cytotoxicity of breast cancer cell lines on administering *Withania somnifera* methanol and ethanol extracts. Wadhwa et al. (2013) studied *Withania somnifera* water extract treatment on MCF-7 breast cell lines, which resulted in the inactivation of p53 and pRB, cyclin B1 reduction and cyclin D1 rise, and downregulation of MMP-3 and 9. Stan et al. (2008a) reported the MDA-MB-231 cells implanted orthotopically in mice treated with Withaferin A for 2.5 weeks. The *in vivo* efficacy of Withaferin A demonstrated a significant reduction in cell proliferation and apoptosis rise in the treated mice. The Withaferin A-treated mice ($P<0.05$) showed average tumor volume ~1.8 fold lesser than control mice. Nagalingam et al. (2014) reported the Withaferin A antitumor effect in MDA-MB-231 cells. Kim et al. (2016) demonstrated Withaferin A inhibitory effect *in vivo* in genetically modified MDA-MB-231 cells with Notch2 knockdown. Thaiparambil et al. (2011) proved the antitumor activity of Withaferin A in the orthotopic mouse. Here, Withaferin A at two doses 2 and 4 mg/kg were given intraperitoneally on alternate days for a month.

Liu et al. (2019) reported Withaferin A efficacy in the MDA-MB-231 xenograft model. Samanta et al. (2017) reported MNU-induced Luminaltype breast cancer in the rat model and Withaferin A administration was done intraperitoneally at two doses 4 and 8 mg/kg from the second week after

N-Methyl-N-nitrosourea induction for five times in a week till the 10th week. Tumor incidence, multiplicity, and weight showed a decrease. Withaferin A treatment in MCF-7 and MDA-MB-231 cells for 24 hours inhibited cell viability with an IC50 between 1.5 and 2.0 µmol/L being reported by Stan et al. (2008a). Thaiparambil et al. (2011) investigated vimnetin cytoskeleton inhibition by Withaferin A. Vimnetin overexpression has a relationship with the induction of EMT. Imaging studies explained the perinuclear vimnetin accumulation preceded by quick vimnetin depolymerization by Withaferin A. Withaferin A at less than/equal to 500 nM retained an effective anti-invasive effect. Structure–activity relationships explained that the expected vimnetin-binding region of Withaferin A is essential to induce ser56 phosphorylation in vimnetin. Withaferin A showed an antimetastatic activity in breast cancer via its effects on vimnetin and vimnetin ser 56 phosphorylation. Hahm et al. (2011a) investigated the MCF-7 cells proliferation by Estrogen stimulation, inhibited by Withaferin A. Stan et al. (2008b) reported the increase in the G2-M fraction in a dose- and time-dependent manner on the treatment of Withaferin A in breast cancer cell lines; concomitantly, there was a decrease in CDC 25B and/or CDC 25 C, cyclin-dependent kinase 1 causing a buildup of Tyrosine15 phosphorylated (inactive) Cdk1. Withaferin A-mediated arrest at the G2-M phase in MDA-MB-231 cells was protected partially by the overexpression of cell division cycle 25 C. Zhang et al. (2011) reported the G2/M arrest in breast cancer cells from the Withaferin A treatment. Antony et al. (2014) showed Withaferin A arrest at the G2 and mitotic phase in MCF-7, SUM 159, and SK-BR-3 cell lines and linked with reduction of β-tubulin levels. Withanone and Withanolide A is C6, C7-epoxy analogs of Withaferin A could not cause mitotic arrest in these cells. Withaferin A dislocated the spindle morphology. In MCF-7 cells, the Withaferin A bounded covalently to Cys303 of β-tubulin. The molecular docking studies explained that the β-tubulin has the binding pocket (Hydrophobic floor & wall with hydrophilic entrance) on its surface for the Withaferin A. Samanta et al. (2018) investigated the cell cycle arrest on the administration of Withaferin A in the MCF-7 cell line. The role of peptidyl-prolyl cis/trans isomerase 1 is cell cycle arrest. Lee et al. (2012) showed a Withaferin A inhibitory effect on cell migration in breast cancer. Withaferin A treatment in MDA-MB-231 cells inhibited invasion ability and there was a decrease in the gene expression profiles of ADAM8, PLAT, uPA (Extracellular matrix-degrading proteases), pro-inflammatory mediators (ANGPTL2, TNFSF12, CSF1R, IL6) and cell adhesion molecules (Integrins, Laminins) (Szarc et al., 2014).

Withaferin could inhibit experimental EMT and cell migration. Treatment of TNF-α and TGF-β could partially reverse the process (Lee et al., 2015).

Withaferin A induces death in breast cancer cell lines by ROS-mediated parap-tosis (Ghosh et al., 2016). MNU induced Luminal-type breast cancer in the rat model. Withaferin A administration was done intraperitoneally at two doses, 4 and 8 mg/kg, from the second week after N-Methyl-N-nitrosourea induction, five times until the 10th week. There was an increase in TUNEL-positive apoptotic cells (Samanta et al., 2017). There was no change in the complex III assembly in MDA-MB-231, but complex III decreased in SUM159 and MCF-7 cells as analyzed by gel electrophoresis. The fusion process involved proteins, their expression was decreased, with the other proteins OPA1, mitofusion 1, mitofusion 2 also after Withaferin A treatment. The protein DRP1 was also decreased after the Withaferin A treatment (Sehrawat et al., 2019). Withaferin A treatment against MDA-MB-231 xenografts and breast cancer cell lines *in vivo* induce apoptosis. ROS production led to suppres-sion of oxidative phosphorylation and complex III activity after Withaferin A treatment. Breast cancer cell lines Rho-0 variants (Mitochondrial DNA-deficient) were resistant to ROS induced by Withaferin A. A new insight, that is, ROS production and activation of Bax/Bak is seen into the mechanism of Withaferin A-induced apoptosis (Hahm et al., 2011b). One of the transcription factors of the forkhead box, FOXO3a, is identified in breast cancer cells as a mechanistic target and its knockdown in MCF-7 cells resulted in protection against Withaferin A-mediated apoptosis involving target Bim (Stan et al., 2008a). The progression of proliferation, metastasis, and chemoresistance in breast cancer has been implicated by overexpression and constitutive activa-tion of STAT3 (Ma et al., 2020).

Withaferin A inhibits the STAT3 transcription factor in breast cancer cells. It is observed that Withaferin A suppresses constitutive and/or inducible activation of STAT3 along with phosphorylation of Janus activated kinase 2 (Upstream regulator) in breast cancer cell lines. Withaferin A administration in breast cancer cells furthermore affected inhibition of a. STAT3 transcrip-tional activity with or without stimulation of IL-6; b. dimerization of STAT3 in any case in MDA-MB-231 cells; and c. nuclear translocation of STAT3 (phosphorylated) in both cells. In MDA-MB-231 cells, facilitated activa-tion of STAT3 by IL-6 conferred limited defense counter to cell invasion inhibition by Withaferin A (Lee et al., 2010). 17 β -estradiol (E2) attenuated the Withaferin A treatment in MCF-7 cells. In the MDA-MB-231 cell line, overexpression of ER-α conferred limited but statistically significant defense against Withaferin A-mediated apoptosis, but not the arrest of G2/M phase (Hahm et al., 2011a). Zhang et al. (2011) investigated downregulation of ER α protein expression in MCF-7 cells after Withaferin A treatment. Hahm et al. (2011b) reported Withaferin A treatment in MCF-7 cells induced p53 activation and Ser 15 phosphorylation. Withaferin A induced apoptosis in MCF cells and protection was due to RNA interference of p53. Withaferin A induces apoptosis and various molecular changes in human breast cancer. Withaferin A facilitated extracellular signal-regulated kinase (ERK) hyper-phosphorylation but not p38 MAPK/c-jun NH-terminal Kinase, and this change was partially protected by manganese superoxide dismutase overexpression. Withaferin A-induced apoptosis in MCF-7 cells was prominently increased by the suppression of p38 MAPK and ERK. On the other hand in MCF-7 cells were partly attenuated by JNK inhibition (Hahm et al., 2014).

6.3   CERVICAL CANCER

Cervical cancer is seen in women, and the Human papillomavirus expressing onco proteins E6 and E7 is the cause. 311,000 approximately died from the disease in 2018 and in women cervical cancer is the fourth common cancer. Withaferin A inhibits the CaSki cervical cancer cells proliferation. The Withaferin A mechanism involved the decrease in the E6 and E7 oncoproteins expression. There were changes like induction of p53 accumulation and change in the p53 levels-mediated apoptotic markers-cleaved PARP, caspase-3, Bax, Bcl2, rise in p21cip1/waf1. Withaferin A interaction with PCNA causes G2/M phase arrest and the variation of cdc2, cyclin B1, p34 and PCNA levels and STAT3 and its phosphorylation at Tyr705 and Ser727 decreases. The *in vivo* studies in athymic nude mice results showed a similar trend to that of *in vitro* in manipulating molecular markers (Munagala et al., 2011).

6.4   OVARIAN CANCER

Ovarian cancer is seen in women and the seventh most common cancer. Ovarian cancer has the highest death rate among gynecologic cancers. In 2018, ovarian cancer deaths percentage constituted 4.4% of total mortality from cancers in women. The efficacy of Withaferin A in the treatment of ovarian cancer is proved. The combination of Withaferin A and Doxorubicin against A2780/CP70, A2780, and CaOV3 epithelial cancer cells expresses a dose- and time-dependent synergistic effect on inhibition and induction of cell death, thus reducing ill effects and dosage requirement of Doxorubicin. ROS generation showed a noticeable improvement, leading to induction of autophagy, massive DNA damage as observed under TEM. Analysis of cleaved caspase 3 explained the increase in LC3B (autophagy marker) expression and led to cell death. The combination therapy was validated in 3 dimension (3D) tumor model and xenograft model of ovarian cancer; the results showed a reduction of 70–80 % in tumor growth compared with control or animals treated with Doxorubicin or Withaferin A alone (Fong et al., 2012). Withaferin A and Cisplatin showed efficacy in the treatment of refractory ovarian cancer. Withaferin A at 1.5 µM concentration prominently reduced the spheroid formation in the A2780 ovarian cancer cell line and not by Cisplatin. Combination of Withaferin A and Cisplatin prominently reduced spheroid formation comparatively to control, Withaferin A only, and Cisplatin alone treatment. In orthotopic ovarian tumors bearing nude mice, Cisplatin treatment increased cancer stem cells. In contrast, Withaferin A administration prominently decreased Aldehyde Dehydrogenase cancer stem cells and the combination resulted in the reduction of Aldehyde Dehydrogenase Cancer stem cells. Withaferin A and Cisplatin alone/in combination treatment in orthotopic ovarian tumors bearing nude mice caused complete metastasis inhibition and 70–80% decline in tumor growth (Kakar et al., 2014).

6.5   PROSTATE CANCER

Second common cancer in men is prostate cancer and worldwide it is the fifth cause of mortality. There were about 358,989 deaths from prostate cancer in 2018. *Withania somnifera* extracts and Withaferin A are used in prostate cancer for the treatment. Moselhy et al. (2017) reported that Withaferin A showed a chemopreventive effect in *Pten* conditional knockout mice with constitutively activated AKT signaling. Das et al. (2016) investigated Withaferin A delaying tumor progression in prostate cancer by activating prostate apoptosis-4. Rah et al. (2015) reported in prostate cancer cells about the switching from autophagy to apoptosis by pro-apoptotic protein PAWR-mediated suppression of BCL-2 after the administration of 3-azido derivative of Withaferin A. Withaferin A administration resulted in cell death in PC-3 and DU-145 androgen-independent cancer cells (Nishikawa et al., 2015). Balakrishnan et al. (2017) said *Withania somnifera* extract in androgen-independent prostate cancer cell lines (PC3) for the inhibitory effects on metastasis. The expression of COX-2 and IL-8 in PC3 cells in 24 hours is prominently suppressed by *Withania somnifera* extract, controlling disease advancement from androgen-dependent to androgen-independent.

6.6　PANCREATIC CANCER

Pancreatic cancer is more common in men than in women and is the seventh leading death-causing cancer worldwide. Yu et al. (2010) reported Withaferin A inhibition efficacy and the mechanism of Hsp90 in pancreatic cancer. It has shown the antiproliferative effect in BxPc3, MiaPaCa2, and Panc-1 pancreatic cancer cell lines with ICs of 2.78, 2.93, and 1.24 µM. Withaferin A induced apoptosis in Panc-1 cell line in a dose-dependent way and is verified by Annexin V staining. The Hsp90 chaperone effect to promote the breakdown of Hsp90 proteins was suppressed by Withaferin A and evidenced by Western blot. Pancreatic Panc-1 xenografts tumor growth was inhibited by administering Withaferin A at 3, 6 mg/kg by 30 and 58%, respectively. Thus, Withaferin A binds to Hsp90, suppresses its chaperone activity, and shows anticancer activity against pancreatic cancer. Oxaliplatin is limited in use for the treatment of pancreatic cancer. Li et al. (2015) reported Oxaliplatin-induced growth inhibition, apoptosis, and its enhancement of this activity by Withaferin A in pancreatic cells. This growth inhibition and apoptosis involve PI3K/AKT pathway inactivation and mitochondrial dysfunction. Combination therapy leads to an antitumor effect better than the single drug in pancreatic cancer treatment.

6.7　LUNG CANCER

Lung cancer is the leading cause of cancer death worldwide. In 2020, the deaths were 1.80 million and were the highest among all cancers. Choudhary et al. (2010) reported the compounds Withaferin A, chlorinated steroidal lactone and diepoxy withanolide inhibited the growth and had cytotoxic activity in NCI-H460 lung cancer cell line. Withaferin A has been the most significant (GI = 0.18 µg/mL and LC = 0.45 µg/mL). Cai et al. (2014) studied the activity of Withaferin A in non-small cell lung cancer A549 and observed the inhibition of cellular growth and apoptosis by PI3K/AKT pathways inactivation. Withaferin A is screened insilico as a significant antilung cancer and antilung cancer stem like cell agent and showed cytotoxicity toward various lung cancer cells, along with induction of apoptosis and autophagy. ROS activation plays an upstream role in mediating Withaferin A-elicited effects. Withaferin A suppressed the growth of lung cancer stem like cell and spheroid forming capacity, reducing side population. Withaferin combined with Cisplatin and Pemetrexed showed synergistic effects on suppressing

EGFR wild-type lung cancer (Hsu et al., 2019). Withaferin A treatment in non-small cell lung cancer (NSCLC) A549 and H1299 NSCLC cells demonstrated time- and dose-dependent cytotoxicity. After this, the cells were treated with ≤ 0.5 µM Withaferin A for ≤ 4 h to decrease cytotoxicity and examine its effects on EMT, cell adhesion, motility, migration, and invasion. The pretreatment of cells with Withaferin A showed inhibition to cell adhesion, migration, and invasion of A549 and H1299 cells. Withaferin A inhibited Epithelial to mesenchymal transition (EMT) in A549 and H1299 NSCLC cells, induced by TGFβ1 and TNFα. The phosphorylation and nuclear translocation of Smad2/3 and NF-κB in A549 and H1299 cells are inhibited by Withaferin A. Hence, Withaferin A is a potential therapeutic agent against metastasis in NSCLC (Kyakulaga et al., 2018).

6.8 COLORECTAL CANCER

Colorectal cancer is second common cancer seen in women, and third in men. There were 935,000 deaths in 2020 from colorectal cancer. Kuttan (1996) reported *Withania somnifera* methanol root extract (withanolide sulfoxide) treatment on cell lines suppressed TNF-induced NF-κB activation. Alfaifi et al. (2016) showed cell cycle arrest and apoptosis in breast cancer cell lines by the administration of *Withania somnifera* methanol leaf extract. Withaferin A induced apoptosis in colorectal cancer cells as verified by Flow cytometry and western blot results. Withaferin A stimulates production of ROS and its accumulation, which causes the mitochondrial membrane potential loss and mitochondrial dysfunction (Xia et al., 2018). Koduru et al. (2010) reported in the colon cancer cells that the Withaferin A suppressed Notch-1 signaling and downregulated prosurvival pathways. There was also downregulation of the expression with p4E-BP1and pS6K, the rapamycin signaling components of mammals. Withaferin A activated apoptosis via c-Jun-NH-kinase in colon cancer cells.

6.9 OTHER CANCERS

6.9.1 *LEUKEMIA*

Mondal et al. (2010) reported in leukemia patients, derived lymphoid and myeloid cells accompanied by primary cells, administered with WithaD, improved the accumulation of ceramide, and induced apoptosis. Senthil et al. (2007) reported in promyelocytic leukemia HL-60 cell lines about the induction of apoptosis by the treatment of *Withania somnifera* methanolic extract of leaf and withanolide. There was caspase 9,8,3 activation and release of cytochrome c.

6.9.2   MYELOID LEUKEMIA

The mechanism of action by Withaferin A causing cytotoxicity in cancer cell lines is not apparent. It induced reactive oxygen species production and mitochondrial membrane potential loss, subsequently the release of cytochrome c, Bax mitochondrial translocation, and factor inducing apoptosis to cell nuclei, concomitantly, caspases 9, 3 activation and cleavage of PARP. The extrinsic pathway activated by Withaferin A is evident from the hike in the levels of caspase 8 activity in a time-dependent way next to overexpression of TNFR-1. Thus in human leukemia cell lines HL-60, Withaferin A induced ROS and mitochondrial dysfunction elicit actions responsible for apoptosis in mitochondrial-dependent and independent pathways (Malik et al., 2007).

6.9.3   LYMPHOBLASTIC AND MYELOID LEUKEMIA

Withaferin A treated lymphoblastic and myeloid leukemia patients' primary cells and leukemic cell lines from humans showed a potent growth inhibitory effect. Withaferin A induced programmed cell death (apoptosis) as evidenced by the activation of p38 MAPK signaling cascade, that is, the rise in phosphorylated p38MAPK. The phosphatidylserine externalization demonstrates apoptosis, mitochondrial transmembrane potential loss, and the release of cytochrome c. Along with a surge in Bax/Bcl-2 ratio on time, caspases 3, 9 activations, DNA fragmentation showed the buildup of cells in sub-G0 (Mandal et al., 2008).

6.9.4   SKIN CANCER

Mayola et al. (2011) reported in human melanoma cells that Withaferin A induced apoptotic cell death. The mechanism involved the Bcl-2 down-regulation, translocation of Bax to mitochondria, the release of cytochrome c transmembrane dissipation, caspase 9,3 activation, and DNA fragmentation. Swiss albino mice induced by 7, 12-dimethylbenz(a)anthracene resulted in skin cancer. The mice have been treated with the *Withania somnifera* hydroalcoholic root extract at 400 mg/kg po (high dose). The extract had a chemopreventive effect against skin cancer (Prakash et al., 2002). Halder et al. (2015) reported *Withania somnifera* root extract was tested in A375 cells and has a potent cytotoxic effect, DNA fragmentation, nuclear blebbing and cell morphology supported by MTT assay, Agarose gel electrophoresis, DAPI staining, and fluorescence microscopy.

6.9.5 GLIOBLASTOMA

Shah et al. (2009) reported the *Withania somnifera* alcoholic extract of leaf, its constituents and their blends to show the ability to induce growth arrest and differentiation in YKG1 and C6 glioma cell lines. The analysis at the molecular level resulted in cell migration delay, glial fibrillary acidic protein and neuronal cell adhesion molecules enhanced expression. Kataria et al. (2011) investigated the Ashwagandha leaf extract treatment in the glioma cells showing antiproliferative and differentiation activity. The presence of these activities in the Ashwagandha leaf extract is supported by the Neural cell adhesion molecule, Mortalin (Hsp 70) and Glial fibrillary acidic protein expression levels observed changes, the MTT assay and wound scratch assay. The Ashwagandha water extract treatment in the orthotopic glioma allograft in the rat model reduced intracranial tumor volumes and inhibited the cyclin D1, Hsp70, pAKT, NF-κB, VEGF, PSA-NCAM. The upregulation of mortalin and NCAM and decrease in GFAP indicate the Ashwagandha water extract antiglioma efficiency *in vivo* (Kataria et al., 2016).

6.9.6 NEUROBLASTOMA

The study revealed Ashwagandha water extract to decrease cell proliferation and induce differentiation in IMR-32 neuroblastoma cells of humans. Induced expression of NCAM and polysialylation decrease explained the Ashwagandha water extract antimigratory activity. Along with MMP 2 and 9 activity downregulation (Kataria et al., 2013), Caputi et al. (2018) reported methanol root extract of *Withania somnifera* treatment on Neuroblastoma cell lines alters gene expression of opioid receptors.

6.9.7   DALTON'S ASCITIC LYMPHOMA

Withania somnifera has shown an anticarcinogenic effect in the Swiss albino mice with Dalton's Ascitic Lymphoma. *Withania somnifera* ethanolic root extract showed a surge in lifespan, a fall in tumor weight and cancer cell number. There was a hematological improvement (Christina et al., 2004).

6.9.8   MOUSE EHRLICH ASCITES

Devi et al. (1995) investigated *Withania somnifera* aqueous root extract treatment on Mouse Ehrlich ascites carcinoma showing a radio sensitizing effect.

6.9.9   RENAL CARCINOMA

Um et al. (2012) investigated the Caki human cancer cells that were administered with Withaferin A. Withaferin A induced apoptosis mediated by the STAT3 signaling pathway (Bcl-xL, Bcl-2, Survivin, and cyclin D1) downregulation.

6.9.10   LIVER CANCER

Alfaifi et al. (2016) reported cell cycle arrest and apoptosis after the administration of *Withania somnifera* methanol leaf extract on breast cancer cell lines. Ashwagandha water extract showed antioxidant and anticancer activity in the HepG2 hepatocellular carcinoma cell line. The levels prominently were increased in Glutathione S-transferase, glutathione reductase and total antioxidant, caspase-3,8 and 9 activities, Fas-ligand and the level of TNF-α decreased significantly. The HepG2 cells treated were arrested at the G/G and G/M phases (Ahmed et al., 2018). Zhou et al. (2016) reported Withaferin A inhibition in Hepatocellular carcinoma cells proliferation. Along with these changes: (1) there was a rise in apoptosis rate and arrest at G1 phase, (2) rise in p53, bax expression, and reduction in bcl-2 expression, and (3) upregulation of p21 and downregulation of CDK2 and cyclin D1.

6.10   CONCLUSIONS

Traditional medicines can be the best alternative medicines as they are reasonably priced and effective. For these herbal medicines to be an alternative, proper studies have to be conducted to look into safety, effectiveness, and superiority. The research studies of *Withania somnifera* and its components in human cancer cells and rodent models have indicated the anticancer activity. Hence, *Withania somnifera* can be used as an additional therapy to decrease the side effects of chemo and radiotherapy and as a combination therapy with conventional treatments to improvise the effects of chemo and radiotherapy. Finally, clinical studies are to be done for the validation before translation into clinical application.

KEYWORDS

- ***Withania somnifera***
- **withaferin A**
- **anticancer effects**
- **mechanism of anticancer activity**

REFERENCES

Ahmed, W.; Mofed, D.; Zekri, A.R; El-Sayed, N.; Rahouma, M.; Sabet, S. Antioxidant Activity and Apoptotic Induction as Mechanisms of Action of *Withania somnifera* (Ashwagandha) Against a Hepatocellular Carcinoma Cell Line. *J. Int. Med. Res.* **2018**, *46* (4), 1358–1369.

Alfaifi, M. Y.; Saleh, K. A.; El-Boushnak, M. A.; Elbehairi, S. E.; Alshehri, M. A.; Shati, A. A. Antiproliferative Activity of the Methanolic Extract of *Withania somnifera* Leaf from Faifa Mountains, Southwest Saudi Arabia, Against Several Human Cancer Cell Lines. *Asian Pac. J. Cancer Preve.;* **2016**, *17* (5), 2723–2726.

Antony, M. L.; Lee, J.; Hahm, E. R.; Kim, S. H.; Marcus, A. I.; Kumari, V.; Ji, X.; Yang, Z.; Vowell, C. L.; Wipf, P.; Uechi, G. T. Growth Arrest by the Antitumor Steroidal Lactone Withaferin A in Human Breast Cancer Cells Is Associated with Down-Regulation and Covalent Binding at Cysteine 303 of β-Tubulin. *J. Biol. Chem.* **2014**, *289* (3), 1852–1865.

Balakrishnan, A. S.; Nathan, A. A.; Kumar, M.; Ramamoorthy, S.; Mothilal, S. K. *Withania somnifera* Targets Interleukin-8 and Cyclooxygenase-2 in Human Prostate Cancer Progression. *Prostate Int.* **2017** *5* (2), 75–83.

Berghe, W. V.; Sabbe, L.; Kaileh, M.; Haegeman, G.; Heyninck, K. Molecular Insight in the Multifunctional Activities of Withaferin A. *Biochem. Pharmacol.* **2012**, *84* (10), 1282–1291.

Cai, Y.; Sheng, Z. Y.; Chen, Y.; Bai, C. Effect of Withaferin A on A549 Cellular Proliferation and Apoptosis in Non-Small Cell Lung Cancer. *Asian Pac. J. Cancer Prev.* **2014**, *15* (4), 1711–1714.

Caputi, F. F.; Acquas, E.; Kasture, S.; Ruiu, S.; Candeletti, S.; Romualdi, P. The Standardized *Withania somnifera* Dunal Root Extract Alters Basal and Morphine-Induced Opioid

Receptor Gene Expression Changes in Neuroblastoma Cells. *BMC Complement. Altern. Med.* **2018**, *18* (1) 1–10.

Choudhary, M. I.; Hussain, S.; Yousuf, S.; Dar, A. Chlorinated and Diepoxy Withanolides from *Withania somnifera* and Their Cytotoxic Effects Against Human Lung Cancer Cell Line. *Phytochemistry* **2010**, *71* (17–18), 2205–2209.

Christina, A. J.; Joseph, D. G.; Packialakshmi, M.; Kothai, R.; Robert, S. J.; Chidambaranathan, N.; Ramasamy, M. Anticarcinogenic Activity of *Withania somnifera* Dunal Against Dalton's Ascitic Lymphoma. *J. Ethnopharmacol.* **2004**, *93* (2–3), 359–361.

Dar, P. A.; Mir, S. A.; Bhat, J. A.; Hamid, A.; Singh, L. R.; Malik, F.; Dar, T. A. An Anti-Cancerous Protein Fraction from *Withania somnifera* Induces ROS-Dependent Mitochondria-Mediated Apoptosis in Human MDA-MB-231 Breast Cancer Cells. *Int. J. Biol. Macromol.* **2019**, *13*, 77–87.

Das, T. P.; Suman, S.; Alatassi, H.; Ankem, M. K.; Damodaran, C. Inhibition of AKT Promotes FOXO3a-Dependent Apoptosis in Prostate Cancer. *Cell Death Dis.* **2016**, *7* (2), e2111.

Devi, P. U.; Sharada, A. C.; Solomon, F. E. In Vivo Growth Inhibitory and Radiosensitizing Effects of Withaferin A on Mouse Ehrlich Ascites Carcinoma. *Cancer Lett.* **1995**, *95* (1–2), 189–193.

Fong, M. Y.; Jin, S.; Rane, M.; Singh, R. K.; Gupta, R.; Kakar, S. S. Withaferin A Synergizes the Therapeutic Effect of Doxorubicin Through ROS-Mediated Autophagy in Ovarian Cancer. *PloS One* **2012**, *7* (7), e42265.

Ghosh, K.; De, S.; Das, S.; Mukherjee, S.; Sengupta Bandyopadhyay, S. Withaferin A Induces ROS-Mediated Paraptosis in Human Breast Cancer Cell-Lines MCF-7 and MDA-MB-231. *PLoS One* **2016**, *11* (12), e0168488.

Hahm, E. R.; Lee, J.; Huang, Y.; Singh, S. V. Withaferin a Suppresses Estrogen Receptor-α Expression in Human Breast Cancer Cells. *Mol. Carcinogenesis* **2011a**, *50* (8), 614–624.

Hahm, E. R.; Moura, M. B.; Kelley, E. E.; Van Houten, B.; Shiva, S.; Singh, S. V. Withaferin A-Induced Apoptosis in Human Breast Cancer Cells Is Mediated by Reactive Oxygen Species. *PloS One* **2011b**, *6* (8), e23354.

Hahm, E. R.; Lee, J.; Singh, S. V. Role of Mitogen-Activated Protein Kinases and Mcl-1 in Apoptosis Induction by Withaferin A in Human Breast Cancer Cells. *Mol. Carcinogenesis* **2014**, *53* (11), 907–916.

Halder, B.; Singh, S.; Thakur, S. S. *Withania somnifera* Root Extract Has Potent Cytotoxic Effect Against Human Malignant Melanoma Cells. *PLoS One* **2015**, *10* (9), e0137498.

Hsu, J. H.; Chang, P. M.; Cheng, T. S.; Kuo, Y. L.; Wu, A. T.; Tran, T. H.; Yang, Y. H.; Chen, J. M.; Tsai, Y. C.; Chu, Y. S.; Huang, T. H. Identification of Withaferin A as a Potential Candidate for Anti-Cancer Therapy in Non-Small Cell Lung Cancer. *Cancers* **2019**, *11* (7), 1003.

Kakar, S. S.; Ratajczak, M. Z.; Powell, K. S.; Moghadamfalahi, M.; Miller, D. M.; Batra, S. K.; Singh, S. K. Withaferin a Alone and in Combination with Cisplatin Suppresses Growth and Metastasis of Ovarian Cancer by Targeting Putative Cancer Stem Cells. *PloS One* **2014**, *9* (9), e107596.

Kim, S. H.; Hahm, E. R.; Arlotti, J. A.; Samanta, S. K.; Moura, M. B.; Thorne, S. H.; Shuai, Y.; Anderson, C. J.; White, A. G.; Lokshin, A.; Lee, J. Withaferin A Inhibits In Vivo Growth of Breast Cancer Cells Accelerated by Notch2 Knockdown. *Breast Cancer Res. Treat.* **2016**, *157* (1), 41–54.

Koduru, S.; Kumar, R.; Srinivasan, S.; Evers, M. B.; Damodaran, C. Notch-1 Inhibition by Withaferin-A: A Therapeutic Target Against Colon Carcinogenesis. *Mol. Cancer Therap.* **2010**, *9* (1), 202–210.

Kuttan G. Use of *Withania somnifera* Dunal as an Adjuvant During Radiation Therapy. *Indian J. Exp. Biol.* **1996**, *34* (9), 854–856.

Kyakulaga, A. H.; Aqil, F.; Munagala, R.; Gupta, R. C. Withaferin a Inhibits Epithelial to Mesenchymal Transition in Non-Small Cell Lung Cancer Cells. *Sci. Rep.* **2018**, *8* (1), 1–14.

Kataria, H.; Shah, N.; Kaul, S. C.; Wadhwa, R.; Kaur, G. Water Extract of Ashwagandha Leaf Limits Proliferation and Migration, and Induces Differentiation in Glioma Cells. *Evid. Based Complement. Altern. Med.* **2011**.

Kataria, H.; Wadhwa, R.; Kaul, S. C.; Kaur, G. *Withania somnifera* Water Extract as a Potential Candidate for Differentiation Based Therapy of Human Neuroblastomas. *PLoS One* **2013**, *8* (1), e55316.

Kataria, H.; Kumar, S.; Chaudhary, H.; Kaur, G. *Withania somnifera* Suppresses Tumor Growth of Intracranial Allograft of Glioma Cells. *Mol. Neurobiol.* **2016**, *53* (6) 4143–4158.

Lee, J.; Hahm, E. R.; Singh, S. V. Withaferin A Inhibits Activation of Signal Transducer and Activator of Transcription 3 in Human Breast Cancer Cells. *Carcinogenesis* **2010**, *31* (11), 1991–1998.

Lee, J.; Sehrawat, A.; Singh, S. V. Withaferin A Causes Activation of Notch2 and Notch4 in Human Breast Cancer Cells. *Breast Cancer Res. Treat.* **2012**, *136* (1), 45–56.

Lee, J.; Hahm, E.R, Marcus, A. I.; Singh, S. V. Withaferin A Inhibits Experimental Epithelial–Mesenchymal Transition in MCF-10A Cells and Suppresses Vimentin Protein 13. Level In Vivo in Breast Tumors. *Mol. Carcinogenesis* **2015**, *54* (6), 417–429.

Li, X.; Zhu, F.; Jiang, J.; Sun, D.; Wang, X.; Shen, M.; Tian, R.; Shi, C.; Xu, M.; Peng, F.; Guo, X. Synergistic Antitumor Activity of Withaferin A Combined with Oxaliplatin Triggers Reactive Oxygen Species-Mediated Inactivation of the PI3K/AKT Pathway in Human Pancreatic Cancer Cells. *Cancer Lett.* **2015**, *357* (1), 219–230.

Liu, X.; Li, Y.; Ma, Q.; Wang, Y.; Song, A. L. Withaferin-A Inhibits Growth of Drug-Resistant Breast Carcinoma by Inducing Apoptosis and Autophagy, Endogenous Reactive Oxygen Species (ROS) Production, and Inhibition of Cell Migration and Nuclear Factor Kappa B (Nf-κB)/Mammalian Target of Rapamycin (m-TOR) Signalling Pathway. *Med. Sci. Monitor* **2019**, *25*, 6855.

Ma, J. H.; Qin, L.; Li, X. Role of STAT3 Signaling Pathway in Breast Cancer. *Cell Commun. Signal.* **2020**, *18* (1), 1–3.

Malik, F.; Kumar, A.; Bhushan, S.; Khan, S.; Bhatia, A.; Suri, K. A.; Qazi, G. N.; Singh, J. Reactive Oxygen Species Generation and Mitochondrial Dysfunction in the Apoptotic Cell Death of Human Myeloid Leukemia HL-60 Cells by a Dietary Compound Withaferin A with Concomitant Protection by N-Acetyl Cysteine. *Apoptosis* **2007**, *12* (11), 2115–2133.

Mandal, C.; Dutta, A.; Mallick, A.; Chandra, S.; Misra, L.; Sangwan, R. S.; Mandal, C. Withaferin A Induces Apoptosis by Activating p38 Mitogen-Activated Protein Kinase Signaling Cascade in Leukemic Cells of Lymphoid and Myeloid Origin Through Mitochondrial Death Cascade. *Apoptosis* **2008**, *13* (12), 1450–1464.

Mayola, E.; Gallerne, C.; Degli Esposti, D.; Martel, C.; Pervaiz, S. Larue, L.; Debuire, B.; Lemoine, A.; Brenner, C.; Lemaire, C. Withaferin A Induces Apoptosis in Human Melanoma Cells Through Generation of Reactive Oxygen Species and Down-Regulation of Bcl-2. *Apoptosis* **2011**, *16* (10), 1014–1027.

Mondal, S.; Mandal, C.; Sangwan, R.; Chandra, S.; Mandal, C. Withanolide D Induces Apoptosis in Leukemia by Targeting the Activation of Neutral Sphingomyelinase-Ceramide Cascade Mediated by Synergistic Activation of c-Jun N-Terminal Kinase and p38 Mitogen-Activated Protein Kinase. *Mol. Cancer* **2010**, *9* (1), 1–17.

Moselhy, J.; Suman, S.; Alghamdi, M.; Chandarasekharan, B.; Das, T. P.; Houda, A.; Ankem, M.; Damodaran, C. Withaferin A Inhibits Prostate Carcinogenesis in a PTEN-Deficient Mouse Model of Prostate Cancer. *Neoplasia* **2017**, *19* (6), 451–459.

Motiwala, H.F; Bazzill, J.; Samadi, A.; Zhang, H.; Timmermann, B. N.; Cohen, M. S.; Aubé, J. Synthesis and Cytotoxicity of Semisynthetic Withalongolide A Analogues. *ACS Med. Chem. Lett.* **2013** *4* (11), 1069–1073.

Munagala, R.; Kausar, H.; Munjal, C.; Gupta, R. C. Withaferin A Induces p53-Dependent Apoptosis by Repression of HPV Oncogenes and Upregulation of Tumor Suppressor Proteins in Human Cervical Cancer Cells. *Carcinogenesis* **2011**, *32* (11), 1697–1705.

Nagalingam, A.; Kuppusamy, P.; Singh, S. V.; Sharma, D.; Saxena, N. K. Mechanistic Elucidation of the Antitumor Properties of Withaferin A in Breast Cancer. *Cancer Res.* **2014**, *74* (9), 2617–2629.

Nishikawa, Y.; Okuzaki, D.; Fukushima, K.; Mukai, S.; Ohno, S.; Ozaki, Y.; Yabuta, N.; Nojima, H. Withaferin A Induces Cell Death Selectively in Androgen-Independent Prostate Cancer Cells But Not in Normal Fibroblast Cells. *PLoS One* **2015**, *10* (7), e0134137.

Prakash, J.; Gupta, S. K.; Dinda, A. K. *Withania somnifera* Root Extract Prevents DMBA-Induced Squamous Cell Carcinoma of Skin in *Swiss albino* Mice. *Nutr. Cancer* **2002**, *42* (1), 91–97.

Rah, B.; Rasool, R. U.; Nayak, D.; Yousuf, S. K.; Mukherjee, D.; Kumar, L.D; Goswami, A. PAWR-Mediated Suppression of BCL2 Promotes Switching of 3-azido Withaferin A (3-AWA)-Induced Autophagy to Apoptosis in Prostate Cancer Cells. *Autophagy* **2015**, *11* (2), 314–331.

Samanta, S, K.; Sehrawat, A.; Kim, S. H.; Hahm, E. R.; Shuai, Y.; Roy, R.; Pore, S. K.; Singh, K. B.; Christner, S. M.; Beumer, J. H.; Davidson, N. E. Disease Subtype–Independent Biomarkers of Breast Cancer Chemoprevention by the Ayurvedic Medicine Phytochemical Withaferin A. *JNCI: J. Natl. Cancer Inst.* **2017** *109* (6).

Samanta, S. K.; Lee, J.; Hahm, E. R.; Singh, S. V. Peptidyl-Prolyl Cis/Trans Isomerase Pin1 Regulates Withaferin A-Mediated Cell Cycle Arrest in Human Breast Cancer Cells. *Mol. Carcinogenesis* **2018**, *57* (7), 936–946.

Sehrawat, A.; Samanta, S. K.; Hahm, E. R.; Croix, C. S.; Watkins, S.; Singh, S. V. Withaferin A-Mediated Apoptosis in Breast Cancer Cells Is Associated with Alterations in Mitochondrial Dynamics. *Mitochondrion* **2019**, *47*, 282–293.

Senthil, V.; Ramadevi, S.; Venkatakrishnan, V.; Giridharan, P.; Lakshmi, B. S.; Vishwakarma, R. A.; Balakrishnan, A. Withanolide Induces Apoptosis in HL-60 Leukemia Cells via Mitochondria Mediated Cytochrome C Release and Caspase Activation. *Chem.-Biol. Interact.* **2007**, *167* (1), 19–30.

Shah, N.; Kataria, H.; Kaul, S. C.; Ishii, T.; Kaur, G.; Wadhwa, R. Effect of the Alcoholic Extract of Ashwagandha Leaf and Its Components on Proliferation, Migration, and Differentiation of Glioblastoma Cells: Combinational Approach for Enhanced Differentiation. *Cancer Sci.* **2009**, *100* (9) 1740–1747.

Srivastava, A. N.; Ahmad, R.; Khan, M. A. Evaluation and Comparison of the In Vitro Cytotoxic Activity of *Withania somnifera* Methanolic and Ethanolic Extracts Against MDA-MB-231 and Vero Cell Lines. *Sci. Pharma.* **2016**, *84* (1), 41–59.

Stan, S. D.; Hahm, E. R.; Warin, R.; Singh, S. V. Withaferin A Causes FOXO3a-and Bim-Dependent Apoptosis and Inhibits Growth of Human Breast Cancer Cells In Vivo. *Cancer Res.* **2008a,** *68* (18), 7661–7669.

Stan, S. D.; Zeng, Y.; Singh, S. V. Ayurvedic Medicine Constituent Withaferin a Causes G2 and M Phase Cell Cycle Arrest in Human Breast Cancer Cells. *Nutr. Cancer* **2008b**, *60* (S1), 51–60.

Subbaraju, G. V.; Vanisree, M.; Rao, C.V; Sivaramakrishna, C.; Sridhar, P.; Jayaprakasam, B.; Nair, M. G. Ashwagandhanolide, a Bioactive Dimeric Thiowithanolide Isolated from the Roots of *Withania somnifera*. *J. Nat. Products* **2006**, *69* (12), 1790–1792.

Szarc vel Szic, K.; Op de Beeck, K.; Ratman, D.; Wouters, A.; Beck, I. M.; Declerck, K.; Heyninck, K.; Fransen, E.; Bracke, M.; De Bosscher, K.; Lardon, F. Pharmacological Levels of Withaferin A (*Withania somnifera*) Trigger Clinically Relevant Anticancer Effects Specific to Triple Negative Breast Cancer Cells. *PloS one* **2014**, *9* (2), e87850

Tao, S.; Tillotson, J.; Wijeratne, E. K.; Xu, Y. M.; Kang, M.; Wu, T.; Lau, E. C.; Mesa, C.; Mason, D. J.; Brown, R. V.; La Clair, J. J. Withaferin A Analogs That Target the AAA+ Chaperone p97. *ACS Chem. Biol.* **2015**, *10* (8), 1916–1924.

Thaiparambil, J. T.; Bender, L.; Ganesh, T.; Kline, E.; Patel, P.; Liu, Y.; Tighiouart, M.; Vertino, P. M.; Harvey, R. D.; Garcia, A.; Marcus, A. I. Withaferin A Inhibits Breast Cancer Invasion and Metastasis at Sub-Cytotoxic Doses by Inducing Vimentin Disassembly and Serine 56 Phosphorylation. *Int. J. Cancer* **2011**, *129* (11), 2744–2755.

Um, H. J.; Min, K. J.; Kim, D. E.; Kwon, T. K. Withaferin A Inhibits JAK/STAT3 Signaling and Induces Apoptosis of Human Renal Carcinoma Caki Cells. *Biochem. Biophys. Res. Commun.* **2012**, *427* (1), 24–29.

Wadhwa, R.; Singh, R.; Gao, R.; Shah, N.; Widodo, N.; Nakamoto, T.; Ishida, Y.; Terao, K.; Kaul, S. C. Water extract of Ashwagandha Leaf Has Anticancer Activity: Identification of an Active Component and Its Mechanism of Action. *Plos One* **2013**, *10, 8* (10), e77189.

Xia, S.; Miao, Y.; Liu, S. Withaferin A Induces Apoptosis by ROS-Dependent Mitochondrial Dysfunction in Human Colorectal Cancer Cells. *Biochem. Biophys. Res. Commun.* **2018**, *503* (4), 2363–2369.

Yu, Y.; Hamza, A.; Zhang, T.; Gu, M.; Zou, P.; Newman, B.; Li, Y.; Gunatilaka, A. L.; Zhan, C. G.; Sun, D. Withaferin A Targets Heat Shock Protein 90 in Pancreatic Cancer Cells. *Biochem. Pharmacol.* **2010**, *79* (4), 542–551.

Zhang, X.; Mukerji, R; Samadi, A. K.; Cohen, M. S. Down-Regulation of Estrogen Receptor-Alpha and Rearranged During Transfection Tyrosine Kinase Is Associated with Withaferin A-Induced Apoptosis in MCF-7 Breast Cancer Cells. *BMC Complement. Altern. Med.* **2011**, *11* (1), 1–10.

Zhou, Y. F.; Yu, X. T.; Yao, J. J.; Xu, C. W.; Huang, J. H.; Wan, Y.; Wu, M. J. Withaferin A Inhibits Hepatoma Cell Proliferation Through Induction of Apoptosis and Cell Cycle Arrest. *Int. J. Clin. Exp. Pathol.* **2016**, *9* (12), 12381–12389.

CHAPTER 7

Camptotheca acuminata Decne

AMOGHA G. PALADHI[1], ANU PANDITA[2], MANOHAR M. V.[3],
BHOOMIKA INAMDAR[3], SUGUMARI VALLINAYAGAM[4],
DEEPU PANDITA[5], and K. M. SRINIVASA MURTHY[6]

[1]Christ (Deemed to be University), Bengaluru, Karnataka, India

[2]Vatsalya Clinic, Krishna Nagar, New Delhi, India

[3]JSS Medical College (Deemed to be University), Mysuru, Karnataka, India

*[4]Department of Biotechnology, Mepco Schlenk Engineering College,
Sivakasi, Tamil Nadu, India*

*[5]Government Department of School Education, Jammu,
Jammu and Kashmir, India*

*[6]Department of Microbiology and Biotechnology, Jnanabharathi Campus
Bangalore University, Bengaluru, India*

ABSTRACT

Camptotheca acuminata Decne belonging to the family Nyssaceae is a
native flora of Asian countries like South China, India, Pakistan, etc. In the
early 1960s, the fruits of *Camptotheca acuminata* Decne were used in folk
medicine against cancer. In the 1970s, curiosity led to an accidental discovery
of a group of alkaloids named Camptothecin from the plant *Camptotheca
acuminata* Decne. Extractions were made to analyze the therapeutic
properties of this alkaloid against tumors. The cause of previously known
anticancerous property of *C. acuminata* is the alkaloid named Camptothecin.
In the late 1980s, SAR studies were made to learn more about *Camptotheca*
which showed a primitive anticancer property in its native form of planar

pentacyclic ring structure. Analogues of this component were prepared in various experiments to improve the property of Camptothecin like solubility and antitumor properties. Efforts were made to extract Camptothecin from various parts of the plant leading to the comparative studies of concentration of Camptothecin in various stages of its life. Camptothecin enters the stages of cell cycle wherein the enzyme DNA TOP1 acts for replication and winding of DNA. Camptothecin inhibits DNA replication by binding itself to DNA TOP1 by forming covalent complexes. The Camptothecin derivatives are created by various groups of scientists to treat a very wide range of malignancies like Adeno-carcinoma of Thyroid, Hepatocellular, Colo-rectal, Ovarian, Lung, Leukemia, and many more. Topotecan, Irinotecan, and Exatecan are widely used analogues in preclinical and clinical trials. To improve the efficiency of Camptothecin, approaches were made with the use of nanostructures like Poly (methacylic acid-co-methyl methacrylate) nano-formulations. Thus, alkaloids like Camptothecin and its analogues from *Camptotheca acuminata* Decne can be used in anticancerous activity in most human malignancies.

7.1 INTRODUCTION

Camptotheca acuminata Decne is one of the native floras of South China, Punjab, Pakistan, and other South Asian countries. *Camptotheca acuminata* Decne is commonly called Chinese Happy Tree (Martino et al., 2017). It is one of the plants used in native medicine therapies as the alkaloids in the plant are of high medicinal value and an interesting genus of ethnobotany. In the late 1980s, one of its alkaloids belongs to a group of drugs known as camptothecins. Camptothecin (CPT) is known to be a chemotherapeutic drug. The natural alkaloid obtained from various parts of *Camptotheca acuminata* Decne is 20-(S)-camptothecin (CPT).

Camptothecin is known to be an anticancer drug for its property to break the strands of DNA in mammalian cells by efficiently inhibiting DNA topoisomerase 1 (Gong et al., 2018). Due to this property, with the goal of increasing its efficiency to treat cancer, various derivatives or analogues have been developed till date (Guo and Yuan, 2016). In the 1960s, revolution in screening of various plants was made in search of steroid compounds. The screenings for steroids were made to test its properties like antitumor, antibacterial, and antiviral activities. One of those plants was *Camptotheca acuminata* of South China, and the alkaloid from this tree was first extracted from its bark. It was found that the alkaloid belonged to the group of Camptothecin which is an anticancer drug with its property of specifically inhibiting the activity of DNA TOP1 enzyme, which was capable of DNA replication. This led to a milestone in the discovery of anticancer drugs as the root cause of cancer that can be treated by camptothecins. The extract of *Camptotheca acuminata* Decne was first tested in vivo to treat mouse leukemia and its first discovery of naturally obtained alkaloid Camptothecin being an antitumor agent (Wani et al., 1987). In the 1970s, it was found that camptothecins are able to inhibit DNA and RNA replication in various organisms (Chabner, 1992). In cancerous tumor, the root cause of cell differentiation and proliferation is replication of its genetic material due to the increased activity of DNA TOP1. As this process increases the number of copies of DNA leading to karyokinesis and cytokinesis, it causes cell division. The increased action of the enzyme TOP 1 is due to the formation of covalent complex between substrates (DNA) and enzymes. The action of TOP1 is found to be winding an unzipping of replication forks in DNA, where these sites of interests are the target for analogues of Camptothecin to act on. Camptothecin action leads to the formation of noncovalent complex that drastically decreases and inhibits DNA replication, and also prevents recombination of double-stranded helix. DNA is found to be stable and its double-helix model, due to the enzyme inhibition action of CPT; the DNA, once divided into single strands, is not capable of winding torsionally by itself leading to the death of cell. Various experiments have shown that the activity of CPT is directly proportional to the concentration of DNA TOP1 (Hsiang et al., 1985; Redinbo et al., 1998). DNA TOP1 is found to be hypersensitive to its CPT-induced cytotoxicity, meaning that the higher the concentration of enzyme, the easier for the CPT to inhibit its action causing cell death (Dancey and Eisenhauer, 1996). In various studies on cancer, where it is found that due to the high number of dividing cells in tumor, the concentration of the enzyme DNA TOP1 is very high, because it is known that cancer cells are highly dividing regardless of requirement (Martino et al., 2017). It is easier for CPT and its analogues to efficiently act on the enzyme due to its availability as a substrate in cancer tissues. Thus, CPT is known to be a leading anticancer drug in recent decades (Champoux, 1978).

In later discoveries, CPT is found to have very less solubility; this is a limitation of CPT to be used as a naturally available form. Thus, discoveries regarding improvisation of solubility of CPT were made leading to the development of its analogues. The analogues are found to be more water soluble which can be clinically used.

SAR studies suggested that the anticancer activity of camptothecin depends on the structural orientation and presence in a number of rings, meaning that the least number of rings that must be present for CPT to act against cancer is 5 (planar pentacyclic ring structure). Further studies showed that lesser number of rings in CPT makes the molecule very unstable and useless. Addition of benzene ring to its structure makes it hexacyclic in nature showing an elevated anticancer activity. As there were developed SAR studies, it is found that the synthesis of CPT analogues with the addition of benzene ring is more efficient; lactone is one of the core molecules from which various derivatives were obtained, thus adding a sixth ring to the planar structure making it hexacyclic (Chabner, 1992). These SAR studies led to the discovery of analogues like topotecan, irinotecan, and exatecan. Topotecan and Irinotecan are approved for clinical use by FDA. In later studies, Exatecans showed very promising and efficient anticancer activities in preclinical studies (Dancey and Eisenhauer, 1996).

7.2 PHYLOGENY

The genus *Camptotheca* is well known as a source for camptothecin (CPT), an anticancer drug. The genus was first termed in 1983 and was placed under the family Cornaceae along with other genus like *Davidia, Nyssa,* and *Cornus,* based on its morphology, anatomy, palynology, embryology, cytology, pollination, phytochemistry, phylogenetic position, and paleobiology. However, some taxonomists placed *Camptotheca* under the family Nyssaceae. The most recent APG (angiosperm phylogeny Group) system of classification that includes molecular phylogenetic analysis confirmed that *Camptotheca* can be placed under the family Nyssaceae along with other genus *Davidia, Nyssa, Cornus, Diplopanax,* and *Mastixia.* Nyssaceae is now placed as a sister clade comprising loasaceae, hydrostachyaceae, and hydrangeaceae (Gong et al., 2018).

7.3 CAMPTOTHECIN (CPT)—AN ANTICANCER DRUG

Camptothecin (CPT) is an alkaloid that is produced by a native tree of South China called *Camptotheca acuminata* Decne. It is a quinoline alkaloid having a pentacyclic ring system consisting of a quinolinic portion, pyrrolidine, a lactam, a stereo center configured on α-hydroxy lactam system with delocalized aromatic moiety (Wall et al., 1966). In the late 1980s, it was found that this group of camptothecin drug can be used to treat various cancerous conditions. Initially, it was used in various clinical trials against colon carcinoma, thus evaluating CPT as an anticancer drug with the approval of FDA (Food and Drug Administration) (Martino et al., 2017). SAR (Structural Activity Relationship Studies) has led to the discovery of various derivatives of CPT that are more eligible and efficient to treat cancer. Analogues of

CPT with a planar pentacyclic system are the only component that helps in anticancer treatment, while other cyclic systems like bi, tri, and tetracyclic are not found to have any biological activity (Wall et al., 1966; Wani et al., 1980).

There are structures of CPT that contains more than five ring planar structures that also show anticancerous activity in a conserved manner. Hexacyclic and heptacyclic show more activities than pentacyclic, thus showing a planar structure of CPT with five rings and more have elevated biological activity while the lesser ones are inert (Wani et al., 1980). The replacement of lactam-containing benzene ring might lead to the complete loss of anticancer activity in CPT (Nicholas et al., 1990) and also the opening of the ring structures of α-hydroxy lactone by hydrolysis shows very low antitumor activity results in the reduction in the activity of CPT molecule, if the spatial arrangement of structures forming the complex is disoriented (Jaxel et al., 1989; Rowinsky et al., 1992; Wani et al., 1987).

7.3.1 CAMPTOTHECIN ANALOGUES

Camptothecin was later modified to increase its therapeutic value due to its low solubility. Efforts were made to develop the derivatives with more stability and solubility than that of the naturally obtained alkaloid. Various synthetic and semisynthetic derivatives associated with anticancerous property were brought into action. Studies have reported that the in vivo and in vitro evaluation resulted in increased activity of camptothecin analogues that were derived. Various solubilizing groups were added to the planar ring structure of CPT to increase its solubility leading to the discovery of highly efficient therapeutic derivatives called Topotecan (Date et al., 2002), Irinotecan (Fukuda et al., 1996), and Exatecan (Pizzolato and Saltz, 2003).

7.3.1.1 TOPOTECAN

Topotecan is one of the first found derivatives of CPT. Topotecan is a water-soluble analogue approved by FDA (Food and Drug Administration) for clinical use. The structure of topotecan is 9-[(dimethylamino)methyl]-10-hydroxycamptothecin. Various in vitro experiments and preclinical trials were performed that showed an excellent anticancer property. Later, Topotecans were put to use in treating adeno-carcinomas of colon, ovary, and breast cancer. The tumors of CNS (central nervous system) and sarcomas were also treated well and cured by the administration of the same (Pizzolato and Saltz, 2003).

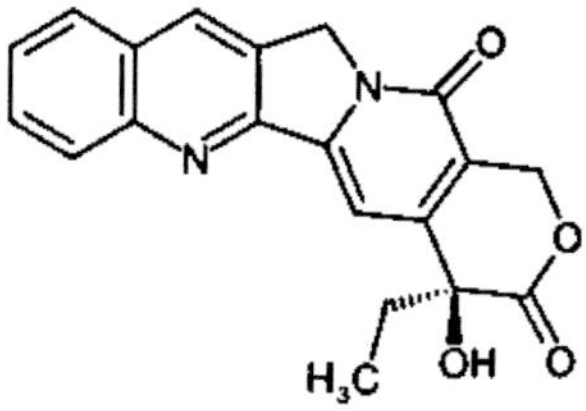

FIGURE 7.1   Structure of camptothecin.

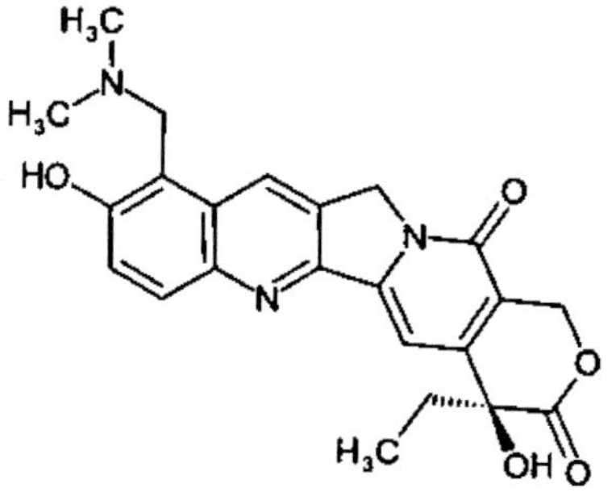

FIGURE 7.2   Structure of topotecan.

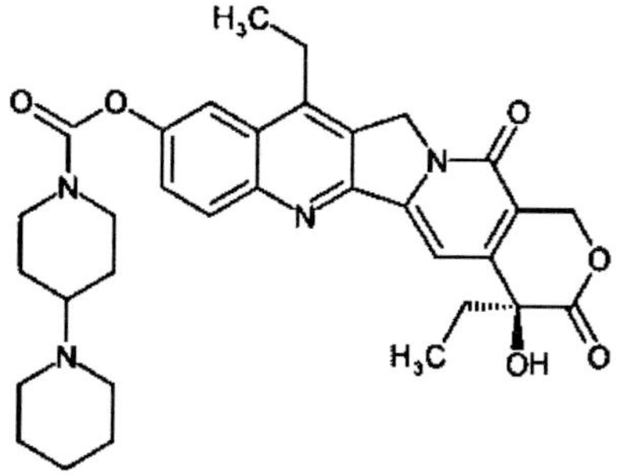

FIGURE 7.3   Structure of irinotecan.

7.3.1.2   IRINOTECAN

Kawoto et al. in 1991 discovered the activity of Irinotecan which is a water-soluble analogue of CPT. It is found that irinotecan is 1000 times more efficient than the native camptothecin. Irinotecan is an active compound that shows anticancerous property in various types of carcinoma cell lines, both in in vitro and in vivo. Irinotecan is otherwise scientifically named as 7-ethyl-10-(4-[1-piperidino]-1-piperidino) methyl-10-hydroxycamptothecin or just CPT-11. SN-38 is the irinotecan analogue of CPT which was used by Kawoto et al. that resulted in the inhibition of topoisomerase- I activity, also showing anticancerous activity in HT-29 human colon carcinoma cells in in vitro. Irinotecan is known to be the most biologically active anticancerous drug found in its array of analogues (Dancey and Eisenhauer, 1996).

7.3.1.3   EXATICAN

Exatican mesylate ($C_{25}H_{26}FN_3O_7S$) is one of the recently known DNA TOP1 inhibitors. Exatican is a derivative analogue of CPT which is water soluble in nature. Clinical trials were known to be experimented on Colo-rectal cancer in earlier days. The studies have shown that Exaticans resulted in the maximum activity on biliary (bile duct and pancreatic) cancer (Dancey and Eisenhauer, 1996).

7.4 EXTRACTION OF CAMPTOTHECIN FROM VARIOUS PARTS OF *CAMPTOTHECA ACUMINATA* DECNE

7.4.1 BARK

As alkaloid Camptothecin is known to exhibit efficient antileukemic activity, Wall et al. considered the extraction of the alkaloid from the stem wood of *Camptotheca acuminata* Decne as they quoted stem wood was extracted by a standard method. They tried to crystallize camptothecin by methane-insoluble material from chloroform to crystallize the alkaloid resulting in the formation of pale yellow needles of camptothecin (Wall et al., 1966).

7.4.2 FRUIT

The alkaloids form *Camptotheca acuminata* was mostly extracted from the fruits of the plant. It was found that folk medicine also used fruits of this plant as therapy against various malignancies. Liu et al. in 2013 extracted camptothecin from the fruit and used it in clinical trials (Y. Liu et al., 2013). An aqueous extract from the fruits of *Camptotheca acuminata* can inhibit the growth of cancer cells and this was experimented by Lin et al. in 2014 (Lin et al., 2014). Guo et al. considered using dried fruits of *Camptotheca acuminata,* 50% ethanol at 80 ℃ was used to extract camptothecin from the coarse powder (He et al., 2019).

7.4.3 LEAF

Wang et al. separated and extracted the alkaloids that were major secondary metabolites in the leaves of *Camptotheca acuminata* (Wang et al., 2015). Experiments have shown that younger leaves contain more camptothecin when compared with older leaves. It was found that the concentration of camptothecin was decreased by 11% as it gets older by every month. The concentration of camptothecin in extraction depends on the way of approach it is extracted. Oven-dried leaves showed 27% less CPT concentration than freeze-dried leaves (Z. Liu et al., 1998).

7.5 STRUCTURE–ACTIVITY RELATIONSHIPS (SAR) OF CAMPTOTHECINS

Various studies have been made to examine the structure of quinoline alkaloid revealing Camptothecin, a pentacyclic planar ring structure that consists of pyrrolo (3,4-β) quinoline moiety and also consists of an α-hydroxy lactone ring within which an asymmetric center is present (Wall et al., 1966). It is suggested that the planar pentacyclic ring is the most capable structure of camptothecin which has the capability to inhibit the activity of DNA TOP1. If this ring is removed or disoriented, the structure of CPT gets distorted and loses the property of its activity. Thus, it is concluded that CPT has a deprived activity in bi, tri, and tetracyclic compounds (Cesare et al., 2000). SAR also showed that the activity of CPT is increased in its structure containing more than five ring compounds suggesting that a pentacyclic ring system is required for anticancerous property of CPT.

Earlier SAR studies have also concluded that an oriented structure of CPT with intact α-hydroxy lactone group is one of the most essential structures (Venditto and Simanek, 2010) that have been identified to denature DNA TOP1 as a target for its antitumor activity (Hsiang et al., 1985; Jaxel et al., 1991, 1989; Kjeldsen et al., 1988; Thomsen et al., 1987).

The interest of cure in cancer has been prioritized in poisoning TOP1 as it is the most efficient enzyme that helps in DNA duplication and other biological activities of DNA. Thus, showing derivatives of camptothecin obtained from *Camptotheca acuminata* Decne with the combination of various other compounds showed an excellent anticancerous property (Hsiang and Liu, 1988; Hertzberg et al., 1989).

7.6 CPT SYNTHESIS, SEMISYNTHETIC DERIVATIVES

CPT was natively extracted from *C. acuminata* which is native species of South China. As this species is found to be endangered and belongs to red line species, if the extraction increases leading to the exploitation then it would result in the extinction of the species. Efforts were made to chemically synthesize this alkaloid synthetically and semisynthetically (Martino et al., 2017). Stork et al. in 1971 were the first to synthesize the synthetic form of CPT (Martino et al., 2017). This procedure consisted of 15 steps in which prior nine were molecules chemically synthesized by Danishefsky in 1993 (Shen et al., 1993). Curran et al. in 1994–1996 developed radicalic pathway (Martino et al., 2017). Comins in 2001 first advanced an optimized procedure consisting of six steps of total synthesis (Comins et al., 2001).

7.7 MODE OF ACTION

Camptothecin (CPT) obtained from *Camptotheca acuminata* Decne is found to be a blocker or inhibitor of an enzyme known as DNA Topoisomerase-I in mammalian cells (Martino et al., 2017). It is a profound anticancer activity where CPT poisons DNA Topoisomerase-I in tumor cells. It has various anticancerous activities, but these are limited due to its poor solubility and high degradation rate (Gigliotti et al., 2017). Researches are conducted to assist the use of camptothecin by manipulating it using the encapsulation of b-cyclodextrin-nanosponges (CN-CPT) that helps in increasing the solubility of CPT. It increases the rate of uptake in tumor or cancer cells and also to reduce the toxicity of the tissue by degrading and inhibiting the growth factors of tumors in both in vitro and in vivo conditions (Gigliotti et al., 2017). Due to its low solubility, experiments were carried out to set up the discovery in the use of camptothecin (CPT) by increasing its property of effectiveness and water solubility by deriving it synthetically and semisynthetically. Structure–activity relationship studies (SAR) were clinically conducted with various numbers of small and large derivatives of CPT. Mainly, topotecan (7-ethyl-10-[4-(1-piperidino)-1-piperidino] carbonyloxycamptothecin) and irinotecan (9-[(dimethylamino)methyl]-10-hydroxy-camptothecin) are the two SAR derivatives of CPT that are in current clinical use (Venditto and Simanek, 2010).

As CPT is well known for its DNA Top1 inhibition activity, the enzyme is mostly involved in various significant biological activities of DNA like repair, recombination, replication, and transcription which are the core stages for cell replication and proliferation that happens in normal cell (Hertzberg et al., 1989; Hsiang et al., 1985). This principle of DNA TOP1 inhibition like CPT is used to damage the enzyme which in turn inhibits the above stages of DNA. This process, when experimented on tumor cells, inhibits cell division and proliferation thus inhibiting the tumor and also helps in curing the existing cancerous conditions. DNA is a stable nucleic acid and also genetic material of all mammals that exist in stable double helix, double strand. When the helix is untangled and the DNA is spilt into single strand, DNA TOP1 helps to maintain and restore the integrity of DNA double helix by rotating each antiparallel single strands together (Chabner, 1992; Pommier, 2006; Stewart et al., 1998).

Core of DNA TOP1 enzyme directly acts on DNA while the other subdomains of enzymes like I and III help in the stabilization of DNA and enzyme complex by the formation of salt bridges. The spatial arrangement of the DNA/enzyme complex allows Tyr123 to form covalent phosphodiester bonds in 3'-end of DNA at active sites of enzymes, thus activating enzyme catalysis (Stewart et al., 1998). The action of camptothecin was discovered in the late 1980s which breaks the interaction of enzyme catalysis. The core target of camptothecin and its derivatives are DNA TOP1 where they act to form a noncovalent bond complex between DNA and Topoisomerase-I enzyme (Hsiang et al., 1985; Redinbo et al., 1998). This results in the breaking of DNA double-stranded helix and leading to the irreversible process causing cell death. Results of various experimental approaches have shown that the toxicity of CPT is directly proportional to the concentration of enzyme in the tumor. This means that higher levels of enzymes present in cancerous cells show hypersensitivity to the cytotoxic activity of CPT. The above explanation also demonstrates the abnormally high availability of enzymes results in the formation of cancer cells, also helping them to replicate and proliferate (Dancey and Eisenhauer, 1996). The equilibrium has to be maintained in between the presence of free TOP and DNA TOP1 complex in the cell. If the equilibrium is disrupted by shifting either of the entities from normal to high, it results in cancer formation. To inhibit this action, a tertiary complex is formed that inverts the free CPT into bound form by the formation of a TOP1-DNA complex (Pizzolato and Saltz, 2003).

The formation of this tertiary complex leads to the shifting of free TOP that results in greater DNA replication leading to the formation of new cells. When such conditions are treated with CPT, it breaks the complex resulting in inhibiting tumor formation.

In contrast, the 3-hydroxy group-induced Camptothecin is used reversibly to interact with an enzyme that unbounds DNA where the cell is damaged due to the inability of DNA to form double helix by the inhibited activity of TOP1 by CPT. 20-hydroxy group is required for DNA TOP1 complex stabilization and also shows anticancerous activity (Hertzberg et al., 1989).

7.7.1   EFFECT OF CAMPTOTHECIN IN VARIOUS INCIDENCES

7.7.1.1   IN VITRO ANTICANCER ACTIVITY

Studies that are made on the treatment of human colon cancer cell-line, Caco2 by using CN-CPT (cyclodextrine nano-sponges) or camptothecin encapsulated with poly-nano-formulations in in vitro to analyze the anticancerous activity. When greater quantities like >1000 µg/ml of camptothecin were encapsulated using poly (methacylic acid-co-methyl methacrylate)

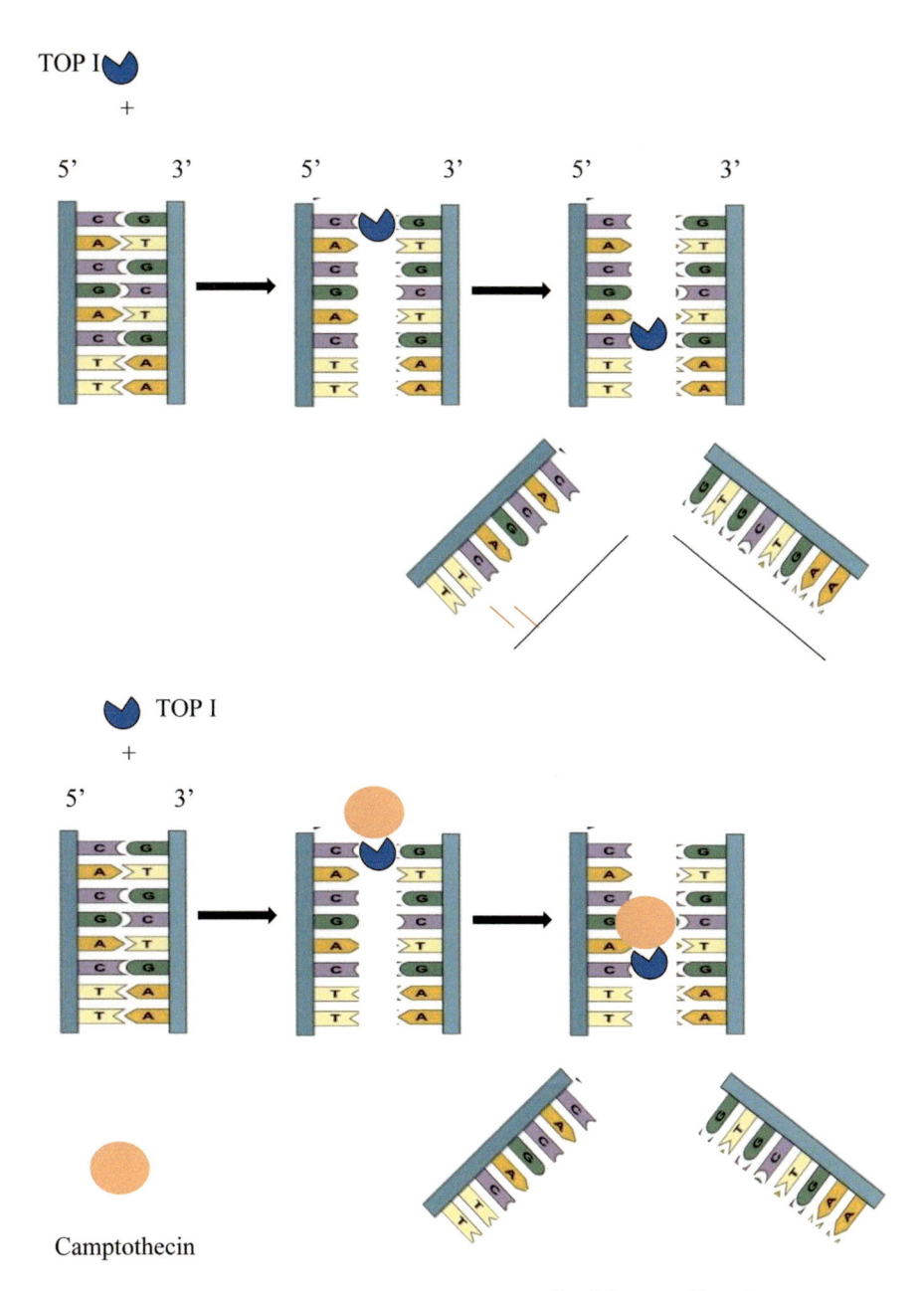

FIGURE 7.4 Mode of action.

nano-formulations resulted in enhanced anticancerous activity of camptothecin on Caco2 cell-line (Manikandan and Kannan, 2017).

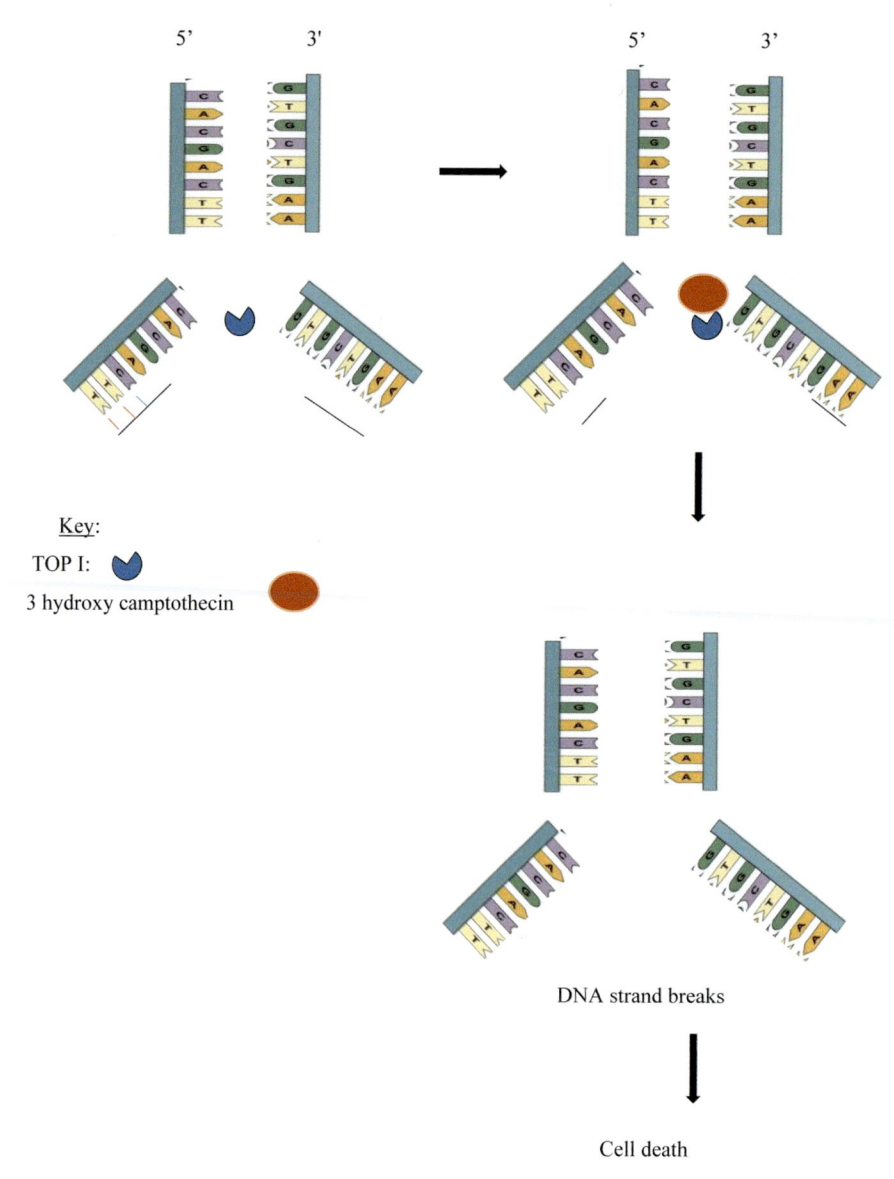

FIGURE 7.5   Inhibition of DNA-TOPI activity.

7.7.1.2 IN VIVO ANTICANCER ACTIVITY

Studies were made on rats considering body weight as a criterion to detect the capability of camptothecin encapsulated with poly (methacylic acid-co-methyl methacrylate) nano-formulations as an anticancerous drug. The above is compared with other batches of rats that were administered with DMH (1,2 dimethylhydrazine). The rats that were treated with camptothecin poly-nano-formulations showed excellent results of increased weight gain denoting increased health of rats. This showed colon carcinoma in rats was cured better by CN-CPT when compared with DMH (Manikandan and Kannan, 2017).

7.7.1.2.1 Effect of Camptothecin on ACF Incidence

Agner et al. conducted an experiment on groups of rats to demonstrate the treatment of CPT to cure colo-rectal carcinoma by considering colonic ACF (aberrant cryptic foci) as a biomarker that shows the intensity of occurring tumor. This experiment demonstrated the use of camptothecin encapsulated poly (methacylic acid-co-methyl methacrylate) nano-formulations significantly decreased the levels of colonic ACF, resulting in 100% cure in rats suffering from colo-rectal carcinoma (Agner et al., 2005).

7.7.1.2.2 Effect of Camptothecin on Tumor Incidence

Hsiang et al. demonstrated an experiment using Burkit's lymphoma cell-line (Kjeldsen et al., 1988), where they reported increase in the concentration of camptothecin by 50 µM in every trial to demonstrate the action of camptothecin that is known to bring the equilibrium of TOP1 and DNA by torsionally relaxing the coiled DNA (Thomsen et al., 1987). Various other experiments also showed the cleavage reaction in the DNA–TOP1 complex (Thomsen et al., 1987). Around 70% of DNA molecules were cleaved at the site of recognition sequences. This shows specificity in the sequence of enzymes in varied ratios of enzyme and substrate. The equilibrium dissociation remains a constant, portraying the action of camptothecin helps in the cure of tumors and also the influence of camptothecin on the cleavage properties of human topoisomerase-I (Thomsen et al., 1987).

7.7.1.3 *CN-CPT INHIBITS CELL PROLIFERATION IN VITRO*

Experiments were conducted to record the anticancerous activity of CN-CPT when compared with naturally occurring form of camptothecin in cell-line studies. In in vitro conditions, animal cancer cell models were cultured in various titrated amounts to obtain a viable number of cells and assessed using MTT (2,3-bis[2-methoxy-4-nitro-5-sulfophenyl]-2H-tetrazolium-5-carboxanilide) assay. The results of the above studies confirm that CN-CPT has a very high efficiency in inhibiting the growth of cultured cell-lines when compared with free CPT. Very little differences were seen when tested for different cell-lines due to the effect of time and concentration. Variable effects of inhibition by effective doses were also observed. To confirm the results, the cells were cultured for more number of days to increase the number of cells. CN-CPT and CPT were again administered where CN-CPT showed the inhibition of multiple colonies being a better derivative of Camptothecin than the free form (Gigliotti et al., 2017).

7.8   TARGET ORGANS

7.8.1   *THYROID CANCER CELLS*

Anaplastic carcinoma of thyroid is a very lethal cancer that is malignant in nature and must be effectively cured by stopping the proliferation (Gigliotti et al., 2017). CN-CPT is used in various in vitro, in vivo experiments and clinical trials. This resulted in inhibiting the proliferation of endothelial cells adhesive to anaplastic carcinoma of Thyroid and also it inhibits the migration of these cells. The CN-CPT activity is studied where it regulates the formation of adhesive cells leading to lamellipodia and increased cell motility (Kim et al., 2013; Occhipinti et al., 2013).

7.8.2   *HEPATOCELLULAR CARCINOMA*

Stotz et al. discovered that hepatocellular carcinoma (HCC) being a malignant tumor condition with high mortality rate can be treated using CPT (Stotz et al., 2015). In the later stages of human hepatocellular carcinoma, the chances of surgical removal of tumor are very less. Simultaneously, an indeed option was found to be chemotherapy (Hu et al., 2013). By this, surgeries can be avoided due to the therapeutic use of CPT in treating human HpG2 and

Hep3B cell-lines which showed more positive results compared with other modes of treatment (He et al., 2019).

7.8.3 COLO-RECTAL CANCER

Tubular adenomas are cancers that are macroscopically sessile that have been induced in a batch of cell-lines; later it was compared with the treatment of DMH and CPT. As CPT does not show much expected results, CPT was encapsulated using poly (methacylic acid-co-methyl methacrylate) nano-formulations and the cell-lines were treated, thus depicting the anticancerous activity of CPT (Manikandan and Kannan, 2017; Pizzolato and Saltz, 2003).

7.8.4 OVARIAN CANCER

Ovarian cancer being a very lethal condition for woman must be treated. The use of CPT derivatives like Topotecans helps in avoiding DNA replication and cell proliferation in Ovary that helps in the maintenance of a static number of cells and also helps in curing the tumor (Pommier, 2006).

7.8.5 LUNG CANCER

Derivatives or analogues of CPT are used in the treatment of lung cancer. Exatecan is intravenously administered in human subjects suffering from lung cancer. Topotecan is also one of the CPT derivatives used in its treatment. The mode of action remains the same in almost every cancer treatment. The motto is to inhibit the binding of DNA TOP 1 enzyme that helps in winding and unwinding of DNA that has torsional rotation, thus helping in chemotherapeutical treatment of lung cancer (Pommier, 2006).

7.8.6 LEUKEMIA

Wall et al., in one of their works, have used leukemia as a subject to be treated using camptothecin. As camptothecin is found to be insoluble, it was standardized using various derivatives and ring structures. The analogues of camptothecin exhibited antileukemic property that cured leukemia in samples. They used X-ray diffraction studies to determine the molecular activity of camptothecin derivatives (Wall et al., 1966).

7.9 LIMITATIONS

Camptothecins and its derivatives despite being beneficial in treatments of cancers also have some limitations that are inevitable. Maintenance of cleavage complexes of DNA and TOP1 is one of the due concerns of CPT administration. The TOP1ccs (cleavage complexes) should be maintained in tumor cells until the camptothecin converts the complex into damaged DNA by its action of inhibition. This leads to a prolonged infusion to maintain the cleavage complexes persistently until DNA is damaged. Another unexpected limitation of camptothecin administration procedure by infusion is that it might cause leucopenia, but this limitation can be resolved by limited administration of camptothecin. Irinotecans when administered have been studied for off-target effects due to the induced diarrhea. The derivatives of camptothecin obtained from lactone like α-hydroxylactone E-ring have a very short shelf-life and been converted into carboxylate that is inactive against DNA TOP1 which binds to serum albumin (Pommier, 2006).

7.10 CONCLUSION

To conclude with, various studies and experiments that are conducted in vitro and in vivo on various animal and human carcinoma modules and cell-lines have shown that camptothecin and its derivatives can efficiently be used as therapeutic drug in cancer. X-ray crystallographic studies have shown that TOP1 is an enzyme that cleaves the DNA and also helps in the replication. In experiments, CPT has shown anti-topoisomerase activity by the action of inhibition. Studies have shown that the camptothecin is more efficient in the presence of Cu (II) and long wavelength UV radiations. Various camptothecin analogues were used in the anticancer treatment and experiments that also showed the similar but elevated activity in forming a noncovalent complex of DNA-TOP1 or inhibiting the enzyme from binding DNA. Free camptothecin and 10-hydroxycamptothecin together are known for its anticancer activity.

Camptothecin has unique properties that act only against the enzyme TOP1 in any given cells with DNA. Studies have confirmed the above, by experiments in which the TOP1 gene is removed from the cells. It is seen that if the cell-lines show single-point mutation in TOP1 gene, such cells show resistance to camptothecin. Thus, SAR studies led to the discovery of analogues of CPT that showed efficient antitumor activities against tumor cells. Tumor cells being highly replicable have increased activity of TOP1 due to which DNA duplication takes place. CPT has a main role in inhibiting this enzyme thus stopping tumor cell proliferation. Studies have shown that Camptothecin and its derivatives have the capability of curing almost any cancer by its inhibition activity that poisons TOP1. Cancers like adenocarcinoma of thyroid, hepatocellular carcinoma, colo-rectal, ovarian, lung, leukemia, etc. increase the activity of camptothecin; the molecules were encapsulated using poly (methacylic acid-co-methyl methacrylate) nano-formulations.

KEYWORDS

- ***Camptotheca acuminata* Decne**
- **Camptothecin**
- **anticancer drug**
- **antitumor activity**
- **Topoteca**
- **Irinotecan**
- **Exatica**
- **Chinese Happy Tree**
- **DNA TOP1**

REFERENCES

Agner, A. R.; Bazo, A. P.; Ribeiro, L. R.; Salvadori, D. M. F. DNA Damage and Aberrant Crypt Foci as Putative Biomarkers to Evaluate the Chemopreventive Effect of Annatto (Bixa orellana L.) in Rat Colon Carcinogenesis. *Mutat. Res.* **2005,** *582* (1–2), 146–154. https://doi.org/10.1016/j.mrgentox.2005.01.009

Cesare, M. De, Zunino, F.; Pace, S.; Pisano, C.; Pratesi, G. Efficacy and Toxicity Profile of Oral Topotecan in a Panel of Human Tumour Xenografts. *Eur. J. Cancer* **2000,** *36*, 1558–1564.

Chabner, B. A. Camptothecins. *J. Clin. Oncol.* **1992,** *10* (1), 3–4. https://doi.org/10.1200/JCO.1992.10.1.3

Champoux, J. J. Mechanism of the Reaction Catalyzed by the DNA Untwisting Enzyme: Attachment of the Enzyme to 3′-Terminus of the Nicked DNA. *J. Mol. Biol.* **1978,** *118* (3), 441–446. https://doi.org/10.1016/0022-2836(78)90238-3

Comins, D. L.; Nolan, J. M.; Carolina, N. *LETTERS A Practical Six-Step Synthesis of (S) -Camptothecin,* 2001; pp 11–13.

Dancey, J.; Eisenhauer, E. A. Current Perspectives on Camptothecins in Cancer Treatment. *Br. J. Cancer* **1996,** *74* (3), 327–338. https://doi.org/10.1038/bjc.1996.362

Date, H.; Kiura, K.; Ueoka, H.; Tabata, M.; Aoe, M.; Andou, A.; Shibayama, T.; Shimizu, N. Preoperative Induction Chemotherapy with Cisplatin and Irinotecan for Pathological N2 Non-Small Cell Lung Cancer. *Br. J. Cancer* **2002,** *86* (4), 530–533. https://doi.org/10.1038/sj.bjc.6600117

Fukuda, M.; Nishio, K.; Kanzawa, F.; Ogasawara, H.; Ishida, T.; Arioka, H.; Bojanowski, K.; Oka, M.; Saijo, N. Synergism Between Cisplatin and Topoisomerase I Inhibitors, NB-506 and SN-38, in Human Small Cell Lung Cancer Cells. *Cancer Res.* **1996,** *56* (4), 789–793.

Gigliotti, C. L.; Ferrara, B.; Occhipinti, S.; Boggio, E.; Barrera, G.; Pizzimenti, S.; Giovarelli, M.; Fantozzi, R.; Chiocchetti, A.; Argenziano, M.; Clemente, N.; Trotta, F.; Marchiò, C.; Annaratone, L.; Boldorini, R.; Dianzani, U.; Cavalli, R.; Dianzani, C. Enhanced Cytotoxic Effect of Camptothecin Nanosponges in Anaplastic Thyroid Cancer Cells In Vitro and In Vivo on Orthotopic Xenograft Tumors. *Drug Delivery* **2017,** *24* (1), 670–680. https://doi.org/10.1080/10717544.2017.1303856

Gong, J. Z.; Li, Q. J.; Wang, X.; Ma, Y. P.; Zhang, X. H.; Zhao, L.; Chang, Z. Y.; Ronse De Craene, L. Floral Morphology and Morphogenesis in Camptotheca (Nyssaceae), and Its Systematic Significance. *Ann. Bot.* **2018,** *121* (7), 1411–1425. https://doi.org/10.1093/aob/mcy041

Guo, Q.; Yuan, Q. A Novel 10-Hydroxycamptothecin-Glucoside from the Fruit of Camptotheca Acuminata. *Nat. Product Res.* **2016,** *30* (9), 1053–1059. https://doi.org/10.1080/14786419.2015.1107057

He, H.; Shang, X. Y.; Liu, W. W.; Zhang, Y.; Song, S. J. Triterpenes from the Fruit of Camptotheca Acuminata Suppress Human Hepatocellular Carcinoma Cell Proliferation Through Apoptosis Induction. *Nat. Product Res.* **2019,** *33* (24), 3527–3532. https://doi.org/10.1080/14786419.2018.1487967

Hertzberg, R. P.; Caranfa, M. J.; Hecht, S. M. On the Mechanism of Topoisomerase I Inhibition by Camptothecin: Evidence for Binding to an Enzyme-DNA Complex. *Biochemistry* **1989,** *28* (11), 4629–4638. https://doi.org/10.1021/bi00437a018

Hsiang, Y. H.; Hertzberg, R.; Hecht, S.; Liu, L. F. Camptothecin Induces Protein-Linked DNA Breaks Via Mammalian DNA Topoisomerase I. *J. Biol. Chem.* **1985,** *260* (27), 14873–14878. https://doi.org/10.1016/s0021-9258(17)38654-4

Hsiang, Y. H.; Liu, L. F. Identification of Mammalian DNA Topoisomerase I as an Intracellular Target of the Anticancer Drug Camptothecin. *Cancer Res.* **1988,** *48* (7), 1722–1726.

Hu, Y.; Wang, S.; Wu, X.; Zhang, J.; Chen, R.; Chen, M.; Wang, Y. Chinese Herbal Medicine-Derived Compounds for Cancer Therapy: A Focus on Hepatocellular Carcinoma. *J. Ethnopharmacol.* **2013,** *149* (3), 601–612. https://doi.org/10.1016/j.jep.2013.07.030

Jaxel, C.; Capranico, G.; Kerrigan, D.; Kohn, K. W.; Pommier, Y. Effect of Local DNA Sequence on Topoisomerase I Cleavage in the Presence or Absence of Camptothecin. *J. Biol. Chem.* **1991,** *266* (30), 20418–20423. https://doi.org/10.1016/s0021-9258(18)54939-5

Jaxel, C.; Kohn, K. W.; Wani, M. C.; Wall, M. E.; Pommier, Y. Structure-Activity Study of the Actions of Camptothecin Derivatives on Mammalian Topoisomerase I: Evidence for a Specific Receptor Site and a Relation to Antitumor Activity. *Cancer Res.* **1989,** *49* (6), 1465–1469.

Kim, Y. S.; Noh, M. Y.; Kim, J. Y.; Yu, H. J.; Kim, K. S.; Kim, S. H.; Koh, S. H. Direct GSK-3β Inhibition Enhances Mesenchymal Stromal Cell Migration by Increasing Expression of β-PIX and CXCR4. *Mol. Neurobiol.* **2013,** *47* (2), 811–820. https://doi.org/10.1007/s12035-012-8393-3

Kjeldsen, E.; Mollerup, S.; Thomsen, B.; Bonven, B. J.; Bolund, L.; Westergaard, O. Sequence-Dependent Effect Of Camptothecin on Human Topoisomerase I DNA Cleavage. *J. Mol. Biol.* **1988,** *202* (2), 333–342. https://doi.org/10.1016/0022-2836(88)90462-7

Lin, C. S.; Chen, P. C.; Wang, C. K.; Wang, C. W.; Chang, Y. J.; Tai, C. J.; Tai, C. J. Antitumor Effects and Biological Mechanism of Action of the Aqueous Extract of the Camptotheca Acuminata Fruit in Human Endometrial Carcinoma Cells. *Evid.-Based Complement. Altern. Med.* **2014,** *2014.* https://doi.org/10.1155/2014/564810

Liu, Y.; Hu, Z.; Lin, X.; Lu, C.; Shen, Y. A New Polyketide from Diaporthe sp. SXZ-19, an Endophytic Fungal Strain of Camptotheca Acuminate. *Nat. Product Res.* **2013,** *27* (22), 2100–2104. https://doi.org/10.1080/14786419.2013.791819

Liu, Z.; Carpenter, S. B.; Bourgeois, W. J.; Yu, Y.; Constantin, R. J.; Falcon, M. J.; Adams, J. C. Variations in the Secondary Metabolite Camptothecin in Relation to Tissue Age And Season in Camptotheca acuminata. *Tree Physiol.* **1998,** *18* (4), 265–270. https://doi.org/10.1093/treephys/18.4.265

Manikandan, M.; Kannan, K. Pharmacokinetic and Pharmacodynamic Evaluation of Camptothecin Encapsulated Poly (Methacylic Acid-Co-Methyl Methacrylate) Nanoparticles. *J. Appl. Pharma. Sci.* **2017,** *7* (3), 9–16. https://doi.org/10.7324/JAPS.2017.70303

Martino, E.; Della Volpe, S.; Terribile, E.; Benetti, E.; Sakaj, M.; Centamore, A.; Sala, A.; Collina, S. The Long Story of Camptothecin: From Traditional Medicine to Drugs. *Bioorg. Med. Chem. Lett.* **2017,** *27* (4), 701–707. https://doi.org/10.1016/j.bmcl.2016.12.085

Nicholas, A. W.; Wani, M. C.; Manikumar, G.; Wall, M. E.; Kohn, K. W.; Pommier, Y. Plant Antitumor Agents. 29. Synthesis and Biological Activity of Ring D and Ring E Modified Analogues of Camptothecin. *J. Med. Chem.* **1990,** *33* (3), 972–978. https://doi.org/10.1021/jm00165a014

Occhipinti, S.; Dianzani, C.; Chiocchetti, A.; Boggio, E.; Clemente, N.; Gigliotti, C. L.; Soluri, M. F.; Minelli, R.; Fantozzi, R.; Yagi, J.; Rojo, J. M.; Sblattero, D.; Giovarelli, M.; Dianzani, U. Triggering of B7h by the ICOS Modulates Maturation and Migration of Monocyte-Derived Dendritic Cells. *J. Immunol.* **2013,** *190* (3), 1125–1134. https://doi.org/10.4049/jimmunol.1201816

Pizzolato, J. F.; Saltz, L. B. The Camptothecins. *Lancet* **2003,** *361* (9376), 2235–2242. https://doi.org/10.1016/S0140-6736(03)13780-4

Pommier, Y. Topoisomerase I Inhibitors: Camptothecins and Beyond. *Nat. Rev. Cancer* **2006,** *6* (10), 789–802. https://doi.org/10.1038/nrc1977

Redinbo, M. R.; Stewart, L.; Kuhn, P.; Champoux, J. J.; Hol, W. G. J. Crystal Structures of Human Topoisomerase I in Covalent and Noncovalent Complexes with DNA. *Science* **1998,** *279* (5356), 1504–1513. https://doi.org/10.1126/science.279.5356.1504

Rowinsky, E. K.; Grochow, L. B.; Hendricks, C. B.; Ettinger, D. S.; Forastiere, A. A.; Hurowitz, L. A.; McGuire, W. P.; Sartorius, S. E.; Lubejko, B. G.; Kaufmann, S. H.; Donehower, R. C. Phase I and Pharmacologic Study of Topotecan: A Novel Topoisomerase I Inhibitor. *J. Clin. Oncol.* **1992,** *10* (4), 647–656. https://doi.org/10.1200/jco.1992.10.4.647

Shen, W.; Coburn, C. A.; Bornmann, W. G.; Danishefsky, S. J. *Drugs fi @ MeA.* **1993,** *9,* 611–617.

Stewart, L.; Redinbo, M. R.; Qiu, X.; Hol, W. G. J.; Champoux, J. J. A Model for the Mechanism of Human Topoisomerase I. *Science* **1998,** *279* (5356), 1534–1541. https://doi.org/10.1126/science.279.5356.1534

Stotz, M.; Gerger, A.; Haybaeck, J.; Kiesslich, T.; Bullock, M. D.; Pichler, M. Molecular Targeted Therapies in Hepatocellular Carcinoma: Past, Present and Future. *Anticancer Res.* **2015,** *35* (11), 5737–5744.

Thomsen, B.; Mollerup, S.; Bonven, B. J.; Frank, R.; Blöcker, H.; Nielsen, O. F.; Westergaard, O. Sequence Specificity of DNA Topoisomerase I in the Presence and Absence of Camptothecin. *EMBO J.* **1987,** *6* (6), 1817–1823. https://doi.org/10.1002/j.1460-2075.1987.tb02436.x

Venditto, V. J.; Simanek, E. E. Cancer Therapies Utilizing the Camptothecins: A Review of the In Vivo Literature. *Mol. Pharma.* **2010,** *7* (2), 307–349. https://doi.org/10.1021/mp900243b

Wall, M. E.; Wani, M. C.; Cook, C. E.; Palmer, K. H.; McPhail, A. T.; Sim, G. A. Plant Antitumor Agents. I. The Isolation and Structure of Camptothecin, a Novel Alkaloidal Leukemia and Tumor Inhibitor from Camptotheca acuminata. *J. Am.erican Chem. Soc.* **1966,** *88* (16), 3888–3890. https://doi.org/10.1021/ja00968a057

Wang, P.; Luo, J.; Wang, X. B.; Fan, B. Y.; Kong, L. Y. New Indole Glucosides as Biosynthetic Intermediates of Camptothecin from the Fruits of Camptotheca Acuminata. *Fitoterapia* **2015,** *103,* 1–8. https://doi.org/10.1016/j.fitote.2015.03.004

Wani, M. C.; Nicholas, A. W.; Wall, M. E. Plant Antitumor Agents. 28. Resolution of a Key Tricyclic Synthon, 5'(rs)-l,5-Dioxo-5'-Ethyl-5'-Hydroxy-2'h,5'h,6'r'-6'-Oxopyrano [3',4'-f]Δ6'8-Tetrahydro-Indolizine: Total Synthesis and Antitumor Activity of 20(s)- and 20(R)-Camptothecin. *J. Med. Chem.* **1987,** *30* (12), 2317–2319. https://doi.org/10.1021/jm00395a024

Wani, M. C.; Ronman, P. E.; Lindley, J. T.; Wall, M. E. Plant Antitumor Agents. 18.1 Synthesis and Biological Activity of Camptothecin Analogs. *J. Med. Chem.* **1980,** *23* (5), 554–560. https://doi.org/10.1021/jm00179a016

CHAPTER 8

Taxus Baccata

ROHIT SAM AJEE[1] and SHUCHI KAUSHIK[2]

[1]*Independent Researcher and Alumni Amity Institute of Biotechnology, Amity University Madhya Pradesh, India*

[2]*State Forensic Science Laboratory, Madhya Pradesh, India*

ABSTRACT

Cancer is the perfect example of an abnormality developed in the cells leading to disastrous effects on overall wellbeing of the organism. It is the result of certain factors which are not always controllable, resulting in a disorder which can afflict almost any part of the human body and also has the ability to spread. Modern science has been looking for a cure to this ailment for several decades with some progress. Current cancer treatment options include the use of toxic compounds or invasive techniques to immobilize and terminate cancerous cells. However, these treatments are also responsible for severe side effects, some of which are almost as deadly as the original disease as well. The desire to treat this debilitating condition with minimal side effects has driven science towards nature to look for an answer. There have been several plants and other biological systems, which have been studied exclusively for their ability to treat cancers. *Taxus baccata* is one such plant. The European Yew, as the plant is commonly known, has been shown to contain some phenolic metabolites which express significant antioxidant as well cytotoxic properties which could aid in tumor treatment. Several studies have also been carried out which used the plant extract as encapsulating material for anticancer nanoparticles as well as the plant for green synthesis of the said nanoparticles. However, the ace in the hole for *Taxus baccata* is its ability to produce the anticancer agent known as Paclitaxel. Although

Potent Anticancer Medicinal Plants: Secondary Metabolite Profiling, Active Ingredients, and Pharmacological Outcomes. Deepu Pandita and Anu Pandita (Eds.)
© 2024 Apple Academic Press, Inc. Co-published with CRC Press (Taylor & Francis)

conventionally, *Taxus brevifolia* is used for the production and harvesting of taxol, recent studies have focused on alternate plants for the same. This study focuses on the characteristic medicinal properties of *Taxus baccata* with special reference to anticancer efficacy.

8.1 BASIC INTRODUCTION TO CANCER, ITS TYPES, AND PLANT-BASED TREATMENT OPTIONS

A rapid, unregulated cellular division that starves the surrounding tissues to death while also occasionally having the ability to infiltrate and traverse to different parts of the body is one way to briefly sum up the biological disorder known as cancer. A lot of research and experiments are done related to this topic for a diverse number of reasons such as to identify the underlying cause of cancer, to discover potential cures, to identify the key features of cancerous cells and tumors, to fully understand the disruptive effects it has on the body, etc. While we have made progress, it is by no means an issue we have completely understood yet. In 2011, Hanahan and Weinberg talked about the Six Hallmarks of Cancer in their report. The hallmarks according to them are (Hanahan and Weinberg, 2011):

- Continuous Proliferative Signaling
- Escaping from the growth Suppressors
- Opposition to Cell Death
- Facilitating Replicative Immortality
- Prompting Angiogenesis
- Stimulating Invasion and Metastasis.

The pair elaborately describes a hypothetical cell which undergoes neoplastic transformations and gradually accrues each of the hallmarks mentioned above that eventually results in malignant tumor mass. In addition to attaining the final neoplastic state, the report also mentions that the tumors seem to become more than just a bunch of rapidly proliferating cells. They express a complex arrangement with the inclusion of multiple, distinct cell types that interact with each other and surrounding tissues to further propagate itself. This revelation was further built upon to create the concept of the Tumor Microenvironment (TME). Earlier studies identified genetic (and eventually epigenetic) changes as the sole factor responsible for cancer development and proliferation. However, research on the matter has revealed that in addition to Genetic and Epigenetics,

Tumor Microenvironment (TME) is just as substantial an element in tumor progression (Baghban et al., 2020). The neoplastic cells within the tumor communicate with surrounding nonmalignant cells and extracellular matrix components like the collagen, laminin, fibronectin, and immune cells like macrophages and lymphocytes via complex signaling pathways. Once control has been obtained, the tumor cells force the surrounding cells to further stimulate tumor growth. The nature of interaction between a tumor and the TME establishes the bond as a critical means of tumor progression. Consequently, the TME can be assumed as an ideal target for disrupting and potentially even terminating tumor development.

To this end, Nia et al. have recently published a study in which the four physical traits associated with most tumors and their TME were discussed upon. As the tumor develops and starts interacting with surrounding noncancerous cells, they induce physical alterations to the microenvironment. These alterations may then affect the biology of the cancer and their response to cures (Nia et al., 2020). One may view the traits mentioned in this review as the physical counterpart (addition even) to the biological hallmarks that were established in Hannagan and Weinberg's review. These physical abnormalities along with the biological hallmarks of tumors provide a more complete understanding of the cellular as well as physical nature of cancer.

- Elevated Solid Stress
- Elevated Interstitial Fluid Pressure (IFP)
- Increased Stiffness and Altered Material Properties, and
- Altered Microarchitecture
  These are the four physical hallmarks of cancer.

Now that we have basic understanding as to the nature of cancer, let us explore some treatment options for it. Classical cancer treatment options generally involve invasive and at times potentially harmful options like chemo and radio therapy and surgery. With time however, science and research has started focusing on more holistic options. Plant-based cancer treatment is one such field that has massive potential. Several plants exist in nature that contains active compounds which can aid curing or treating cancer in one form or another. It is up to the scientific community to identify, extract, and enhance these compounds and put them to use. As indicated by extensive research work, there are numerous anticancer drugs clinically recognized and are suggested for the malignancy therapy (Stark, 2006; Berman et al., 2000).

8.1.1 CURRENT CANCER THERAPY VIA PHYTOCHEMICALS: A NOVEL APPROACH

Since primordial age, use of phytochemicals to treat and cure a number of medical conditions had been a regular practice. Traditional Indian and Chinese medicine systems relied almost extensively on plants and plant-based phytochemical extracts to alleviate sickness. However, due to the lack of written records and general mistrust towards ancient medicine systems, science generally disregarded these cures. All that changed however with the discovery and isolation of morphine from *Papaver somniferum*. This single finding was more than enough to generate curiosity regarding the presence of medicinal compounds in plants, which fleshed out over the years. With time, a wide range of medicinally useful compounds and drugs that were extracted from plants were approved for use (Newmann and Cragg, 2016).

Some of these compounds have been found to possess significant antitumor properties as well, making them suitable for cancer treatment. Based on the property of the compound and the origin of extraction, these complexes show a variety of curative properties and target specific pathways or products involved in the carcinogenic process. While some help in preventing the proliferation of cancerous cells (Yan et al., 2018), others delay the progress of the tumors by neutralizing free radicals (Lee et al., 2013) or interact with tumor activator/suppressor proteins to help in regulating the process (Adams et al., 2010). The following section will briefly talk about a few phytochemical agents with potential anticancerous properties.

A study regarding the efficiency and ability of Allicin, a major biological compound present in garlic, to suppress the proliferation of Cholangiocarcinoma (CCA) indicated that the compound was able to significantly suppress proliferation (Chen et al., 2018). The compound was able to do so by influencing the STAT3 signaling pathway. The organic sulfur compound was found to be capable of inhibiting the STAT3 signaling pathway which then prevents tumor cells from migrating or invading other tissues. The study was also able to identify that allicin induces apoptosis of the tumor cells. These results were obtained from both in vitro as well as in vivo studies. What makes allicin even more attractive is the fact that CCA is largely resistant to conventional cancer therapy treatments like chemotherapy, which necessitates the formulation of effective alternate treatment options.

Apigenin, a phytochemical belonging to the flavonoid class, has been recently targeted for its anticancerous properties. It has potential use in chemotherapy as a substitute to conventionally used agents due to its low toxicity and high antitumor property (Madunic et al., 2018). For cancers that affect the head and neck regions of the body, apigenin was found to promote apoptosis by upregulating the expression of Tumor Necrosis Factor Receptor pathway. It has shown similar activity in the inhibition of breast, prostate, pancreatic, skin, and colorectal cancers. Another factor that works in favor of apigenin is that it is able to positively interact with other drugs and compounds used in conventional chemotherapy. The flavonoid is able to augment the beneficial properties of the drugs while decreasing their harmful effects. For example, apigenin and paclitaxel have a synergistic interaction wherein apigenin promoted the absorption and half-life of the other compound. Apigenin loaded nanoparticles were used to study the effects it had on hepatocellular carcinoma in rats (Bhattacharya et al., 2018). The study was able to conclude that apigenin loaded nanoparticles were able to slow down the progress of hepatocellular carcinoma in both in vivo and in vitro studies. The presence of the carrier increased the accuracy of transmission as well biodistribution of the compound in the tumor-specific site. The controlled release of the drug ensures optimal delivery resulting in higher cytotoxicity and improved cancer suppression as compared to free apigenin.

Resveratrol is yet another phytochemical derived from *Polygonum cuspidatum* which was found to express considerable antitumor properties. When the polyphenolic compound was administered to female Sprague–Dawley rats to test its effects in controlling mammary carcinogenesis, it was observed that the phytochemical was significantly suppressing tumor expression (Banarjee et al., 2002). The compound was able to do so by downregulating the expression of the transcription factor, NF-κB. Over the recent years, the role of NF-κB as a tumor-promoting element has been cemented via multiple studies (de Martin et al., 1999). Resveratrol's ability in inhibiting the expression of this key component in tumorigenesis greatly inhibits overall tumor spread and progression. Furthermore, in vitro studies revealed that resveratrol was able to inhibit the spread of the breast cancer in a concentration-dependent manner.

Gingerol is a type of flavonoid with antioxidant potential and is commonly found in fresh ginger. The compound is shown to exhibit significant anti-inflammatory and anticancerous properties (de Lima et al., 2018). In vitro studies conducted on the action of gingerol on triple negative breast cancer (TNBC) showed that the phytochemical was able to induce concentration-dependent cell death in both human and mouse cell lines (Martin et al., 2017). The study was also able to identify that gingerol initiated cellular apoptosis by influencing the activation of caspase-3 and related pathways.

Quite surprisingly, the study also concludes that the phytochemical is capable of preventing or at the very least, inhibiting the spontaneous metastasis in subjects where the primary tumor was removed.

Taxus baccata has, for quite some time, been under interest because of its alkaloid parts. Despite the fact that its alkaloids share normal primary similitudes with some others, they have a one of a kind taxane ring. These compounds function by advancing and balancing out microtubule arrangement and thus inhibiting mitosis. Microtubules are keys for cell division, and tubulins polymerize to frame microtubules within the sight of microtubule-associated protein (MAP) and GTP. It appears to be conceivable that taxol targets microtubules' dimeric proteins. Subsequently, microtubules become nonfunctional, which meddles cell separating and obstructs cell cycle. Because of strange groups shaped along these lines, nonfunctional microtubules get distributed (Manish et al., 2005). Albeit some atomic instruments are assumed for the activity of taxus species in the disease interaction, it appears to be very conceivable that there may be some different components obscure yet. In this way, further examinations are required.

8.2 BRIEF REVIEW ON *TAXUS BACCATA*: THE WONDERFUL PLANT WITH MULTIFARIOUS ACTIVITIES

The European yew or *Taxus baccata* is found across various parts of the northern side of the equator, mostly in temperate regions. Aside from the organic product, it is a small to medium-sized evergreen tree that generally has been utilized for weapon production and making medicines (Abella, 1996). The Genus *Taxus* has the following classification system:

Class: Pinopsida
Order: Taxales
Family: Taxaceae

Species: *T. baccata* (European or English yew), *T. brevifolia* (Pacific yew or Western yew), *T. canadensis* (Canadian yew), *T. chinensis* (Chinese yew), *T. cuspidata* (Japanese yew), *T. floridana* (Florida yew), *T. globosa* (Mexican yew) and *T. wallichiana* (Himalayan yew)

Cross breeds: Taxus × media = *T. baccata* × *T. cuspidata* and Taxus × hunnewelliana = T. cuspidate × *T. canadensis*.

Because the species are so similar, they are regularly simpler to isolate geologically than morphologically (Cope, 1998).

A group of researchers assessed the potential impacts of *Taxus baccata* extract on D-Adenosine deaminase (ADA) action in harmful and noncancerous human gastric and colon tissues to explain its anticancer potential. D-Adenosine deaminase (ADA) is a chemical (EC 3.5.4.4) associated with purine digestion. It is required for the breakdown of adenosine and for the turnover of nucleic acids in tissues. It is available practically in every single mammalian cell and its essential capability in organism is the development and support of the immune system (Kingston, 2005). However, the full physiological function of ADA isn't yet totally understood (Wilson et al., 1991). ADA affiliation has additionally been seen with epithelial cell separation, neurotransmission, and incubation maintenance (Cristalli et al., 2001). It has likewise been recommended that ADA, notwithstanding adenosine breakdown, animates arrival of excitatory amino acids and is vital for the coupling of A1 adenosine receptors and heterotrimeric G proteins (Wilson et al., 1991).

According to a scientific viewpoint, utilization of ADA inhibitors has helped much in understanding the component activity of adenosine metabolites and analogs. ADA inhibitors have additionally prompted the comprehension of the administrative cycles related to immunodeficiency portrayed by an absence of ADA, and development of the invulnerable response (Glazer, 1980). One of them, pentostatin (Nipent) is a nucleoside simply having the capacity to restrain ADA chemical. Hindrance of ADA impedes the deamination responses in the purine rescue pathway, consequence of which is the restraint of ribonucleotide reductase. Subsequently, this cycle exhausts the nucleotide pool and limits DNA synthesis (Brown et al., 2008).

Furthermore, taxol forestalls spread of metastatic disease cells by repressing cell relocation. Paclitaxel essentially acts at the G-2/M-stage intersection. Nonetheless, docetaxel acts primarily in S-period of mitosis. These compounds are not harmful to non-dividing cells since they don't need the mitotic spindle. Hence, they act just on multiplying cells (Manish et al., 2005). Additionally, as the microtubule framework is fundamental for the arrival of different cytokines, balance of cytokine discharge by this drug might assume a significant part in its anticancer activity (Smith et al., 1995).

8.3 TAXANE AND OTHER ACTIVE COMPOUNDS PRESENT IN *TAXUS BACCATA*

Taxus baccata, or the European Yew which is its common name, has been a subject of interest to the scientific community for quite a while now. These

TABLE 8.1 Pharmacological Activity of Taxanes Against Different Types of Cancer.

Taxane type	Study system	Important outcomes of study
Paclitaxel laden nanostructured lipidic carriers	In vitro: HepG2 liver carcinoma cells In vivo: Wistar rats	In vitro: Enhancement of ROS generation in addition to dose-dependent cell viability decrease. Chromatin condensation, condensed as well as fragmented nuclei were observed all of which points to apoptosis. In vivo: Enhanced absorption of paclitaxel (Harshita et al., 2019)
Docetaxel encapsulated in gold nanoparticles	In vitro: HepG2 liver carcinoma cells	In vitro: The docetaxel nanoparticles lowered cell viability in a concentration-dependent manner. Evidence indicating apoptosis like cell conglomeration and bleb and bulge development were observed.
Apatite carrier	In vivo: Mice	In vivo: When treated with docetaxel nanoparticles, mice were observed to revert to standard live architecture (Wan et al., 2018).
Docetaxel carboxymethylcellulose nanoparticles	In vitro: Hepatic stellate cells, Hep3B, and HLF Human carcinoma cells In vivo: Male C3H/HeNCrNarl mice	In vitro: Dose-dependent decrease in the percentage of cell viability of hepatic stellate cells, HCA-1, Hep3B, and HLF cells. Molecular level: Dose-dependent downregulation of α-smooth muscle actin (α-SMA). Additionally, collagen I at protein and mRNA levels was also observed to be downregulated. In vivo: Inhibition of tumor growth (Chang et al., 2018)
Cabazitaxel	In vitro: SK-hep-1, SM-MC7721, Huh-7, HCC-LM3, Huh-TS-48, and SK- sora-5	In vitro: A time and dose-dependent inhibition of Sk-hep-1, Huh-7, SMMc-7721, and HCC-Lm3 cell lines. Cell cycle arrest: SK-hep-1 and Huh-7 cell lines treated with cabazitaxel showed an increase in the number of cells in the G2-M stage. Molecular level: Decrease in the expression of Cdc2, pCdc2, Cdc25c, and cyclin B1 protein levels. Induction of apoptosis: Increased ratio of apoptotic cells in SK-hep-1 and Huh- 7 cell lines along with a decline in antiapoptotic protein Bcl-2 and a stronger cleavage of poly ADP-ribose polymerase (PARP) (Chen et al., 2018)

TABLE 8.1 *(Continued)*

Taxane type	Study system	Important outcomes of study
Paclitaxel and survinin siRNA encapsulated into polyethyleneimine-block-polylactic acid	In vitro: A549, A549 lung cancer cells.	In vitro: A549 cells showed increased levels of nuclear fragmentation, cytotoxicity, chromosomal abnormalities, and G2/M cell cycle arrest.
	Animal model: BALB/c nude mice	In vivo: Effective tumor inhibition (Jin et al., 2018)
Caffeic acid with paclitaxel	In vitro: Non-small cell lung cancer (NSCLC) H1299 cells and normal Bease-2b cells	Antiproliferative action: NSCLC H1299 showed decreased proliferation. H1299 cells were arrested in the sub G1 phase of the cell cycle and approached apoptosis.
		Molecular level: Bax and Bid were activated, and PARP was cleaved downstream. Phosphorylation of extracellular signal-regulated c-Jun NH2-terminal protein kinase1/2 and kinase1/2 was also observed. MAPK pathway involved in apoptosis was activated (Min et al., 2018).
Micelle delivery system (TP-M) loaded with paclitaxel (PTX) with Plasdone S-630 Copovidone (PVPS630) and vitamin E-TPGS (TPGS) as carriers	In vitro: Caco-2, A549, and Lewis lung cancer cells	In vitro: Due to improved uptake of PTX-TP-M compared to regular PTX, higher toxicity was observed against A549 and Lewis cells.
	In vivo: Male Sprague–Dawley rats and male C57BL/6 mice	In vivo: Treatment with PTX-TP-M yielded higher tumor inhibition. Tumor cell volume and mitotic cells were reduced (Hou et al., 2017).
PEGylated liposomes containing docetaxel with telmisartan	In vitro: A549 lung cancer cell	In vivo: Cotreatment with docetaxel and telmisartan showed significant tumor inhibition and mostly intact lung integrity in mice.
	In vivo: Sprague–Dawley rats, athymic Nu/nu mice	Molecular level: Antiapoptotic marker expression was decreased (survinin) and metastasis markers MMP9 and MMP2 were downregulated (Patel et al., 2016).
Noscapine and paclitaxel	In vitro: LNCaP and PC-3 prostate cancer cells	In vitro: Increased apoptosis in cancerous cell lines coupled with decreased cell viability
		Molecular level: mRNA expression of Bcl-2 was significantly decreased and mRNA expression of Bcl-2-associated X protein Bax, and Bax/Bcl-2 ratios in LNCaP and PC-3 cells were increased. Prostate-specific antigen and androgen receptor expression was also decreased (Rabzia et al., 2017).

TABLE 8.1 *(Continued)*

Taxane type	Study system	Important outcomes of study
Docetaxel with desmopress	In vitro: PC3 prostate cancer cell In vivo: Mice	In vivo: Significant mice tumor volume reduction (Bass et al., 2019)
Combination of Impressic acid and Acankoreanogenin with docetaxel	In vitro: VCaP prostate cancer cell	In vitro: Strong antiproliferative activity was expressed in the form of apoptosis in VCaP cells. Molecular level: Reduced nuclear factor-κB (NF-κB) activity, decreased expression of Bcl-2, NF-κB, p-Akt, and phosphorylated signal transducer and activator of transcription 3 (p-Stat 3) and increased expression of phosphorylated c-Jun N-terminal kinase (p-JNK) (Jiang et al., 2018)
Docetaxel-loaded nanoparticles	In vitro: LnCaP and PC3 prostate cancer cells	In vitro: Increase in lactate dehydrogenase release in the culture medium of prostate cancer cells indicating concentration-dependent cell death (Gallego-Yerga et al., 2017)
Cabazitaxel	In vitro: Human prostate cancer LNCaP and PC-3 cells	In vitro: Reduced proliferation in both the cancer cell lines Molecular level: Androgen receptor and androgen receptor-associated factors HSP90α, HSP40, and HSP70/HSP90 organizing protein levels were lowered. Activity of antiapoptotic factor HSP60 was also suppressed (Rottach et al., 2019).
Cabazitaxel and silibinin coencapsulated cationic liposomes	In vitro: PC-3 and DU-145 prostate cancer cells	In vitro: Dose-dependent inhibition of cell migration, cellular growth, and induction of apoptosis Cell cycle arrest: Upon treatment G2/M cell cycle arrest and prevention of colony formation was observed. Nucleus morphology: Evidence indicating apoptosis like cell blebbing, membrane shrinkage, and nuclear granulation were observed (Mahira et al., 2019).

TABLE 8.1 *(Continued)*

Taxane type	Study system	Important outcomes of study
Bone targeted cabazitaxel nanoparticles	In vitro: PC-3 and C4-2B-luciferase prostate cancer cells	In vitro: Along with high affinity towards bones, the survival percentage of PC-3 and C4-2B prostate cancer cells was determined by the dose and concentration.
		In vivo: Tumor weight was significantly reduced in addition to reduction in the appearance of bone lesions (Gdowski et al., 2017).
Paclitaxel nanoparticles (nab-PTX) bound to albumin and S-nitrosated human serum albumin dimer (SNO-HSA Dimer)	Human pancreatic cancer cell (SUIT2-GLuc)	Antitumor activity in SUIT2 cancer model: Ascites and distant metastasis was efficiently suppressed. Also showed best survival rates for the combination therapy
Docetaxel nanoparticles in combination with radiotherapy	In vitro: AsPC-1 and BxPC-3 pancreatic cancer cells	In vitro: Treatment with docetaxel combined with radiotherapy resulted in decreased cell colonies, increased tubulin polymerization as well as apoptosis.
	In vivo: Mice	Molecular level: Caspase 3 expression levels were boosted
		In vivo: AsPC-1 and BxPC-3 derived tumors were suppressed from growing (Rivera-Franco and Leon-Rodriguez, 2018).
Combination of sorafenib with paclitaxel and radiation therapy	In vitro: MDA-MB-231 breast cancer cells	In vitro: Breast cancer cells proliferation was inhibited with respect to the dose coupled with increased sub-G0-G1 phase.
	In vivo: BALB/c nude mice	Molecular level: p21, CHOP, BAX, Apaf-1, and cleaved caspase-3 levels were heightened and (B-cell lymphoma) Bcl-2 levels were lowered which points to caspase cleavage and inhibition of the Bcl-2 pathway ultimately resulting in cell cycle arrest.
		Cellular level: Cytochrome C levels were heightened in the cytosol which points to a cytochrome c-dependent apoptotic cycle.
		In vivo: Caki1 and MB-231 cell xenograft tumors were suppressed (Foglietta et al., 2018).

TABLE 8.1 *(Continued)*

Taxane type	Study system	Important outcomes of study
Paclitaxel containing cell-penetrating peptide producing nanoliposomes	In vitro: Human breast cancer cell MCF7. In vivo: BALB/c nude mice	In vitro cytotoxicity: Concentration-dependent inhibition of cell proliferation In vivo antitumor efficiency: Reduction in tumor weight indicates that there is significant inhibition of tumor in the MCF7 tumor-bearing mouse model (Choi et al., 2019).
Anti-EGFR anchored immune-nanoparticle bearing paclitaxel	In vitro: MDA-MB-468 breast cancer cell In vivo: Athymic mice	In vitro cytotoxicity: The viability of cancer cells was lowered. In vivo antitumor efficiency: Reduction in tumor and greater accumulation of paclitaxel in the tumor plasma (Zhang et al., 2018)
Nab-Paclitaxel with atezolizumab	Phase III trial: Human females	Nab-Paclitaxel with atezolizumab prolonged the progression-free survival in patients suffering from triple-negative breast cancer (Venugopal et al., 2018).
Paclitaxel and gebcitabine through methoxy-poly (ethyleneglycol)-poly (lactide-coglycolide)-polypeptide nanoparticles	In vitro: 4T1, MCF-7, and MDA-MB-231 breast cancer cells	In vitro: Inhibition of proliferation of 4T1, MCF-7, and MDA-MB-231 in a dose-dependent manner by paclitaxel was observed. A combination of paclitaxel with gebcitabine resulted in a significant reduction in viability of 4T1, MCF-7, and MDA-MB-231 below 50%. Human breast cancer cell resistance to drugs was reversed (Schmid et al., 2018).
Paclitaxel loaded with keratin nanoparticles	In vitro: MCF7 and MDA-MB-231 breast cancer cells	Cell death: In MCF-7, the percentage of late apoptotic cells was increased after 48 h of treatment and in MDA-MB-231, the percentage of early apoptotic cells was increased after 24 h of treatment. Molecular level: Cleaved caspase-3 (CC3) protein and proapoptotic BAX gene expression was heightened (Dong et al., 2018).
Docetaxel loaded in folic acid and thiol-decorated chitosan nanoparticles	In vitro: MDA-MB-231 breast cancer cell	In vitro: Nanoparticles loaded with docetaxel showed better cytotoxicity when compared to docetaxel. Ex vivo: Successful transportation and improved oral bioavailability (Sajjad et al., 2019)

TABLE 8.1 *(Continued)*

Taxane type	Study system	Important outcomes of study
pH-sensitive nanoparticles containing Dihydroartemisinin and docetaxel	In vitro: 4T1 mammary carcinoma cell	In vitro: Reduction in cell viability along with G2/M cell cycle arrest
		Molecular level: Expression of E-cadherin was increased and expression of NF-κB, p-AKT, and p65 was decreased. Additionally, Matrix metalloproteinase-2 (MMP-2) was arrested (Tao et al., 2018).
Docetaxel with ionizing radiation	In vitro: MCF-7 breast cancer cell	In vitro: A concentration and time-dependent decrease in viability of cancer cells upon cotreatment of docetaxel and ionizing radiation suggesting a synergistic effect (Doddapaneni et al., 2016)
Docetaxel with SC-43	In vitro: MDA-MB-231, MDA-MB-468, and HCC-1937 breast cancer cells	In vitro: Dose-dependent increase in antiproliferative activity and treatment with docetaxel and SC-43 induced apoptotic activity.
	In vivo: NCr athymic nude mice	Molecular level: STAT3 downstream effector cyclin D1 and p-stat 3 expressions were lowered. Increased SHP1 expression
		In vivo: Suppression of tumor growth and tumor weight compared to control (Hendrikx et al., 2016)
Noscapine and docetaxel	In vitro: MDA-MBA-231 and MDA-MB-468 breast cancer cells	In vitro: Increased cytotoxic effect and decreased cell viability upon treatment with a combination of noscapine and docetaxel
	In vivo: Mice	Molecular level: Downregulation in the expression of Bcl-2, survivin, α-tubulin, and pAKT
		In vivo: Reduction in tumor collage levels and higher intratumoral uptake of drug-loaded liposomes (Liu et al., 2017)
Ritonavir and docetaxel	In vivo: Cyp3a knockout mice (Cyp3a)	In vivo: Docetaxel and ritonavir cotreatment resulted in reduction to ⅓ of original volume.
	Tumor model: K14cre; Brca1F; p53F mouse model	Histological observation: Mammary tumor cells expressed significant pleomorphism along with expanded stroma, fibrotic changes, and abundance of apoptotic cells in addition to the absence of necrosis (Fard et al., 2017).

TABLE 8.1 *(Continued)*

Taxane type	Study system	Important outcomes of study
Cabazitaxel-loaded poly(2-ethylbutylcyanoacrylate) nanoparticles	In vitro: MDA-MB-231, MDA-MB-468, and MCF-7 breast cancer cells	In vitro: Toxicity of cabazitaxel to MDA-MB-231, MDA-MB-468, and MCF-7 cell lines
	In vivo: Mice	Tumor growth inhibition: Tumor growth was inhibited and was completely remissioned in mice upon treatment. CD206, a marker of M2 macrophages (protumorigenic and anti-inflammatory activity), expression was inhibited (Fusser et al., 2019).
Cabazitaxel and thymoquinone coloaded lipospheres	In vitro: MCF-7 and MDA-MB-231 breast cancer cells	In vitro: MCF-7 and MDA-MB-231 breast cancer cell lines upon treatment with cabazitaxel-thymoquinone-loaded nanoparticles expressed a dose-dependent decrease in viability.
		Cell cycle analysis: Concentration-dependent percent increase in sub G1 population upon treatment with lipospheres loaded with combination drugs
		Apoptosis: Early apoptotic cells were increased in percentage upon treatment with combination drugs (Kommineni et al., 2018).
Hyaluronic acid-coated cabazitaxel-loaded solid lipid nanoparticles	In vitro: MCF-7 breast cancer cells	In vitro: Concentration-dependent decrease in cell viability upon treatment as compared to controls (Zhu et al., 2017)
Cabazitaxel-loaded polymeric micelles	In vitro: 4T1 metastatic breast cancer cells	In vitro: 4T1 cell migration was inhibited upon treatment with cabazitaxel-loaded polymeric micelles.
		In vivo: Tumor growth was inhibited upon treatment with cabazitaxel-loaded polymeric micelles (Zhong et al., 2017).

Source: Reprinted from Sinha (2020). https://creativecommons.org/licenses/by-nc-sa/3.0/#. Open access https://creativecommons.org/licenses/by/4.0/.

conifers belong to the Taxaceae family and are found in parts of Europe, Africa, and Southwest Asia. The reason for the interest lies in the presence of the active compound taxane in all parts of the plant (except for the arils), which undergoes internal enzymatic synthesis to form taxol (Malik et al., 2011). Although the anticancer nature of the compound was discovered initially in *Taxus brevifolia* extracts in 1971, it was eventually figured out that it was better from an economic and efficiency view point to extract the compound from the leaves of *T. baccata* and other *Taxus* species. Additionally, several studies have been conducted which have identified a large amount of taxanes and their subsequent derivatives which were extracted from *T. baccata* (Parmar et al., 1999, Vaishampayan et al., 1999).

As opposed to the medicinal nature of the taxane that ultimately results in the formation of taxol, the seeds and foliage of the yew tree are found to be quite toxic in nature. This is attributed to the presence of the taxine. This complex alkaloid compound exhibits powerful cardiotoxic properties and is readily absorbed in the digestive tract. *Taxus baccata* is also home to a few antibacterial, antifungal, and antioxidant compounds (Erdemoglu and Sener, 2001; Milutinović et al., 2015). A study conducted in 2014 was able to isolate three lignan derivatives from the plant. Of the three only two were shown to have moderate antifungal properties, while the last derivative had antibacterial properties. None of the three lignan derivatives showed any cytotoxic activity (Erdemoglu et al., 2004).

But the focus of this section is going to be on the compound taxane and its antitumor derivative. Taxol, also known as paclitaxel, acts as a mitotic inhibitor and potent antineoplastic agent (Needleman et al., 2005). Basically, taxol interacts with the tubulin dimers of the microtubules and stabilizes them to prevent depolymerization. It does so by binding to the β subunit of tubulin resulting in a hyperstabilized state which prevents the cell from undergoing standard mitotic division. Additionally, studies have also discovered that taxol is also responsible for inducing apoptosis in neoplastic cells by binding to the Bcl-2 protein (Haldar et al., 1996). The Bcl-2 is an apoptosis stopping protein and when taxol binds to it, it becomes nonfunctional resulting in cellular destruction.

8.3.1 PRODUCTION OF TAXOL

Taxol is a secondary metabolite belonging to the diterpenoid variety and as such is derived solely from geranylgeranyl diphosphate (GGPP). The actual process of biosynthesis of taxol involves the actions of various enzymes, cyclization, oxidation, acylation, and hydroxylation. Due to the successful nature of the drug, taxol is constantly in high demand. At present, the main source of paclitaxel is still dependent on the yew tree. If one were to simply depend on naturally extracted taxol for manufacturing the require drugs for global consumption, then we would run out of the viable sources in a couple of years (Ismaiel et al., 2017). This is primarily due to the very low concentrations of taxol in every plant source. Since natural sources can only be used in moderation, lest we eradicate an entire species of plant, the search for alternate modes of taxol production has taken priority. As a result, researchers are attempting to produce taxol using a variety of current approaches, including chemical synthesis, semisynthesis, and plant tissue culture methods. Nonetheless, these procedures offer both benefits and drawbacks (Ismaiel et al., 2017; Shankar Naik, 2019). Although chemical synthesis is an option, the low product yield and highly complex nature of the reaction deters large scale production possibilities (Nicolau et al., 1994). Another approach is the semisynthetic pathway, which focuses on synthesizing taxol from intermediates such as 10-deacetylbaccatin III.

Kusari et al. (2012) in their study revealed that several attempts have been carried out across the world to extract paclitaxel from sources other than the plants, and in this fungal endophytes have been proved to be a good alternate source (Kusari et al., 2012). By and large, isolation and identification of taxol-producing microorganisms is a decent methodology in the creation of taxol (Ismaiel et al., 2017). Along these lines, fostering a less expensive paclitaxel fermentation measure by microorganisms has turned into a maintainable arrangement (Shankar Naik, 2019; Somjaipeng et al., 2015).

Molecular imprinting (MIP) technology is a method of equipping a polymeric material with the ability to detect a specific target molecule. With the progression of science and innovation, molecular imprinting is presently an advanced innovation, and various scientists convey this technology to isolate and get the target product (Li et al., 2017).

Molecular imprinting is generally used for the detachment and advancement of dynamic elements of natural products (Ishkuh et al., 2014). The application of MIP technology to the separation of dynamic elements of natural medicinal resources has attracted an increasingly more consideration recently. For example, using Ultra-spectrophotometry, a few analysts have widely researched the interactions among paclitaxel and some normal functional monomers, such as methacrylic acid (MAA), acrylamide (AM), 2-vinylpyridine (2-VP), and 4-vinylpyridine, in various solvents and tracked down that the most stable communication among paclitaxel and 2-VP in chloroform was seen at a proportion of 1:6 (Li et al., 2013, 2015).

Ishkuh et al. (2014) have used MSP using ethylene glycol dimethacrylate for preparing MIPs for paclitaxel with a genuine degree of crosslinking and found that the most important confining breaking point concerning paclitaxel was 48.4%. However, the particle sizes of the MIPs were generally around 100 nm. Hence, in any case, inspite of its incredible engraving impact, the MIPs couldn't be used further for separation and assessment (Ishkuh et al., 2014).

Contrasted and regularly utilized partition techniques like fluid extraction and column chromatography, MIP innovation partakes in the advantages of economy, speed, and ease (Li et al., 2017). At this point, a gigantic number of studies have set up that paclitaxel shows high anticancer activity. Its antitumor framework incorporates limiting to tubulin, course of action of a consistent cylinder pack, provoking the insufficiency of agreement among dimers, and progression of microtubule gathering polymerization (Li et al., 2017). Consequently, the infection cells are caught in the late G or M stage, mitosis of the danger cells is quelled, augmentation of the harm cells is hindered, and the cells bit by bit contract and in the end die. Notwithstanding, there are relatively few reports on other normal activities of paclitaxel, for instance, prevention of HIV-1 viral replication activity (Shankar Naik, 2019; Wang et al., 2015). Wang et al. (2015) checked out exercises of taxol made by endophytic parasites *Noduli sporium sylviforme* HDFS4-26 with that of taxol extracted from yew bark in controlling turn of events and impelling apoptosis of threatening development of cells (Wang et al., 2015). Cell morphology, cell counting kit (CCK-8), staining (HO33258/PI and Giemsa), DNA agarose gel electrophoresis, and flow cytometry (FCM) examinations were used to choose the apoptosis status of malignant growth of cell lines, for instance, MCF-7 cells, HeLa cells, and ovarian infection HO8910 cells. The parasitic taxol showed cytotoxic activity against HeLa malignant growth cell lines in vitro, and antifungal and antibacterial potential against different pathogenic strains (Das et al., 2017).

However, the most economically viable and large-scale suited alternate for the production of taxol is via plant cell culture. One can further improve the yields obtained via tissue culture by

- Utilizing cell lines with high yield potential
- Selecting and utilizing suitable phytohormones (Khosroushahi et al., 2006)
- Using optimal carbohydrate sources: It has been observed that the presence of fructose stimulates taxol production, while glucose does the opposite.

- Utilizing elicitors, precursors, and other additives that enhance the production process.
- Setting up a two-phase production chamber with in situ product removal is shown to boost yields. This helps in preventing feedback inhibition and the degradation of the metabolite via enzymatic pathways, which ultimately increases overall yield.

8.4   REGULATORY ASPECTS OF HERBAL ANTICANCER DRUGS

Herbal prescriptions have been used for a long time now, with the written records to very nearly since 5000 years ago (Swerdlow, 2000). Till 1890, about 59% of the listing accessible on the US Pharmacopeia was connected somehow to the home grown origins, while till as of late, it has been tracked down that one out of five people in the USA professed to utilize natural products (Barnes et al., 2004; Revathy et al., 2012). The American Botanical Council revealed sales of US$ 5.3 billion in 2011, that increased to US$ 6 billion by the end of 2013 (Lindstrom et al., 2014). Other than these, an enormous extent of the African and Asian populace relies on customary medication and restorative spices for their essential medical care. A part of these were taken forward by the old frameworks of restorative information like Ayurveda, Siddha, and Chinese medication (WHO, 2008; Pan et al., 2014).

As of late, an increase in the inclination of home grown medications along with an adaptable administrative framework and related harmfulness has had many inquiries on the use of these medications and enhancements (Elvin-Lewis, 2001; Cuzzolin et al., 2006; Sahoo et al., 2010). Notwithstanding Herbal medication's usage for a long time, there are no shared view guidelines (Thakkar et al., 2020). While FDA (Food and Drug Administration) is the essential body directing home grown enhancements in the USA, they permit the herbal supplements to be sold as dietary enhancements and not as the endorsed over-the-counter medications. This gives the producer enough freedom to sidestep the tough course of exhibiting security and adequacy as the law characterizes it extensively, and despite the fact that they are not permitted to make direct claims, they can in any case make primary and useful claims on medical advantages (Bent, 2008).

The innovative work plans for new drug are all around spread out and clear, while the equivalent isn't valid for natural medications. As plants seldom have one explicit dynamic pharmacological compound, it makes it hard to report the particular capacities (Sahoo et al., 2010). The overall focus or synthetic profiles of the various mixtures found in a plant could fluctuate a great deal and can altogether impact the outcomes. Other than the previously mentioned heterogeneity, there are different things to be thought about, similar to defilement of one or the other engineered or compound that is normal with the exchange of therapeutic plant is one (Chan, 2003; Yee et al., 2005; Lam et al., 2008), while others are misidentification of plants dependent on morphology (Ramawat and Goyal, 2008; Yap et al., 2008), purposeful or unintentional expansion of harmful substantial metals (Rai et al., 2001; Dargan et al., 2008; Mazzanti et al., 2008; Zhang et al., 2012), and pesticides (Rai et al., 2008; Xue et al., 2018; Xue et al., 2008) or microbial disease (Bugno et al., 2006) might actually influence the helpful qualities and should be regulated.

Phytomedicine improvement is a multistep measure that starts with the assortment of crude material, trailed by identification, adjustment, extractions, quality affirmation, separation of dynamic compound, purging, fractionation, and harmfulness assessment. All through the turn of events, process normalization is a basic advance that relies upon many variables including, however not restricted to, development conditions, circulation, assembling, and advertising. Further, the biochemical profile of a plant and its auxiliary metabolites are likewise answered to be influenced by genetics, moisture, environment, supplement, photoperiod, capacity, collection, extraction, and bundling. To keep the norm of medication adequacy and security, synthetic consistency ought to be routinely checked and kept in ideal reach throughout the process (Sahoo et al., 2010).

Because of innate intricacies and potential varieties at each progression of the interaction, a multimethod approach incorporating plant strategy, synthetic marker, and so on should be taken together for quality control (Mukherjee, 2019; Thakkar et al., 2020). As of late, following methods and steps are utilized for the distinguishing proof and guideline of herbal medications:

- Qualitative investigation is done through moisture content, ash value, extractive worth, pesticides, heavy metals, essential oils, phytochemical constituents, and unrefined fiber (Holst, 1973; Idu et al., 2008; Abdelhadi et al., 2015; Akram et al., 2015; Chen et al., 2015).
- Morphological identification and macroscopic assessment include organoleptic study, micromorphology investigation of roots, bark, stems, rhizomes, woods, leaves, seeds, natural products, and elevated parts (Bauer, 1998; Quettier-Deleu et al., 2000; Grubeši et al., 2005; Smillie and Khan, 2010).
- Chemical investigations, for example, thin layer chromatography (TLC), High-performance thin layer chromatography (HPTLC), and High-performance liquid chromatography (HPLC) are most of the

time utilized for better identification (Lazarowych and Pekos, 1998; Xie et al., 2006; Raclariu et al., 2017).

- Metabolite investigations are typically done through LC/MS, GC-MS, LC-NMR, protein examination, and HPLC/MS (Alali and Tawaha, 2009; Farag et al., 2012; Trivedi et al., 2017; Bendif et al., 2020).
- Bioassay and Herbal medications assessment should be carried out in vitro/in vivo. Moreover, one can likewise consider various proteins, receptors, and possibly can run different high-throughput evaluates for the medication assessment and its parasiticidal, hostile to viral, and antimicrobial impact (Chanda and Nagani, 2013; Tiong et al., 2013; Naz et al., 2015).
- Safety study includes harmfulness study, spice drug connection, enlistment, restraint of CYP450, and pharmacovigilance (Nasri, 2013).
- Quality control is generally directed through the arrangement of specialized rules distributed by WHO throughout the long term. A couple of them are "Quality control strategies for restorative plant material," "Guidelines of good agricultural and collection practices GACP" and "WHO rules for evaluating the nature of herbal drugs concerning impurities and buildups." Other than these, numerous nations like Japan, China, India, and European affiliations have their own arrangements of rules administering the assortment, fabricating, capacity, applications, and engendering of herbal plants (WHO, 1998, 2003, 2007; Liu et al., 2018).

8.5   TRENDS IN TRADITIONAL MEDICINE SYSTEMS AND THE FUTURE OF ANTICANCER PLANT-BASED PRODUCTS

While there is a great amount of studies being conducted on the various avenues for the production of anticancer agents, plant-based compounds probably hold the most promise in the long run. Anticancer agents derived from natural sources in general have two things going for them;

1. Since they are naturally occurring substances, they can be extracted from viable sources, purified and then put to use.
2. Naturally sourced compounds are generally shown to have fewer toxic and adverse effects.

These factors work in tandem to make plant-based and other naturally sourced bioactive compounds a solid contender to their synthetic counterparts.

Over the years, the research and development world has come to the same conclusion and has shifted their attention to these natural products, particularly plant-based anticancer agents. This interest in natural products is only going to increase with time as newer and more advanced technologies for the extraction, production, and purification of these compounds are created (David et al., 2014).

This is not to say that there are no limitations related to the field of plant-based anticancer products. One particularly rampant problem associated with most naturally sourced compounds is the access to biological samples in a day and age, where natural biodiversity is being destroyed at an alarming rate. Additionally, it is quite difficult at times to gain access to specific plants for their bioactive compounds, especially when factors like intellectual property rights and legal supply are thrown around. The other major limitation would be correctly identifying the useful active compounds from natural sources. Due to the lack of communication regarding nonessential compounds present in plant systems, it gets increasingly hard to determine and identify compounds of value. Of course, there are various efforts like the Collective Molecular Activities of Useful Plants, Chinese Medicine Integrated Database, SymMap, Encyclopedia of traditional Chinese medicines, etc. which aim to remedy this situation. By improving the communication thread regarding active plant compounds and related information, we can aim to improve the identification and production of various plant-based compounds, including anticancer agents.

8.6 CONCLUSIONS

Taxus baccata (European yew) has been perhaps the most continuous source of taxol for investigations of the biosynthetic pathway and further developed creation of this anticancer medication. It is by all accounts very clear that the taxanes are continually in upgradation mode both as far as mechanistic viewpoints and clinical perspectives. The roads of up-degree and extemporization of taxanes are totally open for additional investigations. In addition, mix therapies (taxanes alongside some different medications) likewise need to be looked upon to work on the viability and anticancer action. This further opens the way for the investigation of new medications from nature and regular assets plants being the lone retreat for this situation.

As per a characteristic viewpoint, development of *Taxus* sp. for large scale production of taxanes requires more thought and biotechnological approaches and can be done through tissue culture procedures. Treatment of malignant growth would be more refined if the sensibility of the prescriptions is contemplated by the medication business and administrative arrangements. Diminishing the expenses of the prescriptions will not simply make it more accessible to people yet moreover all the while will expect an obvious part as a popular anticancer drug among all the other large number of medicines available in the market. As such, cost assessment of treatment should in like manner outline a huge limit close by drug headway for the overall benefit of humankind, unequivocally with respect to the quantity of patients in underdeveloped countries where the occasion of cancers is increasing to an alarming level, essentially due to contamination and irregular lifestyle. Hence, taxane is among such compounds that has acquired a spot in the medication world for the treatment of a wide bunch of malignancy threat. Its further advancement both to the extent pharmacology and cost viability stay absolutely open for the benefit of patients and mankind. Taxol creation in *T. baccata* suspension cultures has been improved by enriching and improving culture conditions, fundamental media, plant advancement regulators, sugar supplements, etc., and cultures have been increased to a bioreactor level for scaling up. A sensible system might give new comprehension into how the taxol biosynthetic pathway is controlled, with hereditary and metabolic planning strategies, differential hereditary articulation, record factors, and key qualities inciting higher taxol yields. One viewpoint to consider is the taxol release from cells, which could be improved by using a two-stage culture structure, so far not estimated in *T. baccata* cell suspensions. Future perspectives could be focused on the coordinated usage of observational and rational systems and analyzing the two-stage culture structure to make a biotechnological structure for high taxol creation.

KEYWORDS

- **antioxidant**
- **biocompatible**
- **phenolic compounds**
- **secondary metabolites**
- **tumor**

REFERENCES

Abdelhadi, M.; Meullemiestre, A.; Gelicus, A.; Hassani, A.; Rezzoug, S. A. Intensification of *Hypericum perforatum* L. Oil Isolation by Solvent-Free Microwave Extraction. *Chem. Eng. Res. Des.* **2015,** *93*, 621–631.

Abella, I.; La magia de los árboles (Simbolismo, Mitos y tradiciones, Plantacióny cuidados). Barcelona: Ediciones Integral, 1996; pp 99–121.

Adams, L. S.; Phung, S.; Yee, N.; Seeram, N. P.; Li, L.; Chen, S. Blueberry Phytochemicals Inhibit Growth and Metastatic Potential of MDA-MB-231 Breast Cancer Cells Through Modulation of the Phosphatidylinositol 3-Kinase Pathway. *Cancer Res.* **2010,** *70* (9), 3594–3605.

Akram, S.; Najam, R.; Rizwani, G. H.; Abbas, S. A. Determination of Heavy Metal Contents by Atomic Absorption Spectroscopy (AAS) in Some Medicinal Plants from Pakistani and Malaysian origin. *Pak. J. Pharm. Sci.* **2015,** *28* (5), 1781–1787.

Alali, F. Q.; Tawaha, K. Dereplication of Bioactive Constituents of the Genus Hypericum Using LC-(+,−)-ESI-MS and LC-PDA techniques: *Hypericum triquterifolium* as a Case Study. *Saudi Pharma. J.* **2009,** *17* (4), 269–274.

Azmir, J.; Zaidul, I. S. M.; Rahman, M. M.; Sharif, K. M.; Mohamed, A.; Sahena, F.; Omar, A. K. M. Techniques for Extraction of Bioactive Compounds from Plant Materials: A Review. *J. Food Eng.* **2013,** *117* (4), 426–436.

Baghban, R.; Roshangar, L.; Jahanban-Esfahlan, R.; Seidi, K.; Ebrahimi-Kalan, A.; Jaymand, M.; Kolahian, S.; Javaheri, T.; Zare, P. Tumor Microenvironment Complexity and Therapeutic Implications at a Glance. *Cell Commun. Signal.* **2020,** *18*

Banarjee, S.; Bueso-Ramos, C.; Aggarwal, B. B. Suppression of 7,12-dimethylbenz(a) anthracene-Induced Mammary Carcinogenesis in Rats by Resveratrol: Role of Nuclear Factor-Kappa B, Cyclooxygenase 2, and Matrix Metalloprotease 9. *Cancer Res.* **2002,** *62* (17), 4945–4954.

Barnes, P. M.; Powell-Griner, E.; McFann, K.; Nahin, R. L. Complementary and Alternative Medicine Use Among Adults: United States, 2002. *Semin. Integr. Med.* **2004,** *2* (2), 54–71.

Bass, R.; Roberto, D.; Wang, D. Z.; Cantu, F. P.; Mohamadi, R. M.; Kelley, S. O.; Klotz, L.; Venkateswaran, V. Combining Desmopressin and Docetaxel for the Treatment of Castration-Resistant Prostate Cancer in an Orthotopic Model. *Antican. Res.* **2019,** *39* (1), 113–118.

Bauer, R. Quality Criteria and Standardization of Phytopharmaceuticals: Can Acceptable Drug Standards Be Achieved? *Drug Info. J.* **1998,** *32* (1), 101–110.

Bendif, H.; Peron, G.; Miara, M. D.; Sut, S.; Dall'Acqua, S.; Flamini, G.; Maggi, F. Total Phytochemical Analysis of *Thymus munbyanus subsp. coloratus* from Algeria by HS-SPME-GC-MS, NMR and HPLC-MSn Studies. *J. Pharm. Biomed. Ana.* **2020,** *186*, 113330.

Bent, S. Herbal Medicine in the United States: Review of Efficacy, Safety, and Regulation. *J. Gen. Int. Med.* **2008,** *23* (6), 854–859.

Bhattacharya, S.; Modal, L.; Mukherjee, B.; Dutta, L.; Ehsan, I.; Debnath, M. C.; Gaonkar, R. H.; Pal, M. M.; Majumdar, S. Apigenin Loaded Nanoparticle Delayed Development of Hepatocellular Carcinoma in Rats. *Nanomedicine* **2008,** *14* (6), 1905–1917.

Brown, J. B.; Lee, G.; Grimm, G. R.; Barrett, T. A. Therapeutic Benefit of Pentostatin in Severe IL-10 Colitis. *Inflamm. Bowel Disc.* **2008,** *14*, 880–887.

Bugno, A.; Almodovar, A. A. B.; Pereira, T. C.; Pinto, T. D. J. A.; Sabino, M. Occurrence of Toxigenic Fungi in Herbal Drugs. *Braz. J. Microbiol.* **2006,** *37* (1), 47–51.

Chan, K. Some Aspects of Toxic Contaminants in Herbal Medicines. *Chemo* **2003,** *52* (9), 1361–1371.

Chanda, S.; Nagani, K. In Vitro and In Vivo Methods for Anticancer Activity Evaluation and Some Indian Medicinal Plants Possessing Anticancer Properties: An Overview. *J. Pharmacog. Phytochem.* **2013,** *2* (2).

Chang, C. C.; Yang, Y.; Gao, D. Y.; Cheng, H. T.; Hoang, B.; Chao, P. H.; Chen, L. H.; Bteich, J.; Chiang, T.; Liu, J. Y.; Li, S. D.; Chen, Y. Docetaxel-Carboxymethyl Cellulose Nanoparticles Ameliorate CCl_4-Induced Hepatic Fibrosis in Mice. *J. Drug Targ.* **2018,** *26* (5–6), 516–524.

Chen, R.; Cheng, Q.; Owusu-Ansah, K. G.; Chen, J.; Song, G.; Xie, H.; Zhou, L.; Xu, X.; Jiang, D.; Zheng, S. Cabazitaxel, a Novel Chemotherapeutic Alternative for Drug-Resistant Hepatocellular Carcinoma. *Am. J. Can. Res.* **2018,** *8* (7), 1297–1306.

Chen, H.; Zhu, B.; Zhao, L.; Liu, Y.; Zhao, F.; Feng, J.; Jin, Y.; Sun, J.; Geng, R.; Wei, Y. Allicin Inhibits Proliferation and Invasion in Vitro and in Vivo via SHP-1-Mediated STAT3 Signaling in Cholangiocarcinoma. *Cell Physiol. Biochem.* **2018,** *47* (2), 641–653.

Chen, M.; Zhao, Z.; Lan, X.; Chen, Y.; Zhang, L.; Ji, R.; Wang, L. Determination of Carbendazim and Metiram Pesticides Residues in Rapeseed and Peanut Oils by Fluorescence Spectrophotometry. *Measurement* **2015,** *73*, 313–317.

Choi, K. H.; Jeon, J. Y.; Lee, Y. E.; Kim, S. W.; Kim, S. Y.; Yun, Y. J.; Park, K. C. Synergistic Activity of Paclitaxel, Sorafenib, and Radiation Therapy in Advanced Renal Cell Carcinoma and Breast Cancer. *Trans. Onc.* **2019,** *12* (2), 381–388.

Cope, E. A. Taxaceae: The Genera and Cultivated Species. *Bot. Rev.* **1998,** *64*, 291–322.

Cristalli, G.; Costanzi, S.; Lambertucci, C.; Lupidi, G.; Vittori, S.; Volpini, R. et al., Adenosine Deaminase: Functional Implications and Different Classes of Inhibitors. *Med. Res. Rev.* **2001,** *21*, 105–128.

Cuzzolin, L.; Zaffani, S.; Benoni, G. Safety Implications Regarding Use of Phytomedicines. *Eur. J. Clin. Pharma.* **2006,** *62* (1), 37–42.

Dargan, P. I.; Gawarammana, I. B.; Archer, J. R.; House, I. M.; Shaw, D.; Wood, D. Heavy Metal Poisoning from Ayurvedic Traditional Medicines: An Emerging Problem?. *Int. J. Environ. Heal.* **2008,** *2* (3–4), 463–474.

David, B.; Wolfender, J. L.; Dias, D. A. The Pharmaceutical Industry and Natural Products: Historical Status and New Trends. *Phytochem. Rev.* **2014,** *14* (2), 299–315.

de Lima, R. M. T.; Reis, A. C. D.; de Menezes, A. P. M.; Santos, J. V. O.; Filho, J. W. G. O.; Ferreira, J. R. O.; de Alencar, M. C. O. B.; da Mata, A. M. O. F.; Khan, I. N.; Islam, A.; Uddin, S. J.; Ali, E. S.; Islam, M. T.; Tripathi, S.; Mishra, S. K.; Mubarak, M. S.; Melo-Cavalcante, A. A. C. Protective and Therapeutic Potential of Ginger (*Zingiber officinale*) Extract and (6)-Gingerol in Cancer: A Comprehensive Review. *Phytother. Res.* **2018,** *32* (10), 1885–1907.

De Martin, R.; Schmid, J. R.; Hofer-Warbinek, R. The NF-kappaB/Rel Family of Transcription Factors in Oncogenic Transformation and Apoptosis. *Mutat. Res.* **1999,** *437* (3), 231–243.

Doddapaneni, R.; Patel, K.; Chowdhury, N.; Singh, M. Noscapine Chemosensitization Enhances Docetaxel Anticancer Activity and Nanocarrier Uptake in Triple Negative Breast Cancer. *Exp. Cell Res.* **2016,** *346* (1), 65–73.

Dong, S.; Guo, Y.; Duan, Y.; Li, Z.; Wang, C.; Niu, L.; Wang, N.; Ma, M.; Shi, Y.; Zhang, M. Co-Delivery of Paclitaxel and Gemcitabine by Methoxy Poly (Ethylene Glycol)-Poly (Lactide-Coglycolide)-Polypeptide Nanoparticles for Effective Breast Cancer Therapy. *Anti Can. Drug.* **2018,** *29* (7), 637–645.

Elvin-Lewis, M. Should We Be Concerned About Herbal Remedies. *J. Ethnopharm.* **2001,** *75* (2–3), 141–164.

Erdemoglu, N. and Sener, B. Antimicrobial Activity of the Heartwood of *Taxus baccata. Fitoterapia* **2001,** *72* (1), 59–61.

Erdemoglu, N.; Sener, B.; Choudhary, M. I. Bioactivity of Lignans from *Taxus baccata. Zeitschrift Für Naturforschung C* **2004,** *59* (7–8), 494–498.

Farag, M. A.; Porzel, A.; Wessjohann, L. A. Comparative Metabolite Profiling and Fingerprinting of Medicinal Licorice Roots Using a Multiplex Approach of GC–MS, LC–MS and 1D NMR Techniques. *Phytochemistry* **2012,** *76,* 60–72.

Fard, A. E.; Tavakoli, M. B.; Salehi, H.; Emami, H. Synergetic Effects of Docetaxel and Ionizing Radiation Reduced Cell Viability on MCF-7 Breast Cancer Cell. *Appl. Can. Res.* **2017,** *37,* 29.

Foglietta, F.; Spagnoli, G. C.; Muraro, M. G.; Ballestri, M.; Guerrini, A.; Ferroni, C.; Aluigi, A.; Sotgiu, G.; Varchi, G. Anticancer Activity of Paclitaxel-Loaded Keratin Nanoparticles in Two-Dimensional and Perfused Three-Dimensional Breast Cancer Models. *Int. J. Nanomed.* **2018,** *13,* 4847–4867.

Fusser, M.; Øverbye, A.; Pandya, A. D.; Mørch, Ý.; Borgos, S. E.; Kildal, W.; Snipstad, S.; Sulheim, E.; Fleten, K. G.; Askautrud, H. A.; Engebraaten, O.; Flatmark, K.; Iversen, T. G.; Sandvig, K.; Skotland, T.; Mælandsmo, G. M. Cabazitaxel-Loaded Poly (2-Ethylbutyl Cyanoacrylate) Nanoparticles Improve Treatment Efficacy in a Patient Derived Breast Cancer Xenograft. *J. Contr. Rel.* **2019,** *293,* 183–192.

Gallego-Yerga, L.; Posadas, I.; de la Torre, C.; Ruiz-Almansa, J.; Sansone, F.; Ortiz, M. C.; Casnati, A.; García, F. J. M.; Ceña, V. Docetaxel-Loaded Nanoparticles Assembled from β-Cyclodextrin/Calixarene Giant Surfactants: Physicochemical Properties and Cytotoxic Effect in Prostate Cancer and Glio Blastoma Cells. *Front. Pharmacol.* **2017,** *8,* 249.

Gdowski, A. S.; Ranjan, A.; Sarker, M. R.; Vishwanatha, J. K. Bone-Targeted Cabazitaxel Nanoparticles for Metastatic Prostate Cancer Skeletal Lesions and Pain. *Nanomedicine* **2017,** *12* (17), 2083–2095.

Glazer, R. I. Adenosine Deaminase Inhibitors: Their Role in Chemotherapy and Immunosuppression. *Cancer Chemother. Pharmacol.* **1980,** *4,* 227–235.

Grubešić, R. J.; Vuković, J.; Kremer, D.; Valadimir-Knežević, S. Spectrophotometric Method for Polyphenols Analysis: Pre-Validation and Application on *Plantago* L. Species. *J. Pharma. Biomed. Ana.* **2005,** *39* (3–4), 837–842.

Haldar, S.; Chintapalli, J.; Croce, C. M. Taxol Induces BCL-2 Phosphorylation and Death of Prostate Cancer Cells. *Cancer Res.* **1996,** *56* (6), 1253–1255.

Hanahan, D.; Weinber, R. A. Hallmarks of Cancer: The Next Generation. *Cell* **2011,** *144* (5), 646–674.

Harshita; Barkat M. A., Rizwanullah, M.; Beg, S.; Pottoo, F. H.; Siddiqui, S.; Ahmad, F. J. Paclitaxel-Loaded Nanolipidic Carriers with Improved Oral Bioavailability and Anticancer Activity Against Human Liver Carcinoma. *AAPS Pharm. Sci. Tech.* **2019,** *20* (2), 87.

Hendrikx, J. J. M. A.; Lagas, J. S.; Song, J. Y.; Rosing, H.; Schellens, J. H. M.; Beijnen, J. H.; Rottenberg, S.; Schinkel, A. H. Ritonavir Inhibits Intratumoral Docetaxel Metabolismand Enhances Docetaxel Antitumor Activity in an Immunocompetent Mouse Breast Cancer Model. *Int. J. Cancer* **2016,** *138,* 758–769.

Holst, D. O. Holst Filtration Apparatus for Van Soest Detergent Fiber Analyses. *J. Ass. Off. Ana. Chem.* **1973,** *56* (6), 1352–1356.

Hou, J.; Sun, E.; Zhang, Z. H.; Wang, J.; Yang, L.; Cui, L.; Ke, Z. C.; Tan, X. B.; Jia, X. B.; Lv, H. Improved Oral Absorption and Anti-Lung Cancer Activity of Paclitaxel-Loaded Mixed Micelles. *Drug Del.* **2017,** *24* (1), 261–269.

Idu, M.; Omonigho, S. E.; Igeleke, C. L.; Oronsaye, F. E.; Orhue, E. S. Microbial Load on Medicinal Plants Sold in Bini Markets, Nigeria. *Ind. J. Trad. Know.* **2008,** *7* (4), 669–672.

Jeurissen, S. M.; Buurma-Rethans, E. J.; Beukers, M. H.; Jansen-van der Vliet, M.; van Rossum, C. T.; Sprong, R. C. Consumption of Plant Food Supplements in the Netherlands. *Food Funct.* **2018,** *9* (1), 179–190.

Jiang, S.; Zhang, K.; He, Y.; Xu, X.; Li, D.; Cheng, S.; Zheng, X. Synergistic Effects and Mechanisms of Impressic Acid or Acankoreanogein in Combination with Docetaxel on Prostate Cancer. *RSC Adv.* **2018,** *8,* 2768–2776.

Jin, M.; Jin, G.; Kang, L.; Chen, L.; Gao, Z.; Huang, W. Smart Polymeric Nanoparticles with pH-Responsive and PEG-Detachable Properties for Co-Delivering Paclitaxel and Survivin siRNA to Enhance Antitumor Outcomes. *Inter. J. Nanomed.* **2018,** *13,* 2405–2426.

Khosroushahi, A. Y.; Valizadeh, M.; Ghasempour, A.; Khosrowshahli, M.; Naghdibadi, H.; Dadpour, M. R.; Omidi, Y. Improved Taxol Production by Combination of Inducing Factors in Suspension Cell Culture of *Taxus baccata. Cell Bio Inter.* **2006,** *30* (3), 262–269.

Kingston, D. Taxol and Its Analogs. In *Anticancer Agents from Natural Products*; Cragg, G., Kingston, D. G., Newman, D. J., Eds.; Brunner-Routledge Psychology Press, Taylor and Francis Group: Boca Raton, FL, 2005; pp 89–122.

Kinoshita, R.; Ishima, Y.; Chuang, V. T. G.; Nakamura, H.; Fang, J.; Watanabe, H.; Shimizu, T.; Okuhira, K.; Ishida, T.; Maeda, H.; Otagiri, M.; Maruyama, T. Improved Anticancer Effects of Albumin-Bound Paclitaxel Nanoparticle *via* Augmentation of EPR Effect and Albumin-Protein Interactions Using S-Nitrosated Human Serum Albumin Dimer. *Biomaterial* **2017,** *140,* 162–169.

Kommineni, N.; Saka, R.; Bulbake, U.; Khan, W. Cabazitaxel and Thymoquinone Co-Loaded Lipospheres as a Synergistic Combination for Breast Cancer. *Chem. Phys. Lip.* **2018,** *Pii,* 3084.

Lam, Y. H.; Poon, W. T.; Lai, C. K.; Chan, A. Y. W.; Mak, T. W. L. Identification of a Novel Vardenafil Analogue in Herbal Product. *J. Pharma. Biomed. Ana.* **2008,** *46* (4), 804–807.

Lazarowych, N. J.; Pekos, P. Use of Fingerprinting and Marker Compounds for Identification and Standardization of Botanical Drugs: Strategies for Applying Pharmaceutical HPLC Analysis to Herbal Products. *Drug Info. J.* **1998,** *32* (2), 497–512.

Lee, W.; Huang, J.; Shyur, L. Phytoagents for Cancer Management: Regulation of Nucleic Acid Oxidation, ROS, and Related Mechanisms. *Oxid. Med. Cell Longev.* **2013,** 925804.

Lindstrom, A.; Ooyen, C.; Lynch, M.; Blumenthal, M.; Kawa, K. Sales of Herbal Dietary Supplements Increase by 7.9% in 2013, Marking a Decade of Rising Sales: Turmeric Supplements Climb to Top Ranking in Natural Channel. *Herbal Gram* **2014,** *103,* 52–56.

Liu, C. Y.; Chen, K. F.; Chao, T. I., Chu, P. Y.; Huang, C. T.; Huang, T. T.; Yang, H. P.; Wang, W. L.; Lee, C. H.; Lau, K. Y.; Tsai, W. C.; Su, J. C.; Wu, C. Y.; Chen, M. H.; Shiau, C. W.; Tseng, L. M. Sequential Combination of Docetaxel with a SHP-1 Agonist Enhanced Suppression of p-STAT3 Signaling and Apoptosis in Triple Negative Breast Cancer Cells. *J Mol Med* **2017,** *95* (9): 965–975.

Liu, C.; Guo, D. A.; Liu, L. Quality Transitivity and Traceability System of Herbal Medicine Products Based on Quality Markers. *Phytomedicine* **2018,** *44,* 247–257.

Madunic, J.; Madunic, I. V.; Gajski, G.; Popic, J.; Garaj-Vrhovac, V. Apigenin: A Dietary Flavonoid with Diverse Anticancer Properties. *Cancer Lett.* **2018,** *413,* 11–22.

Mahira, S.; Kommineni, N.; Husain, G. M.; Khan, W. Cabazitaxel and Silibinin Co-Encapsulated Cationic Liposomes for CD44 Targeted Delivery: A New Insight into Nanomedicine Based Combinational Chemotherapy for Prostate Cancer. *Biomed. Pharmacother.* **2019,** *110,* 803–817.

Malik, S.; Cusidó, R. M.; Mirjalili, M. H.; Moyano, E.; Palazón, J.; Bonfill, M. Production of the Anticancer Drug Taxol in *Taxus baccata* Suspension Cultures: A Review. *Process. Biochem.* **2011,** *46* (1), 23–34.

Manish, L.; Mangesh, B.; Sabale, V. *Taxus*—The Panacea for Cancer Treatment. *Ancient Sci. Life* **2005,** *24,* 152–159.

Martin, A. C. B. M.; Fuzer, A. M.; Becceneri, A. B.; da Silva, J. A.; Tomasin, R.; Denoyer, D.; Kim, S.; McIntyre, K. A.; Pearson, H. B.; Yeo, B.; Nagpal, A.; Ling, X.; Selistre-de-Araújo, H. S.; Vieira, P. C.; Cominetti, M. R.; Pouliot, N. (10)-Gingerol Induces Apoptosis and Inhibits Metastatic Dissemination of Triple Negative Breast Cancer In Vivo. *Oncotarget* **2017,** *8* (42), 72260–72271.

Mazzanti, G.; Battinelli, L.; Daniele, C.; Costantini, S.; Ciaralli, L.; Evandri, M. G. Purity Control of Some Chinese Crude Herbal Drugs Marketed in Italy. *Food Chem. Tox.* **2008,** *46* (9), 3043–3047.

Milutinović, M. G.; Stanković, M. S.; Cvetković, D. M.; Topuzović, M. D.; Mihailović, V. B.; Marković, S. D. Antioxidant and Anticancer Properties of Leaves and Seed Cones from European Yew (*Taxus Baccata L.*). *Arch. Biol. Sci.* **2015,** *67* (2), 525–534.

Min, J.; Shen, H.; Xi, W.; Wang, Q.; Yin, L.; Zhang, Y.; Yu, Y.; Yang, Q.; Wang, Z. N. Synergistic Anticancer Activity of Combined Use of Caffeic Acid with Paclitaxel Enhances Apoptosis of Non-Small-Cell Lung Cancer h1299 Cells In Vivo and In Vitro. *Cell Physiol. Biochem.* **2018,** *48* (4), 1433–1442.

Mukherjee, P. K. *Quality Control and Evaluation of Herbal Drugs: Evaluating Natural Products and Traditional Medicine*; Elsevier, 2019.

Nasri, H. Toxicity and Safety of Medicinal Plants. *J. HerbMed. Pharmacol.* **2013,** 2.

Naz, S.; Haq, R.; Aslam, F.; Ilyas, S. Evaluation of Antimicrobial Activity of Extracts of In Vivo and In Vitro Grown *Vinca rosea L. (Catharanthus roseus)* against pathogens. *Pak. J. Pharm. Sci.* **2015,** *28* (3), 849–853.

Needleman, D. J.; Ojeda-Lopez, M. A.; Raviv, U.; Ewert, K.; Miller, H. P.; Wilson, L.; Safinya, C. R. Radial Compression of Microtubules and the Mechanism of Action of Taxol and Associated Proteins. *Biophys. J.* **2005,** *89* (5), 3410–3423.

Newmann, D. J.; Cragg, G. M. Natural Products as Sources of New Drugs from 1981 to 2014. *J. Nat. Prod.* **2016,** *79* (3), 629–661.

Ngo, T. V.; Scarlett, C. J.; Bowyer, M. C.; Ngo, P. D.; Vuong, Q. V. Impact of Different Extraction Solvents on Bioactive Compounds and Antioxidant Capacity from the Root of *Salacia chinensis* L. *J. Food Qual.* **2017.**

Nia, H. T.; Munn, L. L.; Jain, R. K. Physical Traits of Cancer. *Science* **2020,** *370* (6516).

Nicolaou, K. C.; Yang, Z.; Liu, J. J.; Ueno, H.; Nantermet, P. G.; Guy, R. K.; Claiborne, C. F.; Renaud, J.; Couladouros, E. A.; Paulvannan, K.; Sorensen, E. J. Total Synthesis of Taxol. *Nature* **1994,** *367,* 630–634.

Organización Mundial de la Salud, World Health Organization, & Światowa Organizacja Zdrowia. *WHO Guidelines on Good Agricultural and Collection Practices (GACP) for Medicinal Plants*; World Health Organization, 2003.

Pan, S. Y.; Litscher, G.; Gao, S. H.; Zhou, S. F.; Yu, Z. L.; Chen, H. Q.; Ko, K. M. Historical Perspective of Traditional Indigenous Medical Practices: The Current Renaissance and Conservation of Herbal Resources. *Evid.-Based Complement. Altern. Med.* **2014.**

Parmar, V. S.; Jha, A.; Bisht, K. S.; Taneja, P.; Singh, S. K.; Kumar, A.; Jain; R., Olsen, C. E. Constituents of the Yew Trees. *Phytochemistry* **1999**, *50* (8), 1267–1304.

Patel, K.; Doddapaneni, R.; Chowdhury, N.; Boakye, C. H.; Behl, G.; Singh, M. Tumor Stromal Disrupting Agent Enhances the Anticancer Efficacy of Docetaxel Loaded PEGylated Liposomes in Lung Cancer. *Nanomedicine* **2016**, *11* (11), 1377–1392.

Quettier-Deleu, C.; Gressier, B.; Vasseur, J.; Dine, T.; Brunet, C.; Luyckx, M.; Trotin, F. Phenolic Compounds and Antioxidant Activities of Buckwheat (*Fagopyrum esculentum* Moench) Hulls and Flour. *J. Ethnopharmacol.* **2000**, *72* (1–2), 35–42.

Rabzia, A.; Khazaei, M.; Rashidi, Z.; Khazaei, M. R. Synergistic Anticancer Effect of Paclitaxel and Noscapine on Human Prostate Cancer Cell Lines. *Iran J. Pharma. Res.* **2017**, *16* (4), 1432–1442.

Raclariu, A. C.; Paltinean, R.; Vlase, L.; Labarre, A.; Manzanilla, V.; Ichim, M. C.; de Boer, H. Comparative Authentication of *Hypericum perforatum* Herbal Products Using DNA Meta-Barcoding, TLC and HPLC-MS. *Sci. Rep.* **2017**, *7* (1), 1–12.

Rai, V.; Kakkar, P.; Khatoon, S.; Rawat, A. K. S.; Mehrotra, S. Heavy Metal Accumulation in Some Herbal Drugs. *Pharma. Bio.* **2001**, *39* (5), 384–387.

Rai, V.; Kakkar, P.; Singh, J.; Misra, C.; Kumar, S.; Mehrotra, S. Toxic Metals and Organochlorine Pesticides Residue in Single Herbal Drugs Used in Important Ayurvedic Formulation–'Dashmoola'. *Environ. Monitor. Assess.* **2008**, *143* (1), 273–277.

Ramawat, K. G.; Goyal, S. The Indian Herbal Drugs Scenario in Global Perspectives. *Bioactive Mol. Med. Plants* **2008**, 325–347.

Revathy, S. S.; Rathinamala, R.; Murugesan, M. Authentication Methods for Drugs Used in Ayurveda, Siddha and Unani Systems of Medicine: An Overview. *Int. J. Pharma. Sci. Res.* **2012**, *3* (8), 2352.

Rivera-Franco, M. M. and Leon-Rodriguez, E. Delays in Breast Cancer Detection and Treatment in Developing Countries. *Breast Cancer* **2018**, *12*, 17752677–11782234.

Rottach, A. M.; Ahrend, H.; Martin, B.; Walther, R.; Zimmermann, U.; Burchardt, M.; Stope, M. B. Cabazitaxel Inhibits Prostate Cancer Cell Growth by Inhibition of Androgen Receptor and Heat Shock Protein Expression. *World J. Urol.* **2019**.

Sahoo, N.; Manchikanti, P.; Dey, S. Herbal Drugs: Standards and Regulation. *Fitoterapia* **2010**, *81* (6), 462–471.

Sajjad, M.; Khan, M. I.; Naveed, S.; Ijaz, S.; Qureshi, O. S.; Raza, S. A.; Shahnaz, G.; Sohail, M. F. Folate-Functionalized Thiomeric Nanoparticles for Enhanced Docetaxel Cytotoxicity and Improved Oral Bioavailability. *AAPS Pharm. Sci. Tech.* **2019**, *20* (2), 81.

Schmid, P.; Adams, S.; Rugo, H. S.; Schneeweiss, A.; Barrios, C. H.; Iwata, H.; Diéras, V.; Hegg, R.; Im, S. A.; Wright, G. S.; Henschel, V.; Molinero, L.; Chui, S. Y.; Funke, R.; Husain, A.; Winer, E. P.; Loi, S.; Emens, L. A. Impassion 130 Trial Investigators. Atezolizumab and Nab-Paclitaxel in Advanced Triple-Negative Breast Cancer. *New Eng. J. Med.* **2018**, *379* (22), 2108–2121.

Schultz, H. Herbal Supplement Sales Hit $5.3 Billion; ABC Report Says, Updated September 5, 2012. http://www.nutraingredientsusa.com/Markets/Herbal-supplement-sales-hit-5.3-billionABC-report-says

Sinha, D. A Review on Taxanes: An Important Group of Anticancer Compound Obtained from Taxus sp. *Int. J. Pharm. Sci. Res.* **2020**, *11* (5), 1969–1985.

Smillie, T. J.; Khan, I. A. A Comprehensive Approach to Identifying and Authenticating Botanical Products. *Clin. Pharmacol. Therap.* **2010**, *87* (2), 175–186.

Smith, R. E.; Thornton, D. E.; Allen, J. A Phase II Trial of Paclitaxel in Squamous Cell Carcinoma of the Head and Neck with Correlative Laboratory Studies. *Semin. Oncol.* **1995**, *22* (6), 41–46.

Swerdlow, J. Nature's Medicine. Plants That Heal, **2000**.

Tao, J.; Tan, Z.; Diao, L.; Ji, Z.; Zhu, J.; Chden, W.; Hu, Y. Co-Delivery of Dihydro-Artemisinin and Docetaxel in pH Sensitive Nanoparticles for Treating Metastatic Breast Cancer *via* the NF-κB/MMP-2 Signal Pathway. *RSC Adv.* **2018**, *8*, 21735–21744.

Thakkar, S.; Anklam, E.; Xu, A.; Ulberth, F.; Li, J.; Li, B.; Tong, W. Regulatory Landscape of Dietary Supplements and Herbal Medicines from a Global Perspective. *Reg. Tox. Pharmacol.* **2020**, *114*, 104647.

Tiong, S. H.; Looi, C. Y.; Hazni, H.; Arya, A.; Paydar, M.; Wong, W. F.; Awang, K. Antidiabetic and Antioxidant Properties of Alkaloids from *Catharanthus roseus* (L.) G. Don. *Molecules* **2013**, *18* (8), 9770–9784.

Trivedi, M. K.; Panda, P.; Sethi, K. K.; Jana, S. Metabolite Profiling in *Withania somnifera* Roots Hydroalcoholic Extract Using LC/MS, GC/MS and NMR Spectroscopy. *Chem. Biodiver.* **2017**, *14* (3), e1600280.

Vaishampayan, U.; Parchment, R. E.; Jasti, B. R.; Hussain, M. Taxanes: An Overview of the Pharmacokinetics and Pharmacodynamics. *Urology* **1999**, *54* (6), 22–29.

Varma, N. Phytoconstituents and Their Mode of Extractions: An Overview. *Res. J. Chem. Environ. Sci* **2016**, *4* (2), 8–15.

Venugopal, V.; Krishnan, S.; Palanimuthu, V. R.; Sankarankutty, S.; Kalaimani, J. K.; Karupiah, S.; Kit, N. S.; Hock, T. T. Anti-EGFR Anchored Paclitaxel Loaded PLGA Nanoparticles for the Treatment of Triple Negative Breast Cancer: In Vitro and In Vivo Anticancer Activities. *PLoS One* **2018**, *13* (11), e0206109.

Wan, J.; Ma, X.; Xu, D.; Yang, B.; Yang, S. and Han, S. Docetaxel-Decorated Anticancer Drug and Gold Nanoparticles Encapsulated Apatite Carrier for the Treatment of Liver Cancer. *J. Photochem. Photobiol.* **2018**, *185*, 73–79.

Wilson, D. K.; Rudolph, F. B.; Quiocho, F. A. Atomic Structure of Adenosine Deaminase Complexed with a Transition-State Analog: Understanding Catalysis and Immunodeficiency Mutations. *Science* **1991**, *252*, 1278–1284.

World Health Organization. *Quality Control Methods for Medicinal Plant Materials*, 1998.

World Health Organization. *WHO Guidelines for Assessing Quality of Herbal Medicines with Reference to Contaminants and Residues*, 2007.

World Health Organization. *Traditional Medicine*, 2008.

Xie, P.; Chen, S.; Liang, Y. Z.; Wang, X.; Tian, R.; Upton, R. Chromatographic Fingerprint Analysis—A Rational Approach for Quality Assessment of Traditional Chinese Herbal Medicine. *J. Chromatogr.* **2006**, *1112* (1–2), 171–180.

Xue, J.; Hao, L.; Peng, F. Residues of 18 Organochlorine Pesticides in 30 Traditional Chinese Medicines. *Chemosphere* **2018**, *71* (6), 1051–1055.

Xue, J.; Liu, D.; Chen, S.; Liao, Y.; Zou, Z. Overview on external contamination sources in traditional Chinese medicines. *World Sci. Tech.* **2008**, *10* (1), 91–96.

Yan, X.; Xie, T.; Wang, S.; Wang. Z.; Li, H.; Ye, Z. Apigenin Inhibits Proliferation of Human Chondrosarcoma Cells via Cell Cycle Arrest and Mitochondrial Apoptosis Induced by ROS Generation-an In Vitro and In Vivo Study. *Int. J. Clin. Exp. Med.* **2018**, *11* (3), 1615–1631.

Yap, K. Y. L.; Chan, S. Y.; Lim, C. S. The Reliability of Traditional Authentication–A Case of Ginseng Misfit. *Food Chem.* **2008**, *107* (1), 570–575.

Yee, S. K.; Chu, S. S.; Xu, Y. M.; Choo, P. L. Regulatory Control of Chinese Proprietary Medicines in Singapore. *Health Policy* **2005,** *71* (2), 133–149.

Zhang, Y.; Ge, Y.; Ping, X.; Yu, M.; Lou, D. and Shi, W. Synergistic Apoptotic Effects of Silibinin in Enhancing Paclitaxel Toxicity in Human Gastric Cancer Cell Lines. *Mol. Med. Rep.* **2018,** *18* (2), 1835–1841.

Zhang, J.; Wider, B.; Shang, H.; Li, X.; Ernst, E. Quality of Herbal Medicines: Challenges and Solutions. *Comp. Ther. Med.* **2012,** *20* (1–2), 100–106.

Zhong, T.; He, B.; Cao, H. Q.; Tan, T.; Hu, H. Y.; Li, Y. P.; Zhang, Z. W. Treating Breast Cancer Metastasis with Cabazitaxel-Loaded Polymeric Micelles. *Acta Pharmacol. Sinca* **2017,** *38* (6), 924–930.

Zhu, C. J.; An, C. G. Enhanced Antitumor Activity of Cabazitaxel Targeting CD44+ Receptor in Breast Cancer Cell Line via Surface Functionalized Lipid Nanocarriers. *Trop. J. Pharma. Res.* **2017,** *16* (6), 1383–1390.

CHAPTER 9

Panax Ginseng

SHALINI GURUMAYUM, SAGAR BARGE, and JAGAT C. BORAH

Chemical Biology Lab I, Institute of Advanced Study in Science and Technology, Paschim Boragaon, Guwahati, Assam, India

ABSTRACT

Panax ginseng C.A. Meyer (deciduous herb belonging to the family Araliaceae), also known as "healing herb," has long been used by the medicinal herbal healers of Far East Asia, including China and Korea, for curing different types of diseases as well in many other purposes to maintain a healthy and long life, and is mostly used as a tonic or adaptogen against hemorrhage, weak constitution, low vitality, fatigue, etc. Apart from curing common ailments, *P. ginseng* was reported to show anticancer activity. Ginsenoside (a triterpene saponin), one of the main compounds present in ginseng, has a steroid rigid four-ringed skeleton, containing sugar residues attaching to a –OH group of glycones having C20 side chain, and the cancer preventing activity gradually increases with respect to the decrease in the number of sugar units present. Metabolomic profiling of UPLC-MS has revealed that bioactive ginsenoside level increases and accumulates in the roots of old ginseng plants, which supports the usage of old roots in medicines. It was also reported that metabolite synthesized from the above-ground portion of the plant shifts underground in the later stage, arising to the elevated levels of ginsenosides biosynthesis across the root area. The ginsenosides are the key players of different signaling cascades. They showed potent antitumor efficacy through cell proliferation inhibition, cell cycle arrest, and induction of apoptosis in tumor cells. Most of the isolated ginsenosides have a unique potency in controlling the progression of different cancer types. A thorough

Potent Anticancer Medicinal Plants: Secondary Metabolite Profiling, Active Ingredients, and Pharmacological Outcomes. Deepu Pandita and Anu Pandita (Eds.)

metabolomic approach to study the variation of metabolites present in the plant according to their growing season and age will be the further perspective to contribute to a proper agronomic as well as quality and quantity specific strategy for effective synthesis of bioactive ginsenosides in the plant.

9.1 INTRODUCTION

Nature has been the primary source of medicine since time immemorial in curing different types of ailments. The traditional herbal healers in other parts of the world have already discovered numerous plant species that are prescribed either solely or in the form of a mixture of different types of medicinal plants (Alyaa, 2019). Mainly, the herbalist uses the plants in various forms like boiling or crushing. So, according to the traditional knowledge we have gained from the herbalist, many natural products have come up in the economy as a substitute for synthetic drugs. Natural derivatives and their structural analogs have historically made a significant contribution to pharmacotherapy, especially for cancer and infectious diseases (Alyaa, 2019).

Panax ginseng belongs to the family of Araliaceae, also known as Asian ginseng, the "king of herb." This plant is mainly grown in Asia, namely China, Japan, and Korea, for thousands of years to treat various diseases, as a health tonic and for the longevity of life (Chang et al., 2003). The government has approved the plant of the said countries as a part of traditional medicine. The herbal healers prescribed this plant as an immune tonic to improve brain function and physical aid performance (Kiefer and Pantuso, 2003; Ellis and Reddy, 2002; Yun, 2001). "*Panax*" was originated from the Greek word meaning "all-healing." Carl Anton von Meyer, the Russian botanist, first used this word (Court, 2000). Among the genus of *Panax*, *P. ginseng* was used primarily as indigenous ginseng, a perennial herb growing up to 60–80 cm tall. Its roots are aromatic, fleshy, and grayish-white to amber yellow in color, grows around 5–6 cm long deep. The ginseng stem is deep red, simple, and erect with oval, thin, and digitate leaves. Ginseng flowers are pink in color and give small red berries (Yun, 2001; Coates et al., 2005).

Medicinal plants are primary source of lead compounds that gave rise to different types of drugs today. These lead compounds are found distributed in various parts of the plant. Similarly, *P. ginseng* leaves, flowers, roots, and berries can be used as medicine or dietary supplement. For example, leaves are being used in the treatment of skin diseases (Lee et al., 2015; Kim et al., 2013). The flowers of *P. ginseng* are used as tea, as a dietary supplement (Hu, 1977), and steamed roots are being consumed for various pharmacological activities (Takagi et al., 1972). Due to its high medicinal value, separation and extraction techniques permitted the isolation of active constituents such as ginsenosides, glycopeptides, polysaccharides, and volatile oils from PG (Choi, 2008).

However, the quality and the quantity of the metabolites in a ginseng plant depends on age, the season of harvest, and the habitat of the plants; as a result, ginsenosides are distributed thoroughly throughout the various parts of the plant in different concentrations (Song et al., 2019). So, a thorough study of the cultivation and harvesting period of the plant to get the maximum quantity of byproducts (ginsenosides) of high quality is required, which is augmented by using different types of metabolomic studies (Yang et al., 2021). This chapter will focus on the characterization of metabolites and metabolite profiling of PG using LCMS, GC–MS, NMR, and HPLC techniques. Metabolic pathways of ginsenoside biosynthesis also depend on the seasonal environmental condition, which in turn affect the secondary metabolite biosynthesis. Also, we will be discussing the mechanism of action of anticancer agents present in the PG ("Ginseng," 2016).

9.2 CHARACTERIZATION OF SECONDARY METABOLITES ISOLATED FROM PG OVER THE YEARS

Recent studies have unveiled that different parts of the plant, originating from a different geographical region with a variation in cultivation status, gives different metabolome. Studies to widen the knowledge of the variation of metabolite concentration in roots and leaves during their growth are a must for exploring a thorough ginseng study (Lee et al., 2019). Metabolomic study has become an emerging trend in the world of science and technology and has given us the insight to take forward experiments on a deeper and more precise level (Shyur and Yang, 2008). The main technique in metabolomic study is to target analysis and profiling of particular metabolite in a sample in place of a wide metabolomic analysis with the help of different types of techniques such as gas chromatography-mass spectrometry (GC–MS) and liquid chromatography-mass spectrometry (LC-MS). Other methods include thin-layer chromatography (TLC), Fourier transforms infrared spectroscopy (FT-IR), Raman spectroscopy, and nuclear magnetic resonance (NMR) (Shyur and Yang, 2008). MS coupled with high field NMR has proven to be a user-friendly as well an efficient system for the analysis of metabolites. Different coupled techniques with NMR, such as HPLC–SPENMR or MALDI- TOF/ TOF MS help in the characterization of the quantity of secondary metabolites present in crude samples as a result of the raised sensitivity of the techniques used (Shyur and Yang, 2008).

9.2.1 METABOLITE PROFILING AND CHARACTERIZATION OF METABOLITES USING LCMS AND NMR

The use of these techniques helps in distinguishing complex chemical structures isolated from the PG. The characterization of chemical constituents present in *P. ginseng* has been widely studied (Chen et al., 2015). Among the different compounds present, ginsenosides (a triterpene saponin) are major bioactive constituents that have a vital role in the pharmaceutical world. The structure of ginsenosides has already been elucidated by Shibata et al. in 1963 (Shibata et al., 1963; Kim, 2012). They have basic similar structural backbone consisting of a nucleus of dammarane steroid of C17 arranged in rings of four (Zhu et al., 2011).

From the year 2000–2019 about 50 new ginsenosides have been isolated where most of them possessed C17 side chain variation (Piao et al., 2020; Liu and Xiao, 1992; Li et al., 2020). With the help of 2D-NMR, 26 metabolites including ginsenosides Rg1, Rg2s, Rg3s, Rb2, Rb1, Re, and Rf, distinguished from among all the isolated metabolites (Chen et al., 2015).

During a study involving metabolomics profiling of *P. ginseng*, with the help of UPLC-Q-TOF, it was seen that *P. ginseng* can also be identified as mountain or garden cultivated according to their ages (Xie et al., 2008; Zhu et al., 2018). It was found that *P. ginseng* constitutes 93 metabolites having surplus signals from other different isotopes, in-source fragmentation as well as HCOO– adduct (Lee et al., 2018; Wang et al., 2017). In another case study, using GC–MS, ginsenosides were found in different concentrations distributed in leaves, stems, and roots in various stages of growth. Ginsenosides Rb1, Rb2, Rc, Rd, Re, and Rg1 make up about 90% of the total ginsenoside content of *P. ginseng* roots which was reported by a ginseng evaluation program. Metabolomic studies with the help of liquid chromatography (LC), gas chromatography-mass spectrometry (GC–MS), and proton nuclear magnetic resonance (1H NMR) showed that the metabolic analysis of ginseng berries varies with respect to the difference in the stages of fruit ripening as well as the different metabolites found in roots according to the cultivation age (Wei et al., 2020; Yang et al., 2012). 1H NMR was also used in the control of quality of ginseng products as well as for the identification of ginseng origin (Lee et al., 2011; Yuk et al., 2013). Using HR-MAS-NMR, ginseng processed products are screened for different secondary metabolites

(Yoon et al., 2020). Abundant levels of ginsenoside have been reported during preharvest stages in ginseng berries (Lee et al., 2020). Furthermore, 35 ginsenosides were identified using HPLC-APCI/MS method; this method allows to identify thermolabile ginsenosides via thermal degradation of ions in both positive and negative ionization, measuring molecular masses of sugar and aglycone structures of ginsenosides (Ma et al., 2005).

The LCMS technique was also used to analyze the quality of ginseng grown in various regions to improve the quality of metabolites and production of ginsenosides (Wu et al., 2018). In this regard, 60 compounds from the ginseng were evaluated for quality upon cultivation from northeast China. GC–MS/GC–TOF–MS was also used for the detection of the longevity of *P. ginseng* seeds, which in turn helps during the sowing season and also for differentiating metabolite differences according to cultivation years and the stage of ripening of ginseng berries (Cui et al., 2015; Kim et al., 2016; Liu et al., 2017; Min et al., 2019; Park et al., 2019). Studies using HPLC have also contributed to the differentiation of different *Panax* species (Pace et al., 2015). During the investigation, it has been observed that environmental conditions and cultivation technology affected the quality of ginseng roots and rhizomes (In et al., 2017).

A thorough investigation of the many constituents in medicinal plants can be achieved through high-throughput metabolomic studies. Metabolomic studies help in the elucidation of components that are essential for the development of pharmaceutical drugs/products that are of exceptional quality. Moreover, this study also aids in the conservation of rare medicinal plants that have been exhausted due to excessive use by mankind in the ever-changing environmental conditions (Shyur and Yang, 2008).

9.3 BIOSYNTHESIS OF SECONDARY METABOLITES OF PG

The biosynthesis of ginsenosides in *P. ginseng* has been studied widely, as ginsenosides are found to have potent activity from the pharmacological point of view. More importance to ginsenoside synthesis, quantity, and quality has been given due to its high demand (Shin et al., 2015).

9.3.1 BIOSYNTHETIC PATHWAY OF GINSENOSIDES

Ginsenosides are mostly amphipathic with an –OH group closely linked to the backbone, which helps to interact with the phospholipid's membrane polar head and the β-OH group in cholesterol. The polar –OH groups are the determinant factors of the physicochemical interactions between ginsenosides and other neighboring lipids, so the difference in the number and sites of these hydroxyl groups gives diversity to ginsenosides. Ginsenosides are reported to be synthesized from 2,3-oxidosqualene. With the help of enzymes cycloartenol synthase, dammarenediol-II synthase, and β-amyrin synthase acting upon 2,3-oxidosqualene, it forms the different types of ginsenosides (Tansakul et al., 2006). With the help of reactions that involves geranyl diphosphate synthase (GPS), farnesyl pyrophosphate synthase (FPS), and squalene synthase (SS), squalene is synthesized via the mevalonate pathway (Liu et al., 2010). Downstream reactions with squalene epoxidase yielded 2,3-oxidosqualene (Deng et al., 2017; Lee et al., 2019). Enzymes like oxidosqualene cyclases (OSCs), dammarenediol-II synthase (PNA), and beta-amyrin synthase (PNY) helps in the cyclization of 2,3- oxidosqualene into dammarenediol and beta-amyrin (Yun-Soo et al., 2009).

Ginsenosides in *P. ginseng* were mainly synthesized through two main biosynthetic pathways: mevalonate (MVA) pathway in the cytosol and 2-C-methyl-D-erythritol 4-phosphate (MEP) pathway in the plastids of the plant (Figure 9.1). In the MVA pathway, in the cytosol, acetyl CoA is converted into HMGCoA with the help of acetyl CoA transferase, and the latter is further reduced to mevalonic acid (MVA). MVA is then phosphorylated to diphosphomevalonate (MVPP) with the help of phosphomevalonate kinase (PMK), which is then decarboxylated to IPP (Isopentenyl phosphate). Similarly, in the plastids of the cell, G3P is converted to 2-C- methyl-D-erythritol-4-phosphate pathway (MEP), which then forms IPP. IPP being the precursor of ginsenosides, with the help of farnesyl diphosphate synthase, forms squalene. Squalene is converted to 2-oxidosqualene and, with the help of two different enzymes: β-amyrin synthase and DDS, helps in the synthesis of oleanane type ginsenoside as well as PPD and PPT type which are the backbone of different ginsenosides found in *P. ginseng* (Figure 9.2) (Han et al., 2012, 2013; Lee et al., 2017).

9.3.2 BIOSYNTHETIC PATHWAY OF OLEANOLIC ACID

Oleanolic acid was known to be synthesized in *P. ginseng* with the help of β-amyrin 28-oxidase encoded by the CYP716A subfamily gene CYP716A52v2 that helps in modifying β-amyrin and converting into oleanolic acid, whose byproduct was confirmed using GC/MS (Han et al., 2013). It is reported that CYP716A52v2 from *P. ginseng* helps the synthesis of oleanolic acid. The CYP716A52v2 gene present in yeast confirmed the potency to synthesize oleanolic acid, which was further confirmed in vitro. Moreover, it was also found that change in Ro ginsenoside biosynthesis was also regulated by CYP716A52v2 overexpression and RNA interference (RNAi) as shown by studies in transgenic ginseng roots. Experimental results showed that CYP16A52v2 helps in the catalyzation of critically important steps during the synthesis of ginsenoside in *P. ginseng* (Han et al., 2013).

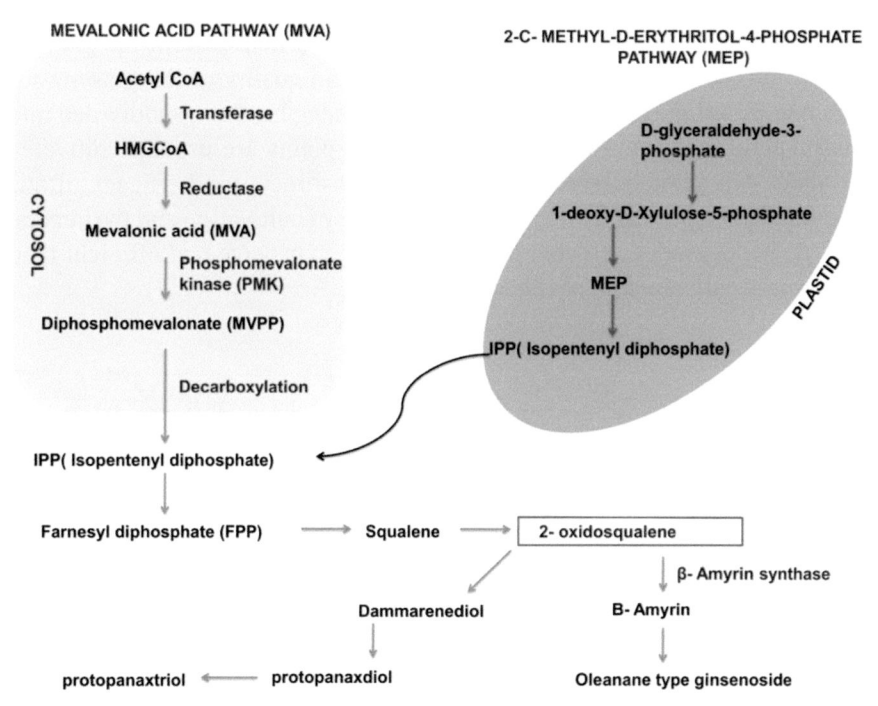

FIGURE 9.1 The biosynthetic pathway of ginsenosides in *P. ginseng* plants where IPP (isopentenyl phosphate), the precursor of ginsenoside, was synthesized from the mevalonate pathway (MVA) and 2-C- methyl-D-erythritol-4-phosphate pathway (MEP), which downstream the synthesis of oleanane type ginsenoside along with protopanaxdiol (PPD) and protopanaxtriol (PPT) type ginsenosides.

Source: Adapted from Lee et al. (2017). https://creativecommons.org/licenses/by-nc/3.0/

9.3.3 SEASONAL AND ENVIRONMENTAL EFFECTS ON THE BIOSYNTHESIS OF SECONDARY METABOLITES IN PG

P. ginseng is widely famous for its myriad health benefits. The harvesting season and time of ginseng has always been the key to its bioactivity since ancient times, and mainly the roots are being used to prepare herbal concoctions (Liu et al., 2017). The main reasons for these particularities have been explained through metabolomics studies of the plant grown in different seasons at different timelines (Yang et al., 2021). The harvesting season, along with the age of the plant, was analyzed using different techniques. Moreover, the berries of the ginseng plant also have a certain amount of ginsenosides that make it a potent medicine. The distribution of ginsenosides across various parts of the plant according to their growing season and age is one of the most interesting studies that can help benefit the growth, production, and maintenance of the quantity and quality of the ginseng for commercial value. The traditional use of *P. ginseng* has been subdivided into hairs, lateral, and main roots. The *P. ginseng* plants are usually cultivated for about 4–6 years before harvesting. The constituents and efficacy of this medicinal plant are highly dependent on the age of cultivation and the harvest season. So, 6-years cultivated ginseng is more expensive and efficient than the 4-years cultivated plant (Shan et al., 2014).

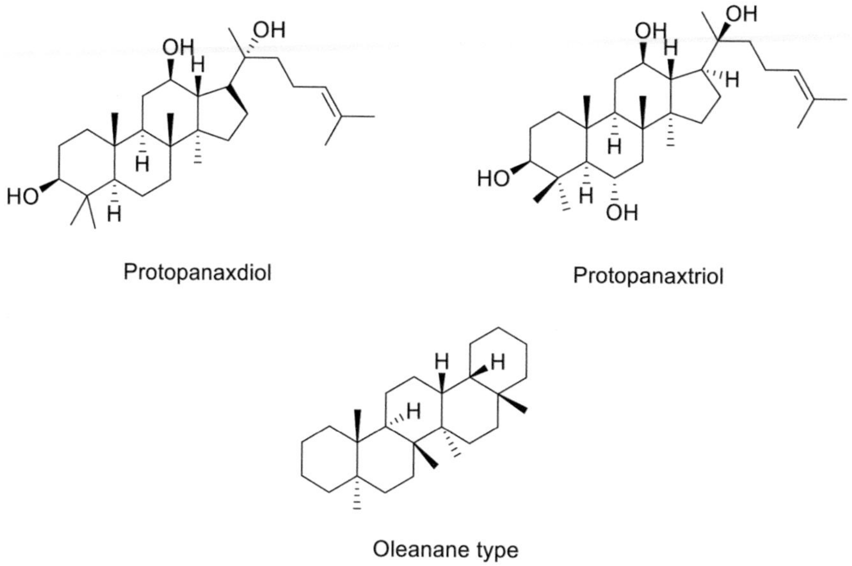

Protopanaxdiol   Protopanaxtriol

Oleanane type

FIGURE 9.2   Different types of ginsenosides (backbone) isolated from *Panax ginseng.*

The quality and quantity of main root ginsenosides were distinguishably changed during the whole growing season and were found to be spiked till the month of May. With the help of orthogonal partial least squares discriminant analysis (OPLS-DA) it was identified that the main roots of ginseng, which were harvested in the month of April, have elevated levels of secondary metabolites but lower levels of glucose and sucrose as compared to those which were harvested in the month of March. Glucose and phenylalanine levels were higher in main ginseng roots that were harvested in June and were found to be decreased as compared to those harvested in May; sucrose levels were significantly elevated in main ginseng roots harvested in June. Lateral ginseng roots have higher contents of ginsenoside compounds, FAs, sterol, and arginine as compared to the main roots, with higher sucrose level as compared to that in lateral roots. It was also found that less exposure time in the sun with a high level of rainfall after May also contributed to the inhibition of ginsenoside synthesis in ginseng roots due to a decline in photosynthesis (Lee et al., 2019).

9.4 ANTICANCER ACTIVITY OF *P. GINSENG* AND ITS CONSTITUENTS

Cancer can be defined as the uncontrolled growth or proliferation of cells defying the natural agenda of contact inhibition as the normal cells usually do. There are two types of cancer: benign, where it is localized at a particular place, and malignant, where the cancer cells get detached and move along the lymph nodes or blood and get attached at another site and establish new tumors (GM, 2000). The chromosomes of cancer cells are highly unstable and prone to mutations like deletions and duplications and have more nucleus content as compared to normal cells (US, 2007). Cancer, as studies showed, arises due to both environmental factors as well as of genetic origin. With the progress in evolution, the genetic pool also acquired different types of recessive or dominant mutations. In recent findings, plant-based anticancer drugs have gained considerable attention worldwide. Since ancient times, plants have always been a source of medicine for different human ailments, and many cancer drugs commercially available today have been derived from plant products (Greenwell and Rahman, 2015; Rahman et al., 2020). Currently used anticancer drugs, although effective to some point, have their own share of side effects, and most of the cancer drugs are not effective when it comes to higher disease progression. The drugs circulate and not only target the cancer cells but also cause effects on other parts of the body like rapidly dividing normal cells, including hair follicles and gastrointestinal cells (Iqbal et al., 2017). In another different case, there arise drug-resistant cancer cells where due to a high rate of mutation, the cancer cells develop its own immunity against a particular drug. So, the single-target drug has also become ineffective during such conditions and requires a combination of drugs (Iqbal et al., 2017; Rahman et al., 2020). Plant-based medicines have been circulating on the table as they help in providing multiple-target drugs with lesser side effects (Greenwell and Rahman, 2015; Iqbal et al., 2017; Rahman et al., 2020).

One of the most famous anticancer plants is *Panax ginseng* or Asian ginseng (Kamei et al., 2000). Since ancient times, this plant has been used as a cure for different types of cancer, and studies revealed that ginsenosides are the main active ingredient that possesses anticancer activity apart from the polysaccharides, flavanols, etc. (Ahuja et al., 2018; Alyaa, 2019; Kamei et al., 2000; Park, 2019; Shareef et al., 2016). The activity of ginsenoside against cancer, as already stated, differs with respect to the amount of sugar units present; so, lesser the sugar moieties higher the anticancer activity. PPD and PPT type (no sugar moieties) shows the highest activity in different types of cancer (Nag et al., 2012). Ginsenosides play multiple roles in inhibiting cancer growth following different pathways. The common pathway includes inhibition of metastasis, cell cycle arrest, inhibition of angiogenesis, and apoptosis (Yue et al., 2007; Ahuja et al., 2018; Chen et al., 2020) (Figure 9.3).

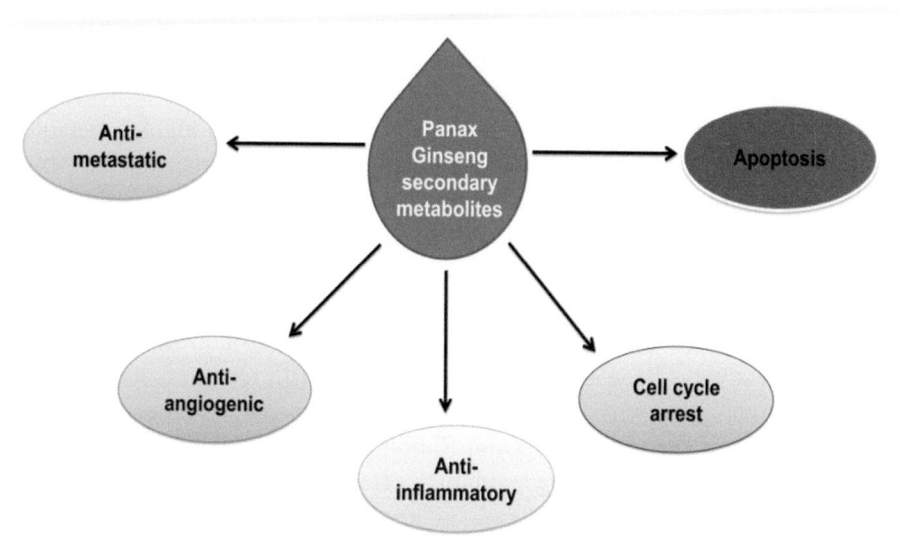

FIGURE 9.3   Anticancer properties of *Panax ginseng*.

Source: Adapted from Nag et al. (2012). https://creativecommons.org/licenses/by-nc/3.0/

The different types of ginsenosides found in *P. ginseng* follow different signaling pathways that help in the inhibition of cancer progression. It helps in the activation of both intrinsic (through a mitochondrial pathway involving the formation of apoptosomes) as well as the extrinsic apop-totic pathways through caspase activation (Chen et al., 2018). They also act as antimetastatic as well as antiangiogenic through NF-κβ inhibition, which helps in the downregulation of vascular endothelial growth factor (VEGF) as well as fibroblast growth factor (FGF), which helps promote angiogenesis (Chen et al., 2020; Chen et al., 2018; Kashfi, 2009; Lee et al., 2019) (Figure 9.5). Ginsenosides also inhibit cancer cell growth via cell cycle arrest by activation of tumor suppressor genes like p53, p21, and p27 (Figure 9.4) which in turn helps in the activation of cyclin proteins that interrupts cell cycle progression and stops further growth and division of cells (Chen et al., 2018, 2020).

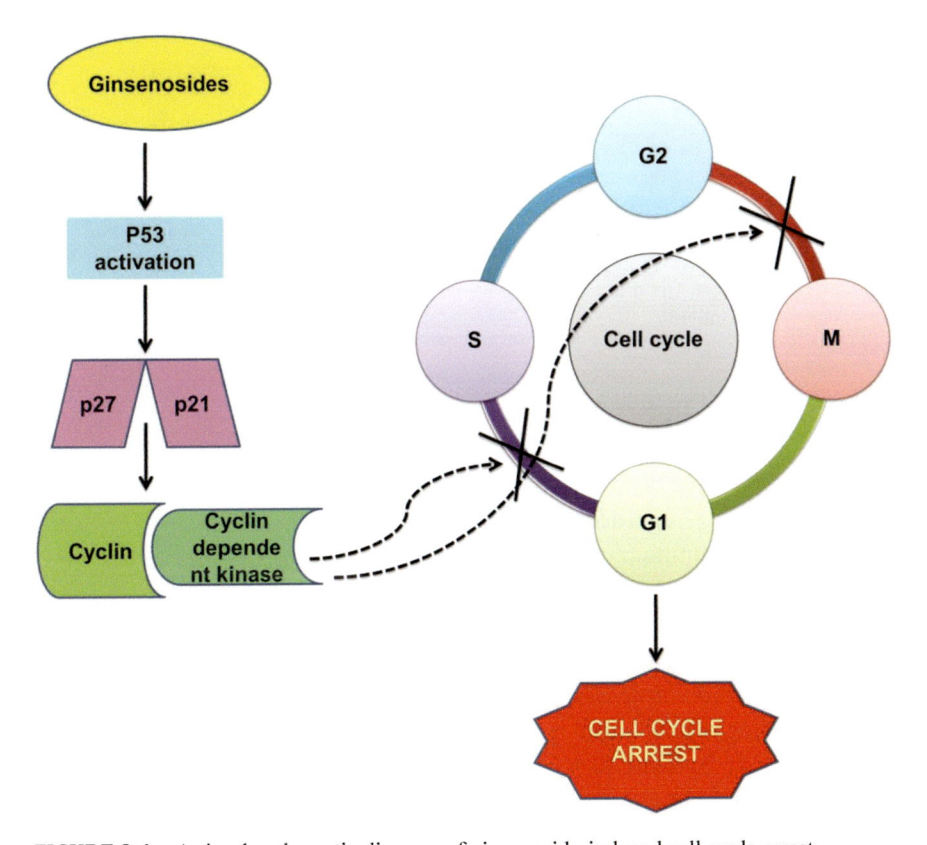

FIGURE 9.4   A simple schematic diagram of ginsenoside-induced cell cycle arrest.

Source: Reprinted from Mohanan et al. (2018). https://creativecommons.org/licenses/by-nc-nd/4.0/

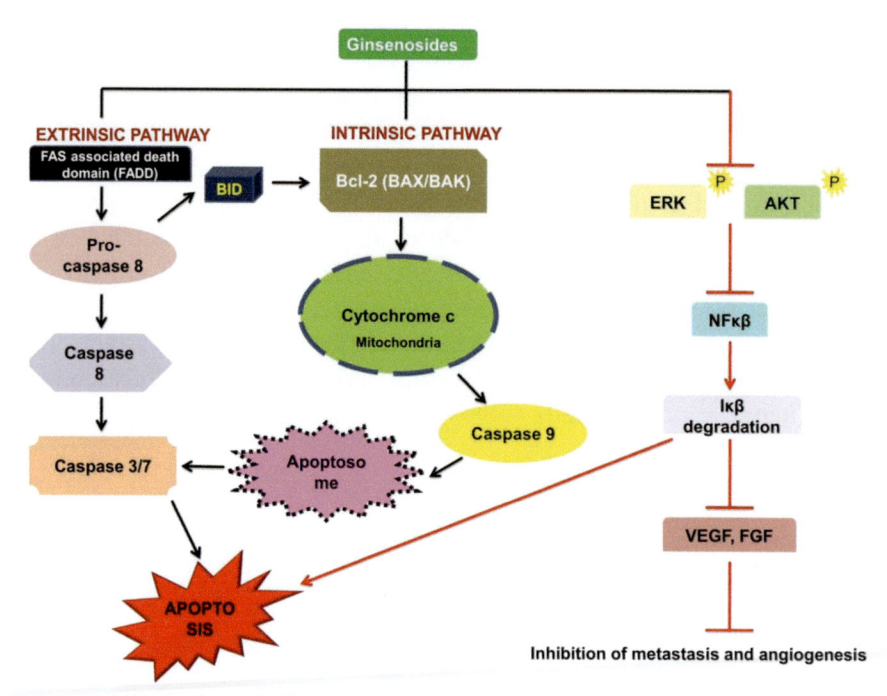

FIGURE 9.5   A schematic diagram of anticancer activity signaling pathway of ginsenosides.

Source: Adapted from Chen et al. (2018); Mohanan et al. (2018). https://creativecommons.org/licenses/by-nc-nd/4.0/

9.5   CONCLUSIONS

Several studies have been carried out using *P. ginseng* both in vitro as well as in vivo. *P. ginseng*, regarded as the "all healing herb," has lived up to its name as many of the compounds from this plant have different roles in different types of diseases (Park et al., 2018). Due to high demand in economic market, the ginseng plant has been regarded as a rare plant, and conservation of this plant has also been a major challenge among ginseng experts. By using high throughput metabolomic devices and software a thorough screening of the plant's growing season, age, type of cultivation, habitat as well as mode of biosynthesis of its secondary metabolites have been closely monitored to get a higher yield of desired plant products, which are of high quality and quantity. The origin and the quality of the plant itself were also chosen to benefit the evergrowing agroeconomic conditions and ensure a high quality adulteration-free product.

TABLE 9.1 The Antitumor/Anticancer Activity of *P. ginseng* and Its Constituents.

Sl.no	Extract/compound	Antitumor activity and signaling pathway	References
1	Protopanaxdiol	Inhibited growth of human lung cancer cells A549 and H460, downregulate levels of β-catenin, cyclin D1, CDK4, and c-myc	Bi et al. (2014)
2	(20R)-12b-O- (L-chloracetyl)-dammarane-3b, 20, 25-triol (xl), (20R)-3b-O- (L-alanyl)-dammarane-12b, 20, 25-triol (1c), and (20R)-3b-O- (Boc-L arginyl)-dammarane-12b, 20, 25-triol (8b)	Decrease growth of cancer cells, induces apoptosis through caspase signaling	Xia et al. (2014)
3	a1-a4, b1-b4 (first class, a5, a6 b5, and b6 (second class), a7, a8, and b7 (third class)	a5, a7, b5, and b7 show anticancer activity	Qu et al. (2017)
4	Water-soluble ginseng oligosaccharides (WGOS)	Hinder tumor formation, helps in TNF-production, and stimulates TNF-α production	Jiao et al. (2014)
5	Ginsenoside Rk1	Inhibits telomerase activities, cell growth, and proliferation, and induces apoptosis through executioner caspases	Kim et al. (2008)
6	Ginsenoside Ro (Ro)	Helps in suppression of tumor growth	Zheng et al. (2019)
7	Panaxadiol from roots of *Panax ginseng*	Helps in suppression of hypoxia-inducible factor (HIF)-1α and regulates the expression of phosphoinositide 3-kinase (PI3K) and mitogen-activated protein kinase (MAPK). It also shows antiproliferative activity.	Wang et al. (2020)
8	Neoginseng	Inhibits tumor progression due to its anti-inflammatory as well as antioxidative activity	Keum et al. (2000)
9	Ginsenoside Rg3	Suppresses cancer cell proliferation through inhibition of C/EBPβ/NF-κB signaling	Yang et al. (2017)

TABLE 9.1 *(Continued)*

194

Sl.no	Extract/compound	Antitumor activity and signaling pathway	References
10	Ethanolic extract of Asian ginseng, EAG (*Panax ginseng*)	Suppresses tumor growth along with downregulation of PCNA proliferative marker, shows cancer cell cytotoxicity, induces MAPK and p53 signaling through suppression of cyclin B–cdc2 complex, and induces G2–M arrest leading to apoptosis	Wong et al. (2010)
11	Ginseng polysaccharide (GPS)	Antiproliferative and induce apoptotic pathway, transcription of p38 and JNK were elevated while that of ERK was downregulated while it suppressed phosphorylation of ERK, NF-κB, and cyclin D1	Xiong et al. (2017); Zhao et al. (2019)
12	Ginsenoside Rg3	Antitumor activity	Keum et al. (2003)
13	Ginsenoside Rg1	Inhibited JAK2/STAT5 pathway activity, helps in caspase-3 and C-PAPR expression, it also shows anticancer activity in drug-resistant nasopharyngeal cancer cells via activation of autophagy, cell cycle arrest at S phase, and inhibition of PI3K/AKT pathway	Li et al. (2014); Li et al. (2019)
14	compound K (CK),	Reduces the metastatic growth by reducing colony formation, adhesion, and invasion, decreases nuclear NF-κB p65 and increases cytosolic NF-κB p65 suppressing the NF-κB pathway	Ming et al. (2011)
15	Ginsenoside Rg3	Synergistic effect is shown by the combination of cisplatin and Rg3 through alteration of G2/M phase and intrinsic apoptotic pathway.	Lee et al. (2014)
16	Ginsenoside Rb2, 20(R) and 20(S)	Shows inhibition of metastasis of lung cancer cells by inhibiting cell adhesion and invasion along with antiangiogenic effect	Mochizuki et al. (1995)

TABLE 9.1 *(Continued)*

Sl.no	Extract/compound	Antitumor activity and signaling pathway	References
17	Ginsenoside Rb1	Shows antiangiogenic effect via suppression of PEDF	Leung et al. (2007)
18	Compound K	Induced apoptosis through caspases 8, 9, and 3, and of Bid protein	Cho et al. (2009)
19	Ginsenoside Rh2	Induces apoptosis in nephrotoxicity via caspase mediated pathway	Qi et al. (2019)
20	Ginsenoside Rd	Shows antitumor activity through G0/G1 phase	Tian (2020)
21	Ginseng proteins	Shows antitumor activity and induces intrinsic apoptotic pathway	Wan et al. (2019)
22	Ginsenoside Rg5	Helps in the inhibition of cancer cell proliferation via intrinsic apoptotic pathway and induction of autophagy	Liu and Fan (2020)

KEYWORDS

- *Panax ginseng*
- **herbal medicines**
- **pharmacological uses**
- **metabolites**
- **active compounds**
- **metabolomics**
- **antitumor activity**

REFERENCES

(US), N. I. o. H. *Understanding Cancer*, 2007.

Ahuja, A.; Kim, J. H.; Kim, J.-H.; Yi, Y.-S.; Cho, J. Y. Functional role of Ginseng-Derived Compounds in Cancer. *J. Ginseng Res.;* **2018**,*42* (3), 248–254.

Alyaa, M. Panax ginseng—A review. *Univ. Thi-Qar J. Sci.;* **2019**, *7* (1).

Bi, X.; Xia, X.; Mou, T.; Jiang, B.; Fan, D.; Wang, P.; Liu, Y.; Hou, Y.; Zhao, Y. Anti-Tumor Activity of Three Ginsenoside Derivatives in Lung Cancer Is Associated with Wnt/β-Catenin Signaling Inhibition. *Eur. J. Pharmacol.;* **2014**, *742*, 145–152.

Chang, Y. S.; Seo, E. K.; Gyllenhaal, C.; Block, K. I. Panax Ginseng: A Role in Cancer Therapy? *Integr Cancer Ther*, **2003**, *2* (1), 13–33.

Chen, H.; Yang, H.; Fan, D.; Deng, J. The Anticancer Activity and Mechanisms of Ginsenosides: An Updated Review. *eFood*, **2020**, *1*.

Chen, T.; Li, B.; Qiu, Y.; Qiu, Z.; Qu, P. Functional Mechanism of Ginsenosides on Tumor Growth and Metastasis. *Saudi J. Biol. Sci.* **2018**, *25* (5), 917–922.

Chen, X.-p.; Lin, Y.-p.; Hu, Y.-z.; Liu, C.-x.; Lan, K.; Jia, W.; Phytochemistry, Metabolism, and Metabolomics of Ginseng. *Chinese Herbal Med.* **2015**, *7* (2), 98–108.

Cho, S. H.; Chung, K. S.; Choi, J. H.; Kim, D. H.; Lee, K. T.; Compound K, a Metabolite of Ginseng Saponin, Induces Apoptosis via Caspase-8-Dependent Pathway in HL-60 Human Leukemia Cells. *BMC Cancer* **2009**, *9*, 449.

Choi, K. T. Botanical Characteristics, Pharmacological Effects and Medicinal Components of Korean *Panax ginseng* C A Meyer. *Acta Pharmacol. Sin* **2008**, *29* (9), 1109–1118.

Coates, P. B.; Blackman, M.; Crag, G. *Asian Ginseng, (Panax ginseng) in Encyclopedia of Dietary Supplements*, 2005.

Court, W. E. Ginseng, the Genus Panax, 2000.

Cui, S.; Wang, J.; Yang, L.; Wu, J.; Wang, X. Qualitative and Quantitative Analysis on Aroma Characteristics of Ginseng at Different Ages Using E-Nose and GC-MS Combined with Chemometrics. *J. Pharm. Biomed. Anal.* **2015**, *102*, 64–77.

Deng, B.; Zhang, P.; Ge, F.; Liu, D.-Q.; Chen, C.-Y. Enhancement of Triterpenoid Saponins Biosynthesis in Panax Notoginseng Cells by Co-Overexpressions of 3-Hydroxy-3-Methyl-glutaryl CoA Reductase and Squalene Synthase Genes. *Biochem. Eng. J.* **2017**, *122*, 38–46.

Ellis, J. M.; Reddy, P. Effects of Panax Ginseng on Quality of Life. *Ann. Pharmacother.* **2002**, *36* (3), 375–379.

GM, C. *The Cell: A Molecular Approach*, 2000.

Greenwell, M.; Rahman, P. K. S. M. Medicinal Plants: Their Use in Anticancer Treatment. *Int. J. Pharma. Sci. Res.* **2015**, *6* (10), 4103–4112.

Han, J. Y.; Hwang, H. S.; Choi, S. W.; Kim, H. J.; Choi, Y. E. Cytochrome P450 CYP716A53v2 Catalyzes the Formation of Protopanaxatriol from Protopanaxadiol During Ginsenoside Biosynthesis in *Panax ginseng*. *Plant Cell Physiol.* **2012**,*53* (9), 1535–1545.

Han, J. Y.; Kim, M. J.; Ban, Y. W.; Hwang, H. S.; Choi, Y. E. The Involvement of β-amyrin 28-Oxidase (CYP716A52v2) in Oleanane-Type Ginsenoside Biosynthesis in *Panax ginseng*. *Plant Cell Physiol.* **2013**,*54* (12), 2034–2046.

Hu, S. Y. A Contribution to Our Knowledge of Ginseng. *Am. J. Chin. Med. (Gard City N Y)*, **1977**, *5* (1), 1–23.

In, G.; Seo, H. K.; Park, H.-W.; Jang, K. H. A Metabolomic Approach for the Discrimination of Red Ginseng Root Parts and Targeted Validation. *Molecules* **2017**, *22* (3), 471.

Iqbal, J.; Abbasi, B. A.; Mahmood, T.; Kanwal, S.; Ali, B.; Shah, S. A.; Khalil, A. T.; Plant-Derived Anticancer Agents: A Green Anticancer Approach. *Asian Pac. J. Trop. Biomed.* **2017**, *7* (12), 1129–1150.

Jiao, L.; Zhang, X.; Li, B.; Liu, Z.; Wang, M.; Liu, S. Anti-Tumour and Immunomodulatory Activities of Oligosaccharides Isolated from *Panax ginseng* C. A. Meyer. *Int. J. Biol. Macromol.* **2014**, *65*, 229–233.

Kamei, T.; Kumano, H.; Iwata, K.; Nariai, Y.; Matsumoto, T. The Effect of a Traditional Chinese Prescription for a Case of Lung Carcinoma. *J. Altern. Complement. Med.* **2000**, *6* (6), 557–559.

Kashfi, K. Anti-Inflammatory Agents as Cancer Therapeutics. In Enna, S. J., Williams, M., Eds.; *Adv. Pharmacol.* **2009**, *57*, 31–89.

Keum, Y. S.; Han, S. S.; Chun, K. S.; Park, K. K.; Park, J. H.; Lee, S. K.; Surh, Y. J. Inhibitory Effects of the Ginsenoside Rg3 on Phorbol Ester-Induced Cyclooxygenase-2 Expression, NF-Kappa B Activation and Tumor Promotion. *Mutat. Res.* **2003**, *523–524*, 75–85.

Keum, Y.-S.; Park, K.-K.; Lee, J.-M.; Chun, K.-S.; Park, J. H.; Lee, S. K.; Kwon, H.; Surh, Y.-J. Antioxidant and Anti-Tumor Promoting Activities of the Methanol Extract of Heat-Processed Ginseng. *Cancer Lett.* **2000**, *150* (1), 41–48.

Kiefer, D.; Pantuso, T. *Panax ginseng. Am. Fam. Phys.* **2003**, *68* (8), 1539–1542.

Kim, D.-H. Chemical Diversity of *Panax ginseng*, *Panax quinquifolium*, and *Panax notoginseng*. *J. Ginseng Res.* **2012**, *36* (1), 1–15.

Kim, S. W.; Gupta, R.; Lee, S. H.; Min, C. W.; Agrawal, G. K.; Rakwal, R.; Kim, J. B.; Jo, I. H.; Park, S. Y.; Kim, J. K.; Kim, Y. C.; Bang, K. H.; Kim, S. T. An Integrated Biochemical, Proteomics, and Metabolomics Approach for Supporting Medicinal Value of Panax ginseng Fruits. *Front. Plant Sci.* **2016**, *7*, 994.

Kim, W. K.; Song, S. Y.; Oh, W. K.; Kaewsuwan, S.; Tran, T. L.; Kim, W. S.; Sung, J. H. Wound-Healing Effect of Ginsenoside Rd from Leaves of *Panax ginseng* via Cyclic AMP-Dependent Protein Kinase Pathway. *Eur. J. Pharmacol.* **2013**, *702* (1–3), 285–293.

Kim, Y. J.; Kwon, H. C.; Ko, H.; Park, J. H.; Kim, H. Y.; Yoo, J. H.; Yang, H. O. Anti-Tumor Activity of the Ginsenoside Rk1 in Human Hepatocellular Carcinoma Cells Through Inhibition of Telomerase Activity and Induction of Apoptosis. *Biol. Pharm. Bull.* **2008**, *31* (5), 826–830.

Lee, A. R.; Gautam, M.; Kim, J.; Shin, W. J.; Choi, M. S.; Bong, Y. S.; Hwang, G. S.; Lee, K. S. A Multianalytical Approach for Determining the Geographical Origin of Ginseng Using

Strontium Isotopes, Multielements, and 1H NMR Analysis. *J. Agric. Food Chem.* **2011**, *59* (16), 8560–8567.

Lee, D.-Y.; Cha, B.-J.; Lee, Y.-S.; Kim, G.-S.; Noh, H.-J.; Kim, S.-Y.; Kang, H. C.; Kim, J. H.; Baek, N.-I. The Potential of Minor Ginsenosides Isolated from the Leaves of Panax ginseng as Inhibitors of Melanogenesis. *Int. J. Mol. Sci.* **2015**, *16* (1).

Lee, D.-Y.; Park, C. W.; Lee, S. J.; Park, H.-R.; Kim, S. H.; Son, S.-U.; Park, J.; Shin, K.-S. Anti-Cancer Effects of Panax ginseng Berry Polysaccharides via Activation of Immune-Related Cells [Original Research]. *Front. Pharmacol.* **2019**, *10* (1411).

Lee, H.-J.; Jeong, J.; Alves, A. C.; Han, S.-T.; In, G.; Kim, E.-H.; Jeong, W.-S.; Hong, Y.-S. Metabolomic Understanding of Intrinsic Physiology in *Panax ginseng* During Whole Growing Seasons. *J. Ginseng Res.* **2019**, *43* (4), 654–665.

Lee, J. W.; Ji, S.-H.; Choi, B.-R.; Choi, D. J.; Lee, Y.-G.; Kim, H.-G.; Kim, G.-S.; Kim, K.; Lee, Y.-H.; Baek, N.-I.; Lee, D. Y. UPLC-QTOF/MS-Based Metabolomics Applied for the Quality Evaluation of Four Processed Panax ginseng Products. *Molecules (Basel, Switzerland)* **2018**, *23* (8), 2062.

Lee, M. Y.; Seo, H. S.; Singh, D.; Lee, S. J.; Lee, C. H. Unraveling Dynamic Metabolomes Underlying Different Maturation Stages of Berries Harvested from *Panax ginseng. J. Ginseng Res.* **2020**,*44* (3), 413–423.

Lee, Y. J.; Lee, S.; Ho, J. N.; Byun, S. S.; Hong, S. K.; Lee, S. E.; Lee, E. Synergistic Antitumor Effect of Ginsenoside Rg3 and Cisplatin in Cisplatin-Resistant Bladder Tumor Cell Line. *Oncol. Rep.* **2014**, *32* (5), 1803–1808.

Lee, Y. S.; Park, H.-S.; Lee, D.-K.; Jayakodi, M.; Kim, N.-H.; Koo, H. J.; Lee, S.-C.; Kim, Y. J.; Kwon, S. W.; Yang, T.-J. Integrated Transcriptomic and Metabolomic Analysis of Five Panax ginseng Cultivars Reveals the Dynamics of Ginsenoside Biosynthesis [Original Research]. *Front. Plant Sci.* **2017**, *8*, 1048.

Leung, K. W.; Cheung, L. W. T.; Pon, Y. L.; Wong, R. N. S.; Mak, N. K.; Fan, T. P. D.; Au, S. C. L.; Tombran-Tink, J.; Wong, A. S. T. Ginsenoside Rb1 Inhibits Tube-Like Structure Formation of Endothelial Cells by Regulating Pigment Epithelium-Derived Factor Through the Oestrogen Beta Receptor. *Br. J. Pharmacol.* **2007**, *152* (2), 207–215.

Li, J.; Wei, Q.; Zuo, G. W.; Xia, J.; You, Z. M.; Li, C. L.; Chen, D. L. Ginsenoside Rg1 Induces Apoptosis Through Inhibition of the EpoR-Mediated JAK2/STAT5 Signalling Pathway in the TF-1/Epo Human Leukemia Cell Line. *Asian Pac. J. Cancer Prev.* **2014**, *15* (6), 2453–2459.

Li, K.-K.; Li, S.-S.; Xu, F.; Gong, X.-J. Six New Dammarane-Type Triterpene Saponins from *Panax ginseng* Flower Buds and Their Cytotoxicity. *J. Ginseng Res.* **2020**, *44* (2), 215–221.

Li, W.; Li, G.; She, W.; Hu, X.; Wu, X. Targeted Antitumor Activity of Ginsenoside (Rg1) in Paclitaxel-Resistant Human Nasopharyngeal Cancer Cells Are Mediated Through Activation of Autophagic Cell Death, Cell Apoptosis, Endogenous ROS Production, S Phase Cell Cycle Arrest and Inhibition of m-TOR/PI3K/AKT Signalling Pathway. *J. Buon.* **2019**, *24* (5), 2056–2061.

Liu, C. X.; Xiao, P. G. Recent Advances on Ginseng Research in China. *J. Ethnopharmacol.* **1992**, *36* (1), 27–38.

Liu, J.; Liu, Y.; Wang, Y.; Abozeid, A.; Zu, Y. G.; Zhang, X. N.; Tang, Z. H.; GC-MS Metabolomic Analysis to Reveal the Metabolites and Biological Pathways Involved in the Developmental Stages and Tissue Response of Panax ginseng. *Molecules* **2017**, *22* (3).

Liu, L.; Zhu, X. M.; Wang, Q. J.; Zhang, D. L.; Fang, Z. M.; Wang, C. Y.; Wang, Z.; Sun, B. S.; Wu, H.; Sung, C. K. Enzymatic Preparation of 20(S, R)-Protopanaxadiol by Transformation of 20(S, R)-Rg3 from Black Ginseng. *Phytochemistry* **2010**, *71* (13), 1514–1520.

Liu, Y.; Fan, D.; The Preparation of Ginsenoside Rg5, Its Antitumor Activity against Breast Cancer Cells and Its Targeting of PI3K. *Nutrients* **2020**, *12* (1), 246.

Liu, Z.; Wang, C.-Z.; Zhu, X.-Y.; Wan, J.-Y.; Zhang, J.; Li, W.; Ruan, C.-C.; Yuan, C.-S.; Dynamic Changes in Neutral and Acidic Ginsenosides with Different Cultivation Ages and Harvest Seasons: Identification of Chemical Characteristics for Panax ginseng Quality Control. *Molecules (Basel, Switzerland)* **2017**, *22* (5), 734.

Ma, X. Q.; Liang, X. M.; Xu, Q.; Zhang, X. Z.; Xiao, H. B.; Identification of Ginsenosides in Roots of *Panax ginseng* by HPLC-APCI/MS. *Phytochem. Anal.* **2005**, *16* (3), 181–187.

Min, J. E.; Hong, J. Y.; Kwon, S. W.; Park, J. H. Integrated Metabolomics Signature for Assessing the Longevity of *Panax ginseng* Seeds. *J. Sci. Food Agric.* **2019**, *99* (13), 6089–6096.

Ming, Y.; Chen, Z.; Chen, L.; Lin, D.; Tong, Q.; Zheng, Z.; Song, G. Ginsenoside Compound K Attenuates Metastatic Growth of Hepatocellular Carcinoma, Which Is Associated with the Translocation of Nuclear Factor-κB p65 and Reduction of Matrix Metalloproteinase-2/9. *Planta Med.* **2011**, *77* (5), 428–433.

Mochizuki, M.; Yoo, Y. C.; Matsuzawa, K.; Sato, K.; Saiki, I.; Tono-oka, S.; Samukawa, K.; Azuma, I. Inhibitory Effect of Tumor Metastasis in Mice by Saponins, Ginsenoside-Rb2, 20 (R)- and 20(S)-Ginsenoside-Rg3, of Red Ginseng. *Biol. Pharm. Bull.* **1995**, *18* (9), 1197–1202.

Mohanan, P.; Subramaniyam, S.; Mathiyalagan, R.; Yang, D.-C. Molecular Signaling of Ginsenosides Rb1, Rg1, and Rg3 and Their Mode of Actions. *J. Ginseng Res.* **2018**, *42* (2), 123–132.

Nag, S. A.; Qin, J.-J.; Wang, W.; Wang, M.-H.; Wang, H.; Zhang, R. Ginsenosides as Anticancer Agents: In Vitro and In Vivo Activities, Structure-Activity Relationships, and Molecular Mechanisms of Action. *Front. Pharmacol.* **2012**, *3*, 25.

Pace, R.; Martinelli, E. M.; Sardone, N.; E, D. E. C. Metabolomic Evaluation of Ginsenosides Distribution in Panax Genus (*Panax ginseng* and *Panax quinquefolius*) Using Multivariate Statistical Analysis. *Fitoterapia* **2015**, *101*, 80–91.

Park, J. D. Antitumor Activities of Korean Ginseng as a Health Food Based Upon Underlying Mechanisms of Ginsenosides. In *Chemistry of Korean Foods and Beverages*; *Am. Chem. Soc.* **2019**, *1303*, 149–168.

Park, S.-E.; Seo, S.-H.; Kim, E.-J.; Park, D.-H.; Park, K.-M.; Cho, S.-S.; Son, H.-S. Metabolomic Approach for Discrimination of Cultivation Age and Ripening Stage in Ginseng Berry Using Gas Chromatography-Mass Spectrometry. *Molecules (Basel, Switzerland)*, **2019**, *24* (21), 3837.

Park, S.-Y.; Park, J.-H.; Kim, H.-S.; Lee, C.-Y.; Lee, H.-J.; Kang, K. S.; Kim, C.-E.; Systems-Level Mechanisms of Action of *Panax ginseng*: A Network Pharmacological Approach. *J. Ginseng Res.* **2018**, *42* (1), 98–106.

Piao, X.; Zhang, H.; Kang, J. P.; Yang, D. U.; Li, Y.; Pang, S.; Jin, Y.; Yang, D. C.; Wang, Y. Advances in Saponin Diversity of *Panax ginseng*. *Molecules (Basel, Switzerland)*, **2020**, *25* (15), 3452.

Qi, Z.; Li, W.; Tan, J.; Wang, C.; Lin, H.; Zhou, B.; Liu, J.; Li, P. Effect of ginsenoside Rh(2) on Renal Apoptosis in Cisplatin-Induced Nephrotoxicity In Vivo. *Phytomedicine* **2019**, *61*, 152862.

Qu, F.-Z.; Zhao, C.; Cao, J.-Q.; Zhang, Y.; Zhao, Y.-Q. One-Pot Synthesis, Anti-Tumor Evaluation and Structure-Activity Relationships of Novel 25-OCH(3)-PPD Derivatives. *MedChemCommun.* **2017**, *8* (9), 1845–1849.

Rahman, M. A.; Bulbul, M. R. H.; Kabir, Y. 13—Plant-Based Products in Cancer Prevention and Treatment. In Kabir, Y., Ed.; *Functional Foods in Cancer Prevention and Therapy*; 2020; pp 237–259.

Shan, S. M.; Luo, J. G.; Huang, F.; Kong, L. Y. Chemical Characteristics Combined with Bioactivity for Comprehensive Evaluation of *Panax ginseng* C. A. Meyer in Different Ages and Seasons Based on HPLC-DAD and Chemometric Methods. *J. Pharm. Biomed. Anal.* **2014**, *89*, 76–82.

Shareef, M.; Ashraf, M. A.; Sarfraz, M. Natural Cures for Breast Cancer Treatment. *Saudi Pharma. J.* **2016**, *24* (3), 233–240.

Shibata, S.; Tanaka, O.; Sado, M.; Tsushima, S. On Genuine Sapogenin of Ginseng. *Tetrahedron Lett.* **1963**, *4* (12), 795–800.

Shin, B. K.; Kwon, S. W.; Park, J. H. Chemical Diversity of Ginseng Saponins from *Panax ginseng*. *J. Ginseng Res.* **2015**, *39* (4), 287–298.

Shyur, L.-F.; Yang, N.-S. Metabolomics for Phytomedicine Research and Drug Development. *Curr. Opin. Chem. Biol.* **2008**, *12* (1), 66–71.

Song, S. Y.; Park, D. H.; Seo, S. W.; Park, K. M.; Bae, C. S.; Son, H. S.; Kim, H. G.; Lee, J. H.; Yoon, G.; Shim, J. H.; Im, E.; Rhee, S. H.; Yoon, I. S.; Cho, S. S. Effects of Harvest Time on Phytochemical Constituents and Biological Activities of Panax ginseng Berry Extracts. *Molecules* **2019**, *24* (18).

Takagi, K.; Saito, H.; Nabata, H. Pharmacological Studies of Panax Ginseng Root: Estimation of Pharmacological Actions of Panax Ginseng Root. *Jpn. J. Pharmacol.* **1972**, *22* (2), 245–259.

Tansakul, P.; Shibuya, M.; Kushiro, T.; Ebizuka, Y. Dammarenediol-II Synthase, the First Dedicated Enzyme for Ginsenoside Biosynthesis, in *Panax ginseng*. *FEBS Lett.* **2006**, *580* (22), 5143–5149.

Tian, Y.-Z.; Liu, Ya-Peng, Tian, San-Chun, Ge, Su-Yin, Wu, Yong-Jie, Zhang, Bao-Lai Antitumor Activity of Ginsenoside Rd in Gastric Cancer via Up-Regulation of Caspase-3 and Caspase-9. *Int. J. Pharma. Sci.* **2020**, *75* (4), 147–150.

Wan, X.; Jin, X.; Ren, Y.; Xiu, Y.; Li, Y.; Chen, C.; Liu, S. Antitumor Effects and Mechanism of protein from *Panax ginseng* C. A. Meyer on Human Breast Cancer Cell Line MCF-7 [Original Article]. *Pharmacogn. Magaz.* **2019**, *15* (65), 715–721.

Wang, H.-p.; Liu, Y.; Chen, C.; Xiao, H.-b. Screening Specific Biomarkers of Herbs Using a Metabolomics Approach: A Case Study of *Panax ginseng*. *Sci. Rep.* **2017**, *7* (1), 4609.

Wang, Z.; Li, M. Y.; Zhang, Z. H.; Zuo, H. X.; Wang, J. Y.; Xing, Y.; Ri, M.; Jin, H. L.; Jin, C. H.; Xu, G. H.; Piao, L. X.; Jiang, C. G.; Ma, J.; Jin, X. Panaxadiol Inhibits Programmed Cell Death-Ligand 1 Expression and Tumour Proliferation via Hypoxia-Inducible Factor (HIF)-1α and STAT3 in Human Colon Cancer Cells. *Pharmacol. Res.* **2020**, *155*, 104727.

Wei, G.; Yang, F.; Wei, F.; Zhang, L.; Gao, Y.; Qian, J.; Chen, Z.; Jia, Z.; Wang, Y.; Su, H.; Dong, L.; Xu, J.; Chen, S. Metabolomes and Transcriptomes Revealed the Saponin Distribution in Root Tissues of *Panax quinquefolius* and *Panax notoginseng*. *J. Ginseng Res.* **2020**, *44* (6), 757–769.

Wong, V. K.; Cheung, S. S.; Li, T.; Jiang, Z. H.; Wang, J. R.; Dong, H.; Yi, X. Q.; Zhou, H.; Liu, L. Asian GINSENG EXTRACT INHIBITS In Vitro and In Vivo Growth of Mouse Lewis Lung Carcinoma via Modulation of ERK-p53 and NF-κB Signaling. *J. Cell Biochem.* **2010**, *111* (4), 899–910.

Wu, W.; Jiao, C.; Li, H.; Ma, Y.; Jiao, L.; Liu, S. LC-MS Based Metabolic and Metabonomic Studies of *Panax ginseng*. *Phytochem. Anal.* **2018**, *29* (4), 331–340.

Xia, X.; Jiang, B.; Liu, W.; Wang, P.; Mou, Y.; Liu, Y.; Zhao, Y.; Bi, X. Anti-Tumor Activity of Three Novel Derivatives of Ginsenoside on Colorectal Cancer Cells. *Steroids* **2014**, *80*, 24–29.

Xie, G.; Plumb, R.; Su, M.; Xu, Z.; Zhao, A.; Qiu, M.; Long, X.; Liu, Z.; Jia, W. Ultra-Performance LC/TOF MS Analysis of Medicinal Panax Herbs for Metabolomic Research. *J. Sep. Sci.* **2008**, *31* (6–7), 1015–1026.

Xiong, W.; Li, J.; Jiang, R.; Li, D.; Liu, Z.; Chen, D. Research on the Effect of Ginseng Polysaccharide on Apoptosis and Cell Cycle Of Human Leukemia Cell Line K562 and Its Molecular Mechanisms. *Exp. Ther. Med.* **2017**, *13* (3), 924–934.

Yang, S. O.; Shin, Y. S.; Hyun, S. H.; Cho, S.; Bang, K. H.; Lee, D.; Choi, S. P.; Choi, H. K. NMR-Based Metabolic Profiling and Differentiation of Ginseng Roots According to Cultivation Ages. *J. Pharm. Biomed. Anal.* **2012**, *58*, 19–26.

Yang, X.; Zou, J.; Cai, H.; Huang, X.; Yang, X.; Guo, D.; Cao, Y. Ginsenoside Rg3 Inhibits Colorectal Tumor Growth via Down-Regulation of C/EBPβ/NF-κB Signaling. *Biomed. Pharmacother.* **2017**, *96*, 1240–1245.

Yang, Y.; Ju, Z.; Yang, Y.; Zhang, Y.; Yang, L.; Wang, Z. Phytochemical Analysis of Panax Species: A Review. *J. Ginseng Res.* **2021**, *45* (1), 1–21.

Yang, Y.; Yang, Y.; Qiu, H.; Ju, Z.; Shi, Y.; Wang, Z.; Yang, L. Localization of Constituents for Determining the Age and Parts of Ginseng Through Ultraperfomance Liquid Chromatography Quadrupole/Time of Flight-Mass Spectrometry Combined with Desorption Electrospray Ionization Mass Spectrometry Imaging. *J. Pharma. Biomed. Analys.* **2021**, *193*, 113722.

Yoon, D.; Shin, W. C.; Lee, Y.-S.; Kim, S.; Baek, N.-I.; Lee, D. Y. A Comparative Study on Processed Panax ginseng Products Using HR-MAS NMR-Based Metabolomics. *Molecules (Basel, Switzerland)*, **2020**, *25* (6), 1390.

Yue, P. Y. K.; Mak, N. K.; Cheng, Y. K.; Leung, K. W.; Ng, T. B.; Fan, D. T. P.; Yeung, H. W.; Wong, R. N. S. Pharmacogenomics and the Yin/Yang Actions of Ginseng: Anti-Tumor, Angiomodulating and Steroid-Like Activities of Ginsenosides. *Chin. Med.* **2007**, *2* (1), 6.

Yuk, J.; McIntyre, K. L.; Fischer, C.; Hicks, J.; Colson, K. L.; Lui, E.; Brown, D.; Arnason, J. T.; Distinguishing Ontario Ginseng Landraces and Ginseng Species Using NMR-Based Metabolomics. *Anal. Bioanal. Chem.* **2013**, *405* (13), 4499–4509.

Yun, T. K. Brief Introduction of *Panax ginseng* C. A. Meyer. *J. Korean Med. Sci.* **2001**, *16*, S3–S5.

Yun-Soo, K.; JungYeon, H.; Soon, L.; YongEui, C. Ginseng Metabolic Engineering: Regulation of Genes Related to Ginsenoside Biosynthesis. *J. Med. Plants Res.* **2009**, *3*, 1270–1276.

Zhao, B.; Lv, C.; Lu, J. Natural Occurring Polysaccharides from *Panax ginseng* C. A. Meyer: A Review of Isolation, Structures, and Bioactivities. *Int. J. Biol. Macromol.* **2019**, *133*, 324–336.

Zheng, S. W.; Xiao, S. Y.; Wang, J.; Hou, W.; Wang, Y. P. Inhibitory Effects of Ginsenoside Ro on the Growth of B16F10 Melanoma via Its Metabolites. *Molecules* **2019**, *24* (16).

Zhu, G. Y.; Li, Y. W.; Hau, D. K.; Jiang, Z. H.; Yu, Z. L.; Fong, W. F.; Acylated protopanaxadiol-type ginsenosides from the root of Panax ginseng. *Chem Biodivers*, **2011**, *8*(10), 1853–1863.

Zhu, H.; Lin, H.; Tan, J.; Wang, C.; Wang, H.; Wu, F.; Dong, Q.; Liu, Y.; Li, P.; Liu, J. UPLC-QTOF/MS-Based Nontargeted Metabolomic Analysis of Mountain- and Garden-Cultivated Ginseng of Different Ages in Northeast China. *Molecules* **2018**, *24* (1).

CHAPTER 10

Tinospora cordifolia (Thunb.) Miers.

LEPAKSHI M. D. BHAKSHU[1], PULALA RAGHUVEER YADAV[2],
K. VENKATA RATNAM[3], ANU PANDITA[4], DEEPU PANDITA[5], and
K. M. SRINIVASA MURTHY[6]

[1]*Department of Botany, PVKN Government College (Autonomous), Chittoor, Andhra Pradesh, India*

[2]*Department of Biotechnology, Indian Institute of Technology Hyderabad, Kandi, Telangana State, India*

[3]*Department of Botany, Rayalaseema University, Kurnool, Andhra Pradesh, India*

[4]*Vatsalya Clinic, Krishna Nagar, New Delhi, India*

[5]*Government Department of School Education, Jammu, Jammu and Kashmir, India*

[6]*Department of Microbiology and Biotechnology, Jnanabharathi Campus Bangalore University, Bengaluru, India*

ABSTRACT

Tinospora cordifolia (Thunb.) Miers, is a woody twiner, belongs to family Menispermaceae, with characteristic lenticels on stems. It is commonly used in Ayurvedic medicine and popularized as Gaduchi or Amrit (heavenly elixir). The stem, stem bark, and leaves were appraised for the human ailments like anti-diabetic, anti-oxidant, anti-stress, anti-malarial, anti-spasmodic, anti-inflammatory, anti-arthritic, anti-leprotic, hepatoprotective, anti-allergic, immune-modulatory, and anti-neoplastic activities. In addition, the plant is also scientifically proved for the treatment of various types of cancers such as IMR

32 human neuroblastoma, C6 glioma cells, and IMR-32 cells. The extracts or compounds such as berberine, were potentially inhibited the cancer cells and have been proved the molecular mechanism including the genetic expression of homeostasis and senescence indicating proteins-HSP70 and Mortalin. This plant contains glycosides like cordiosides, tinocordifolioside, tinocordiside, palmitosides, and steroids such as ecdysterone and β-sitosterol. Alkaloids like berberine, palmatine, 11-hydroxymustakone magnoflorine, and some diterpenoid lactones along with aliphatic compounds as pharmacologically active constituents responsible for its medicinal properties. The present review focused on the utilization of extracts and their principles in the cancers as well the leading path for the apoptosis and involvement of cytokines expression studies in relation to cancer therapies supports the extensive use with safe and cost-effective.

10.1 INTRODUCTION

Mother nature is blessed with curative properties of many human ailments by gifting medicinal plants with plethora of secondary metabolites which have to be scientifically investigated and properly utilized in human welfare. The changing human culture might be a reason for leading to dreadful illness and Cancer is one among them. The medicaments developed for the treatment of Cancer, have to supplement with the herbal boosters for the improvement of overall health during the course. In this connection, the Ancient systems of medicines were providing such medicaments and *Tinospora cordifolia* (Thunb.) Miers. (Figure 10.1) belongs to Menispermaceae family, is one among the lifesaving herbs and believed to be safe, non-toxic, and cost-effective. It is most commonly used as a potent herbal recipe in the Ayurveda system of medicine, known as Amrita or heavenly elixir (Giloy).

The scientific world have appreciations and reported several bioactive components from *T. cordifolia* such as alkaloids, steroids, lactones, glycosides, aliphatic compounds sesquiterpenoid, diterpenoid and polar compounds, polysaccharides, and phenolics with varied pharmacological properties. Amrit has been reported as a remedy for jaundice. The plant has been reported as hepato-protective, immunomodulatory and immunostimulatory, potential for diabetes and related disorders in addition to adaptogenic, cardiotonic, anti-oxidant, anti-inflammatory, and anti-psychotic activities. Interestingly, many of the reports were demonstrated its effectiveness on cancerous cells.

FIGURE 10.1 *Tinospora cordifolia* (Thunb.) Miers.

The recent studies have evidenced for the inhibition role of the extracts of *T. cordifolia* on the anti-proliferative activity in various *in vitro/in vivo* cancer models supporting anti-neoplastic, cytotoxic, anti-angiogenesis, and anti-metastatic activities (Mishra et al., 2013; Sharma et al., 2019) were being discussed. The modern tools were provided the characterization for its anti-cancer properties with possible mechanisms even providing evidences at a molecular level and the present study able to focus on the sequential roles of the *T. cordifolia* on the effectiveness of Cancer.

10.2   ACTIVE COMPONENTS

The extracts of *T. cordifolia* have tremendous medicinal potential and the pharmacological activities are ascribed to the presence of secondary metabolites such as terpenoids, for example, furanolactone, tinosporide, tinosporaside, and ecdysterone, etc., glucosides like poly acetate, phenylpro- pene disaccharides, tinocordioside (Figure 10.2), cordifoliosides, cordioside,

Alkaloids such as palmatine (Figure 10.3), berberine (Figure 10.4), tinosporine, magnoflorine, Choline, Jatrorrhizine, lignans, steroids, and others were characterized (Sharma et al., 2019).

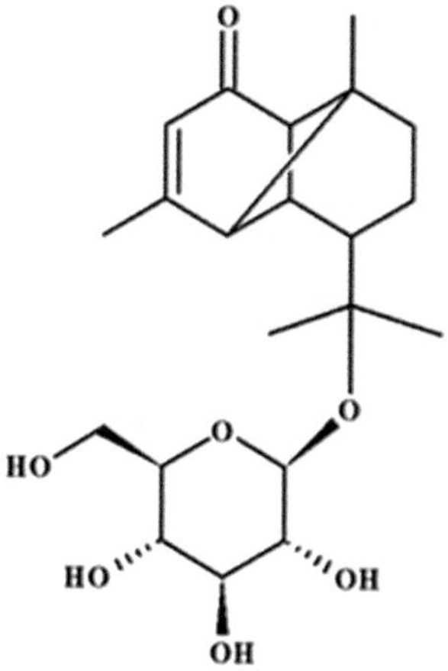

FIGURE 10.2   Tinocordiside.

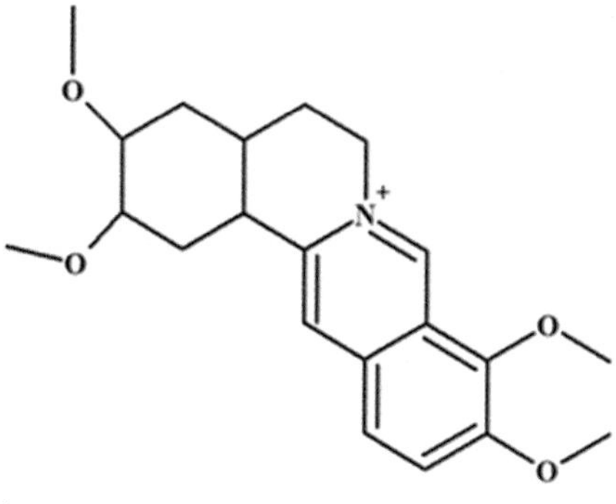

FIGURE 10.3   Palmatine.

10.3   ANTI-CANCER ACTIVITIES

The extracts of *T. cordifolia* (TCEX) obtained from the methanol and water and methylene chloride when tested on the survival of HeLa cells at the dose of 50 and 100 µg/mL exhibited significant killing effect when compared to non-treated cells whereas the methylene chloride extract is more effective (up to 6.8 times) as compared to the standard (doxorubicin) in a linear pattern and suggested as anti-neoplastic agent for further studies (Jagetia et al., 1998).

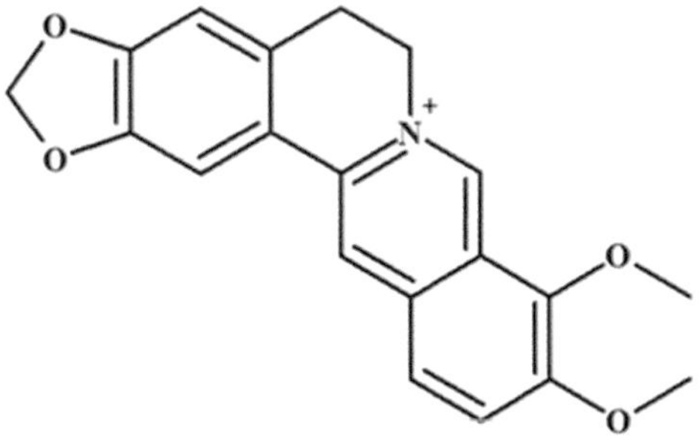

FIGURE 10.4   Berberine.

10.3.1   EFFECT ON SKIN CANCER

Anti-cancer efficiency of Immu-21, an Ayurvedic polyherbal formulation containing amla, holy basil, ashwagandha, and TCEX extracts on B/T cell mitogens, phytohemagglutinin, LPS, and concanavalin-induced cell proliferation in splenic leukocytes using *in vitro* and K 562 cells in mice. Cells treated with Immu-21 enhanced the division capacity of splenic leukocytes exposed with LPS and B-cell mitogens. Further, the formulation exhibited notable cytotoxicity in K 562 cells in mice (Nemmani et al., 2002).

The effect of TCEX on 7,12-dimethylbenz(a)anthracene (DMBA) induced skin cancer and evaluated at two-stage levels and pretreated with 100 mg/kg bw (TCEX) and tumor formation for 16 days was found significant reduction in incidence and weight of the tumor when compared to the untreated mice besides declined tumor burden, papillomas, the elevation of detoxifying enzymes, reduction on lipid peroxides in liver and skin during the studies strongly supported its anti-tumor effect (Chaudhary et al., 2008). Further, the studies of Ali and Dixit (2011), supported the effect of the alkaloid obtained from *T. cordifolia* on the skin cancer in the DMBA-effected skin tumor model using response surface methodology (RSM). The cytogenic effect of the mixed plant extracts of *T. cordifolia,* ashwagandha, and holy basil was assessed on the blood samples collected from healthy volunteers, revealed that the extracts significantly declined colchicine-induced chromosomal abnormalities (Sharma et al., 2011).

10.3.2 EFFECT ON MULTI DRUG RESISTANT SKIN CANCER CELLS

The anti-tumor potential of ethanolic extracts of TCEX and ashwagandha on the side population of drug-resistant cancer cells showed that TCEX exhibited a noteworthy reduction in side population and still no effect was observed in ashwagandha extract-treated cells. Flow cytometry results revealed that petroleum ether and dichloromethane extracts of TC strongly suppressed the expression of ABC-G2 and ABC-B1 genes, which play an important role to sensitize the cancer cells to the chemotherapeutic drug (Maliyakkal et al., 2015).

10.3.3 EFFECT ON ORAL CANCER

The anti-cancer effect of an herbal regimen made from *Emblica officinalis* (EO) and *T. cordifolia* was evaluated in oral cancer patients (114) with buccal and tongue cancer, revealed that the formulation showed promising effect on reducing mucosa formation on cheeks and back of the lips and tongue (Dias et al., 2018). TCEXAQ reported the induction of apoptosis in AW13516 oral cancer line attenuating its potential for epithelial–mesenchymal transition in a dose-dependent pattern and exhibited potential effect at 5 µg/mL (Patil et al., 2021).

10.3.4 EFFECT ON PROSTRATE CANCER

The anti-cancer actions of ethanolic extract of *T. cordifolia* against human prostate cancer cell lines (LNCaP cells) for a period of 48 h experimental schedule, demonstrated that extracts showed concentration variant stimulation and proliferation of prostate cancer cells. The anti-androgen flutamide reversed the ethanol extract and stimulated the secretion of prostate-specific antigen (PSA) in cancer cells, which is similar to the standard dihydrotestosterone (DHT) and suggested TCEX may contain androgenic compounds and stimulated the androgen receptor proving endocrine modulation (Kapur et al., 2009).

10.3.5 ANTI-TUMOR POTENTIALITY

The polysaccharide fraction from *T. cordifolia* was reported significant effects on the suppression of the passage of melanoma cells (B16F-10) and strongly suppressed the growth of tumors in the lungs of C57BL/6 mice and the biochemical markers of neoplastic cancer like lung collagen hydroxyproline uronic acids secretions and hexosamines, were decreased significantly in the TCEX-treated mice compared with the control mice in addition to affecting serum γ-glutamyl-transpeptidase (γ-GT) and sialic acid levels (Leyon and Kuttan, 2004).

Pretreatment of alcoholic TCEX to mice with T cell lymphoma, resulted in a significant reduction in tumor growth in thymus. Further, TCEX enhanced the proliferation rate of thymocytes and reduced its apoptosis. The extract also reduced Hassal's corpuscles number. TCEX showed its mechanism by reducing thymocyte's apoptosis and enhancing the secretion of growth-promoting cytokines such as IL-2 and IF-γ in thymocytes (Singh et al., 2005a).

Cytotoxic efficiency of the water extract of *T. cordifolia* stem (TCEXAQ) and bioactive polysaccharides such as arabinogalactan (AG), were assessed on pulmonary cancer cells and its associated cytokines in mice-induced tumorogenesis with benzo(a)pyrene (B(a)P) and dose-dependent treatments in different groups. The B (a) P-induced serum or plasma markers like lactate dehydrogenase, circulating tumor DNA, tumor necrosis factor, and carcinoembryonic antigen significantly attenuated in the TCEX treatments which depict TCEX's protective role and even enhanced the apoptosis in cancer-affected cells. The apoptosis is more effective in treatments with TCEXAQ than AG in modulating lung carcinogenesis (Mohan and Koul, 2017).

The radio-sensitizing activity of *T. cordifolia* dichloromethane extract was investigated in gamma rays induced radiation in mice with Ehrlich ascites carcinoma at five different concentrations. The results demonstrated that mice treated with extract at 30 mg/kg b.wt. expressed dose-dependent tumor suppression and radio-sensitizing activity and enhanced the life span of irradiated mice. The extract showed its radio-sensitizing mechanism by reducing glutathione S-transferase and glutathione concentrations and increased the levels of lipid peroxidation and DNA fragmentation in cancer cells (Rao et al., 2008).

Anti-tumor efficiency of the alcoholic extract of *T. cordifolia* was assessed in *in vivo* mice models with bone marrow cells and DL-bearing (Daltons Lymphoma) mice, revealed that *T. cordifolia* extract significantly increased the formation of colonies of granulocyte-macrophages in mice with bone marrow cells and enhanced the bone marrow cells proliferation in DL-bearing mice. The treated mice showed an improved response to

LPS-activated production of IL-1 and TNF indicating the *T. cordifolia* role in the differentiation of precursor of myeloid bone marrow progenitor cells and the involvement of the macrophages (evidenced by increased number during the TCEX treatment) in response to tumor growth *in situ* are directly related to anti-tumor effect, particularly on myelopoiesis (Singh et al., 2006).

10.3.6   EFFECT ON BONE CANCER CELLS

Anti-cancer and anti-mitotic activity of methanolic extract of *T. cordifolia* at three different concentrations was investigated *in vivo* models, that is, Swiss albino mice and C57 Bl mice, demonstrated that the extract significantly reduced tumor size and increased life span of animals in both the tested animal models (Verma et al., 2011).

10.3.7   CYTOTOXICITY ON FEMININE CANCER CELL

The cytotoxic effect of various organic fractions of *T. cordifolia* was evaluated *in vitro* on three human breast (MCF-7), ovary (IGR-OV-1), prostrate (DU-145), and the tested cancer cell lines affected by the fractions with moderate cytotoxic potential. The chloroform, acetone, and aqueous extracts exhibited cytotoxic activity against the breast cancer cell lines with >50% inhibition, while other extracts showed <50% inhibition on breast cancer cells. The extracts demonstrated <40% cytotoxic effects against the prostate and ovarian cancer cell lines. Paclitaxel, mitomycin-C, and adriamycin were used as standard anti-cancer drugs as reference drugs (Mishra et al., 2013). Further, the potent anti-tumor activity of TCEX was reported in mice with Dalton's lymphoma ascites tumors (Adhvaryu et al. (2007). However, the exact mechanism suggests that decline in clonogenicity, protection from DNA damage, suppression of glutathione-S-transferase and topoisomerase-II activities, and induction of tumor-associated profile involved in the cytotoxic activity (Jagetia and Rao, 2013).

The methanol and aqueous extracts of WS (WSMEX and WSAQEX) and TC (TCEXM and TCEXAQ) were tested and exhibited the cytotoxic and apoptotic effects on MDA MB 231and MCF7 in the concentration-dependent pattern except aqueous extracts. Further studies revealed hallmark patho-cascade of programmed cell death like blebbing of membrane, condensation of nuclear material, and fragmentation of DNA with the TCEXM and enhanced the sub-G content, however. In addition to that WS extracts seized the cell cycle progression in the G/M phase and the extracts not affected the normal cells (HaCaT) while affected the breast cancer cells (Maliyakkal et al., 2013).

10.3.8 EFFECT OF BERBERINE AT GENETIC LEVEL IN CANCER CELLS

A comparative anti-cancer activity of commercial and isolated berberine was investigated on HCA-7, human colon carcinoma cell lines *in vitro* methods. The results demonstrate that out of 44 oncogenes tested in HCA-7 cell lines, 33 genes were significantly suppressed in a time and concentration-dependent manner. The results suggest that the presence of berberine in *T. cordifolia* may be attributed to its anti-tumor activity (Palmieri et al., 2019).

10.3.9 EFFECT OF NANOPARTICLES OF TCEX ON CANCER CELLS

Cytotoxic effects of silver nanoparticles prepared in the combination of *T. cordifolia* leaves, defined *via* numerous techniques like scanning microscopy, Fourier Transformed infrared, X-ray diffraction, transmission microscopy, and energy dispersive X-ray analysis were assessed on A549 cell lines. The size and morphology of nanoparticles were confirmed with transmission microscopy and exhibited very strong anti-tumor activity and *T. cordifolia* leaves are found to be toxic against human respiratory organ glandular cancer cell line A549 (Mittal et al., 2020).

The role of nano-formulation of TCEX stem extract in glandular carcinoma cells reported, that there's no caspase-mediated cell death induction, however, stops proliferation in prostate cancer cells without apoptosis and can be regarded as an anti-proliferative agent against prostate cancer (Karuppath et al., 2016). Anti-proliferative activity was in cervical malignant neoplastic disease in HeLa cell lines by demonstrating that the dichloromethane and ethyl alcohol fraction of TCEX exhibited potent cytotoxic impact within the effective concentrations (Polu et al., 2017).

10.3.10 TCEX IMPACT ON NERVE FIBER CELLS

T. cardifolia showed a broad spectrum of immune-augmentary effects and tumor-associated macrophages which in turn upregulates the anti-tumor property. The differentiation of macrophages to dendritic cells was enhanced by the TCEX. The treatment of alcoholic extract (200 mg/kg bw) for two days, showed improvement in post-tumor transplantation through tumor cytotoxicity and NO, TNF, and IL-1 production. The survival was extended in tumor-bearing mice after the adoptive transfer of dendritic cells obtained from tumor-associated macrophages into mice with Dalton's lymphoma (Singh et al., 2003, 2004, 2005b).

The cytotoxicity of bone marrow cells obtained by dendritic cells (BMDC) enhances maturity by arabinogalactan (G1-4A) by *T. cardifolia* was due to Nitric oxide synthase and apocynin-inhibited killing. BMDC phagocytosed killed cells and activates cytotoxic T-cells (Pandey et al., 2014).

The *T cardifolia* stem extracts were used to isolate bioactive constituents with the aid of chromatography and characterized by NMR and MS. Eight compounds were isolated tinocordiside, cordifolioside, yangambin, magnoflorine, palmatine, jatrorrhizine, and 11-hydroxymustakone, these are from different classes. The tested phytoconstituents were effective on CHOK-1, KB, SiHa, HT-29, and murine primary cells differentially of which some compounds possess anti-cancer and while others exhibited immunomodulatory properties (Bala et al., 2015).

HeLa cells were exposed to 2-Gy- γ-radiation, and were treated with *T. cardifolia* dichloromethane extract for 4 h before exposure to radiation. This led to cell death, nearly approximately 50%. There is a downfall in the viability of HeLa cells, depending on the doses of γ-radiation. The concentration of *T. cardifolia* dichloromethane fraction was also important along with the concentration of γ-radiation, this combination led to a further decrease in the cell viability. A total of 1–4 Gy γ-radiation was radiated on Hela cells and later treated with different doses of *T. cardifolia* dichloromethane fraction demonstrated a significant decrease in cell viability. A dose-dependent decrease in the surviving fraction was observed with the increasing dose before irradiation. Among different doses, 4 µg/mL *T. cardifolia* dichloromethane extract showed the lowest survival fraction. There was an increase in LDH and a decrease in GST activity in HeLa cells treated with *T. cardifolia* dichloromethane fraction during post-irradiation times. Lipid peroxidation multiplied in 4 h of post-irradiation and slowly declined by 12 h of post-irradiation (Rao and Rao, 2010).

T. cordifolia and *Zingiber officinale* were treated in MCF-7 cells, MTT assay resulted in IC_{50} 509 and 1 mg/mL, respectively. Combination of *T. cordifolia* and *Zingiber officinale* administered in MCF-7 cells showed IC_{50} of 2 µg/mL from MTT assay which indicates that combination treatment is more effective than alone. Combination treatment of *T cordifolia* and *Zingiber officinale* induced apoptosis and arrest at the G_0/G_1 phase. *T.*

cordifolia and *Zingiber officinale* anti-cancer effect are due to MMP2, 9, and ALOX5 major gene targets. Pharmacological network and *in vitro* studies prove the synergistic effect of *T. cordifolia* and *Zingiber officinale* combination in MCF-7 cells (Javir and Joshi, 2019).

10.3.11 CORRELATION WITH HEAT SHOCK PROTEINS (HSP) EXPRESSION AND CANCER

Rashmi et al. (2017) reported the inhibitory property and regulation of heat shock proteins and angiogenesis in certain tumor models such as Ehrlich ascites MDA-MB-231 cell lines through the inactivation of PI3K or AKT and related cassette of gene expression by the isolated pyrrole derivative from the leaves and this activity was demonstrated in the *in vivo* and *ex vivo* models. Clerodane furano diterpene glycoside was a new molecule isolated from the stems of *T. cardifolia* (hydro-alcoholic extract), shown anti-cancer activity by inducing apoptosis and autophagy in HCT116 (human colorectal cancer) cells. Besides this, N-nitroso-diethylamine-induced liver cancer in male Wister albino rats were treated with *T. cordifolia* ethanolic extract showed a prominent increase in LPO and declining of enzymic and nonenzymic anti-oxidants (Sharma et al., 2018).

The TCEX affected the Bcl-2 or Bax protein expression which impacts on "mitochondrial membrane depolarization," MPTP, and peroxidation of cardiolipin. It evoked the cytochrome unharness of the cytoplasm, activation of protease, leading to deoxyribonucleic acid fragmentation. The initiation of apoptosis and modifications in cell structure was evident from the "phosphatidylserine externalization" and increases the "subG" population.

The Ehrlich neoplasm (EAT) in mouse (*in vivo*) demonstrated the effectiveness of TCEX in declining the neoplasm burden and enhanced the chance of ~2 folds in the survival of mice reporting marginal "hepato-renal damage." *T. cordifolia* extracts reported for the effect on the ROS and restore p50 activity and mitochondrial-mediated cell death in "MDA-MB-231 cells" besides the activation of EAT necrobiosis, which inhibited the neoplasm proliferation (Rashmi et al., 2019).

10.3.12 EFFECT ON CELL CYCLE

The inhibition effect of TCEX of proliferation of "KB cells" has been linked with the arrest of G0/G1-phase in cell cycle and proved to be safe on the normal peripheral blood mononuclear cells (PBMC) indicates its specific effect might be beneficial in the treatment of cancer (Bansal et al., 2017).

10.3.13   EFFECT ON GLIOBLASTOMA

The extracts of *T. cordifolia* showed anti-cancer effect on neuroblastoma cells (IMR-32) along with its metastasis effecting on cell cycle at G0/G1 and modulated the expression of polymer clamp slippery macromolecules (PCNA) and cyclin D1. Further, TCEX-treated cells showed differentiation as unconcealed by their morphology and also the expression of neural cell-specific differentiation markers NF200, MAP-2, and NeuN in metastatic tumor cells. The differentiated composition was related to the induction of senescence and pro-apoptosis pathways by enhancing the expression of senescence marker mortalin and Rel-A fractional monetary unit of nuclear factor kappa beta (NFkB) beside decreased expression of anti-apoptotic marker, Bcl-xl. TCEX exhibited anti-metastatic activity and considerably reduced cell migration within the scratched space beside the downregulation of neural cell adhesion molecule (NCAM) polysialylation and secretion of matrix metalloproteinases (MMPs). Mishra and Kaur (2013) showed the anti-brain cancer potential, 500th ethanolic extract effects C6 brain tumor cells considerably elicited differentiation in C6 brain tumor cells, and reduced the cell-proliferation (Mishra and Kaur, 2015).

10.3.14   EFFECT ON NEUROBLASTOMA

The anti-cancer effect of TCEX on Neuroblastoma (IMR-32 cells) was established with different doses and found effective at 200 μg/mL (IC$_{50}$) and reported nontoxic on normal cells. Based on the studies of Mishra and Kaur (2015) and the following mechanism has been provided on the Glioblastoma.

1. TCEX-arrested cell cycle at G0/G1 phase.
2. It is associated with the growth retardation and expression of regulatory proteins in cell cycle.
3. Verified the stimulatory role in the cell cycle process from the G1/S phase.
4. TCEX treatment inhibited the expression of cyclin-D1.
5. TCEX treatment affected the PCNA protein expression of the majority of IMR-32 cells.

6. TCEX induces differentiation of morphology in IMR-32 Cells with multiple and elongated processes.

7. TCEX-treatment-proceeded IMR-32 cells induce differentiation through enhanced expression of cytoskeleton proteins of neurons ("NF-200" and "MAP-2").

8. Elevated NeuN only at the translational and transcriptional level.

9. TCEX-affected cells showed enhanced mitochondrial activity in IMR-32 cells and indicated that the cell survival and differentiations were dose-dependent.

10. TCEX induces senescence in neuroblastoma cells through the production of mortalin, a chaperone of senescence marker which exerts prominently at 0.3 mg/mL. In addition, the HSP70 also reported as a senescence indicator in TCEX-treated cells in association with the phosphorylated form of Akt-1 (pSER473).

11. The suppression of Bcl-xl production significantly induces the pro-apoptotic/apoptotic pathways.

12. The TCEX affected the cell survival and differentiation and induced senescence and preceded the apoptosis besides significant anti-migratory potential of cells from scratched areas.

Based on the above-mentioned properties, clearly emphasizing its efficacy on the neuroblastoma and found to safe usage in a combination of rational therapies of neuroblastomas (Mishra and Kaur, 2013). The fractions obtained from Chloroform and hexane of TCEX reported for the significant inhibition of proliferation and differentiation in glioblastoma and neuroblastoma.

The non-polar fractions obtained from hexane and Chloroform affected significantly and reduced the rate of proliferation and induced cell differentiation in glioblastoma and neuroblastoma, respectively, which are correlated through the stress markers such as Mortalin and HSP-70 which were upregulated in treated cells in addition to metastasis of glioblastoma (U87MG and IMR-32) or neuroblastoma and declined the production of NCAM found to be an important event in the treatment of neural carcinomas (Anuradha et al., 2019).

10.4 MECHANISM OF ACTION OF PHYTOCOMPONENTS ON CANCER CELLS

The phytocomponents of TCEX affected the iatrogenic malignant tumor, have demonstrated through mitochondrial-mediated caspase-mediated cell death, cytotoxic effect, tumor size reduction, formation of ROS, attenuate the cell cycle, and gene expression resulting in significant inhibition of cancer cell proliferation. The mechanism of action of TCEX depends on the phyto-components in which the "clerodane furano-diterpene" organic compound proved for inhibition of malignant tumor through the programmed cell death by initiating reactive chemical element agents (ROS) besides autophagy. Additionally, the phenolics of the TCEX have geno-protective against cancer cells (Singh et al., 2006). The ethanolic extracts showed iatrogenic apoptosis by increasing the "sub-G0 phase" without affecting the cell cycle (Maliyakkal et al., 2013). Arabinogalactans of aqueous TCEX initiate the immunological and cytotoxic effects. The chemoprotective role has attributed to flavonoids in cancer besides pyrrole-phytoconstituents induce programmed cell death and cytotoxic effects (Rashmi et al., 2019).

Palmatine showed an increased inhibition effect by enhancing the levels of anti-oxidant enzymes and conjointly exhibited detoxifying effect in addition to declining of lipid peroxidation. The berberine and hexane fractions proved the tumor inhibition property programme cell death through "capsase 3-activated DNase," respectively. The epoxy-clerodane diterpene of TCEX polysaccharides resulted in anti-neoplastic effect (Leyo and Kuttan, 2004) besides it involved in the blockage of the metabolic activation and encouraged in the detoxification of carcinogens. Singh et al. (2006) reported that the TCEX showed direct tumoricidal effect. Thus, TCEX proved the anti-carcinogenic effects through the stimulation of DNA injury, clonogenicity, inhibition of topoisomerase-II, glutathione S enzyme activity, apoptosis, and enhancing lipid oxidase activity.

The berberine of TCEX has demonstrated the cell cycle inhibition and differentiation, HEP2 human speech organ neoplastic cells. The clerodane-furano-diterpene-glycoside proved for vital cytotoxic effect and iatrogenic programmed cell death in human prostate cancer (PC-3, SF-269 [CNS], skin cancer [MDA-MB-435], pulmonary cancer [A549], carcinoma [MCF-7] cells, and carcinoma [HCT-116]). Singh et al. (2005a, 2005b) demonstrated with the known synthetic resin compounds from a fungus extract of endo-phytic flora *Cladosporium velox* TN-9S which are isolated from the stems of TCEX, showed a marginal geno-protective effect against DNA injury on Chinese hamster ovary cell lines when the treatment with non-ionic water nonylphenol. It had been conjointly noted that the endophyte's capability to synthesize phytocomponent was just like the host plant. Anti-cancer potency of chemical constituents such as cordifolioside A, tinocordiside, mangoflorine, jatrorrhizine, palmatine, N-formylannonain, yangambin, and

11-hydroxymustakone isolated from *T. cardifolia* using chromatography technique was evaluated against four types of cancer cell lines. It revealed that among the tested constituents palmatine supressed oral and colon cancer cell lines, tinocordiside active on the ovarian and oral cancer cells and yangambin inhibited the growth of oral cancer cells. The components 11-hydroxymustakone and N-formylannonain exhibited the highest immu-nomodulatory activity in the tested cancer cell lines (Bala et al., 2015). In addition, the TCEX and fractions were compared the malignant tumor growth inhibition and had been proposed that the effect is magnified due to combinations and synergistic (Bala et al., 2015).

Rashmi et al. (2017), determined apoptotic induction using different apoptotic markers, generation of ROS, activity of proteinase, and cell cycle analysis and found phytocomponents of TCEX proved its anti-cancer effect and conjointly inhibited the tumor proliferation. Though thorough studies demonstrated with TCEX, the exact mechanism has to derive. The flavonoids such as rutin and quercetin from TCEX showed anti-proliferative activity of human carcinoma neuroblastoma cells through ignition of programmed cell death, intracellular ROS and altered gene expression.

Anti-tumor efficiency of aqueous extract and arabinogalactan, bioac-tive polysaccharide of *T. cardifolia* stem was assessed in benzo(a)pyrene-induced tumors in lungs and its associated biochemical markers in animal model, demonstrated that the extract and arabinogalactan significantly attenuated benzo(a)pyrene-induced tumor markers such as TNF-alpha, carcinoembryonic antigen, lactate dehydrogenase, and ctDNA in animal models. The results indicated that extract was more potent that polysac-charide in reducing lung tumors in animal models (Mohan and Koul, 2018).

The phenolics of TCEX such as ellagic acid and kaempferol as geno-protective property on the fish model and found the reduction in nuclear abnormalities (Mishra et al., 2013; Bakrania et al., 2017). Palmatine rein-forces the inhibitor catalyst levels and conjointly inhibits carcinogenesis on singe administration (Ali and Dixit, 2013) besides berberine showed growth remission on Swiss anomaly mice. Jagetia et al. (1998; Jagetia and Rao, 2006), reported that the malignant tumor suppression was dose-dependent and opined that the combinatory result of the alkaloids inhibits the malignant tumor. Leyon and Kuttan (2004) studied carbohydrate fraction from TCEX and reported the pathologic process result on C55BL/6 mice and, therefore, the extremely vital inhibition effect on skin cancer cell line B16F-10 has attributed to its immune modulation.

10.5 MOLECULAR MECHANISM OF ANTI-CANCER EFFICACY

The details of the mechanism of anti-tumor activity of the Amruth and its phytoconstituents were established on certain cell lines and were still under the progression. However, TCEX roles have clearly proved on the certain anti-tumor and its metastasis effects of cancers. Moreover, the TCEX triggers of ROS (reactive oxygen species) which elevates the inhibition effect of growth cells through DNA damage causing mortality of cancer cells besides reduced anti-oxidant enzymes activity, and exaggerated oxidation of lipids and DNA damage might lead to cell death. The binding of TCEX components with DNA topoisomerases plays a prominent role in DNA replication especially Topoisomerase I/II may induce cytotoxic effects and cell proliferation. Berberine of TCEX, demonstrated to suppress the DNA topoisomerase I/II activity and stabilizes the catalyst mediated-DNA "cleavable complex," which can additionally trigger DNA damage and mortality. In addition, Berberine arrests end elongation causing toxicity in carcinoma cells. However, suppression of COX-II, NF–κB, and Nrf2 might have additionally extended the vital role in neoplastic cell death by *T. cordifolia* where the berberine and the extracts involved in the declining of anti-inflammatory markers (Kapil and Sharma, 1997; Gao and Cai, 2008; Kapur et al., 2010; Birla et al., 2019).

Berberine, a nitrogenous compound isolated from aerial parts of *T. cordifolia* has been found to seize the progression of the G2/M phase of the cell cycle, over the expression of cytotoxic markers Bax, CDk2, and p53. In addition to this, it activated the caspase enzyme complex and other markers such as PARP, involved in inducing programmed cell death (Palmeri et al., 2019).

10.6 CONCLUSIONS

The extracts/principles of TCEX have proven for its beneficial effects on the distinguished cancer cell lines and in the animal models (Figure 10.5). Moreover, the selected plant possesses numerous medicinal properties of which anti-cancer effect is appreciated by many scientific studies and berberine, magnoflorine, palmatine, rutin, quercetin, and arabinogalactans may be an important principle during the cancer treatment. More insightful efforts have been made to develop this plant for cancer treatment since it has proved as a best anti-diabetic, anti-inflammatory, and anti-cancer besides many physiological or pathogen induced ailments.

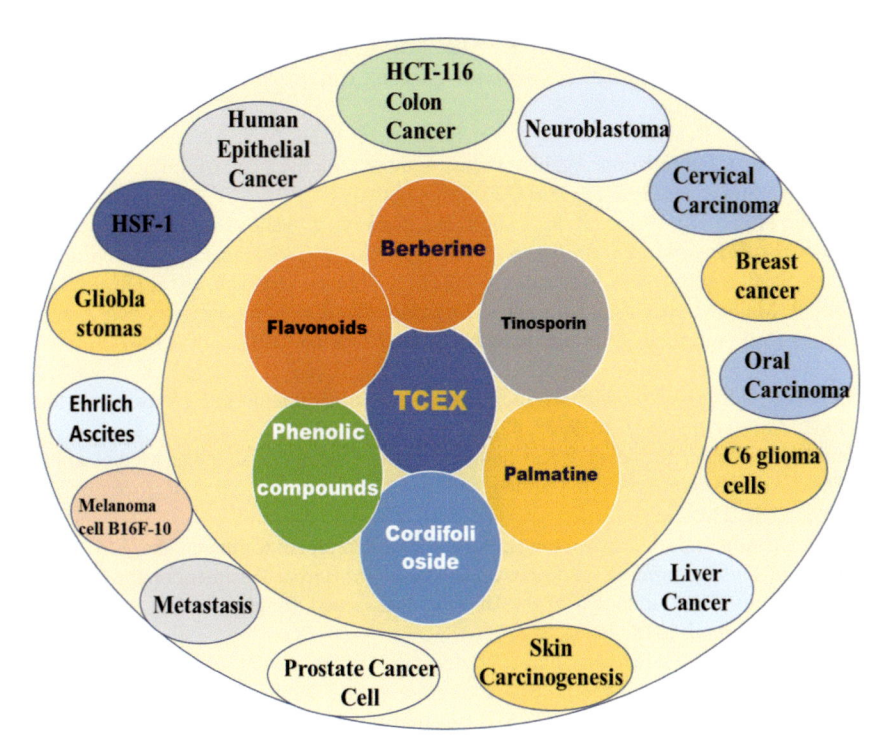

FIGURE 10.5 Effect of *T.cordifolia* on cancer cells.

KEYWORDS

- ***Tinospora cordifolia***
- **anti-cancer properties**
- **mechanism of anti-cancer properties**

REFERENCES

Adhvaryu, M. R.; Reddy, N.; Parabia, M. H. Effects of Four Indian Medicinal Herbs on Isoniazid, Rifampicin and Pyrazinamide-Induced Hepatic Injury and Immunosuppression in Guinea Pigs. *World J. Gastroenterol.* **2007,** *13* (23), 3199–3205.

Ali, H.; Dixit, S. Extraction Optimization of *Tinospora cordifolia* and Assessment of the Anticancer Activity of Its Alkaloid Palmatine, *Sci. World J.* **2013,** *28*, 1–10.

Ansari, J. A.; Rastogi, N.; Ahmad, M. K.; Mahdi, A. A., et al., ROS Mediated Pro-Apoptotic Effects of *Tinospora cordifolia* on Breast Cancer Cells. *Frontiers in Bioscience—Elite,* **2017,** *9* (1), 89–101.

Bakrania, A. K.; Nakka, S.; Variya, B. C.; Shah, P. V.; Patel, S. S. Antitumor Potential of Herbomineral Formulation Against Breast Cancer: Involvement of Inflammation and Oxidative Stress. *Indian J. Exp. Biol.* **2017,** *55,* 680–687.

Bala, M.; Pratap K.; Verma, P. K.; Singh, B., Padwad, Y. Validation of Ethnomedicinal Potential of *Tinospora cordifolia* for Anticancer and Immunomodulatory Activities and Quantification of Bioactive Molecules by HPTLC. *J. Ethnopharmacol.* **2015,** *175,* 131–137.

Bansal, P.; Malik M. A.; Das S. N.; Kaur, J. *Tinospora Cordifolia* Induces Cell Cycle Arrest in Human Oral Squamous Cell Carcinoma Cells. *Gulf J. Oncol.* **2017,** *1* (24),10–14.

Birla, H.; Rai, S. N.; Singh, S. S.; Zahra, W.; Rawat, A.; Tiwari, N.; Singh, R. K.; Pathak, A.; Singh, S. P. *Tinospora cordifolia* Suppresses Neuroinflammation in Parkinsonian Mouse Model. *Neuro. Mol. Med.* **2019,** *21,* 42–53.

Chaudhary, R.; Jahan, S.; Goyal, P. K. Chemopreventive Potential of an Indian Medicinal Plant (*Tinospora Cordifolia*) on Skin Carcinogenesis in Mice. *J. Environ. Pathol. Toxicol. Oncol.* **2008,** *27* (3), 233–243.

Dias, D. K.; Atukorala, C.; Amaratunga, N. A. D. S.; Perera, B.; Karunathilake, L. P. A. The Effects of *Emblica officinalis* and *Tinospora cordifolia* Herbal Treatment on the Prognosis of Squamous Cell Carcinoma of the Buccal Mucosa and Tongue. *J. Oral Maxillofacial Surg. Med. Pathol.* **2018,** *30* (1), 21–29.

Gao, L.; Cai, G. Beta-Ecdysterone Induces Osteogenic Differentiation in Mouse Mesenchymal Stem Cells and Relieves Osteoporosis. *Biol. Pharm. Bull.* **2008,** *31,* 2245–2249.

Jagetia, G. C.; Nayak V.; Vidyasagar M. S. Evaluation of the Antineoplastic Activity of Guduchi (*Tinospora cordifolia*) in Cultured HeLa Cells. *Cancer Lett.* **1998,** *127* (1–2), 71–82.

Jagetia, G. C.; Rao, S. K. Evaluation of the Antineoplastic Activity of guduchi (*Tinospora cordifolia*) in Ehrlich Ascites Carcinoma Bearing Mice. *Biol. Pharma. Bull.* **2006,** *29* (3), 460–466.

Javir, G.; Joshi, K. Evaluation of the Combinatorial Effect of *Tinospora cordifolia* and *Zingiber officinale* on Human Breast Cancer Cells. *3 Biotech* **2019,** *9* (11), art. no. 428.

Jayaprakash, R.; Ramesh, V.; Sridhar, M. P.; Sasikala, C. Antioxidant Activity of Ethanolic Extract of *Tinospora cordifolia* on N-Nitrosodiethylamine (Diethylnitrosamine) Induced Liver Cancer in Male Wister Albino Rats. *J. Pharm. Bioallied Sci.* **2015,** *7,* S40–S45.

Kapil, A.; Sharma, S. Immunopotentiating Compounds from *Tinospora cordifolia. J. Ethnopharmacol.* **1997,** *58,* 89–95.

Kapur, P.; Pereira, B. M. J.; Wuttke, W.; Jarry, H. Androgenic Action of *Tinospora cordifolia* Ethanolic Extract in Prostate Cancer Cell Line LNCaP. *Phytomedicine* **2009,** *16* (6–7), 679–682.

Kapur, P.; Wuttke, W.; Jarry, H.; Seidlova, D. W. Beneficial Effects of Beta-Ecdysone on the Joint, Epiphyseal Cartilage Tissue and Trabecular Bone in Ovariectomized Rats. *Phytomedicine* **2010,** *17,* 350–355.

Karuppath, S.; Snima, K. S.; Ravindranath, K. C.; Nair, S. V.; Lakshmanan, V. K. Anti-Proliferative Effect of *Tinospora cordifolia* Nano Particles in Prostate Cancer Cells. *J. Bionanosci.* **2016,** *10* (2), 127–133.

Leyon, P. V.; Kuttan, G. Inhibitory Effect of a Polysaccharide from *Tinospora cordifolia* on Experimental Metastasis. *J. Ethnopharmacol.* **2004,** *90* (2–3), 233–237.

Maliyakkal, N.; Appadath B. A.; Balaji, S. A.; Udupa, N.; Ranganath, P. S.; Rangarajan A. Effects of *Withania somnifera* and *Tinospora cordifolia* Extracts on the Side Population

Phenotype of Human Epithelial Cancer Cells: Toward Targeting Multidrug Resistance in Cancer. *Integr. Cancer Therap.* **2015,** *14* (2), 156–171.

Maliyakkal, N.; Udupa N.; Pai, K. S. R.; Rangarajan, A. Cytotoxic and Apoptotic Activities of Extracts of *Withania somnifera* and *Tinospora Cordifolia* in Human Breast Cancer Cells. *Int. J. Appl. Res. Nat. Products* **2013,** *6* (4)1–10.

Mishra, A.; Kumar, S.; Pandey, A. K. Scientific Validation of the Medicinal Efficacy of *Tinospora cordifolia*. *Sci. World J.* **2013,** *1–8.*

Mishra, R.; Kaur, G. Aqueous Ethanolic Extract of *Tinospora cordifolia* as a Potential Candidate for Differentiation-Based Therapy of Glioblastomas, *PLoS One 8*, **2013,** e78764.

Mishra, R.; Kaur, G. *Tinospora cordifolia* Induces Differentiation and Senescence Pathways in Neuroblastoma Cells. *Mol. Neurobiol.* **2015,** *52* (1), 719–733.

Mittal, J.; Pal, U.; Sharma, L.; Verma, A. K.; Ghosh, M.; Sharma, M. M. Unveiling the Cytotoxicity of Phytosynthesised Silver Nanoparticles using *Tinospora cordifolia* Leaves Against Human Lung Adenocarcinoma A549 Cell Line. *IET Nanobiotechnol.* **2020,** *14* (3), 230–238.

Mohan, V.; Koul, A. Anticancer Potential of *Tinospora cordifolia* and Arabinogalactan Against Benzo(a)pyrene Induced Pulmonary Tumorigenesis: A study in Relevance to Various Biomarkers *J. Herb Med. Pharmacol.* **2018,** *7* (4), 225–235.

Nemmani, K. V. S.; Jena, G. B.; Dey, C. S.; Kaul, C. L.; Ramarao, P. Cell Proliferation and Natural Killer Cell Activity by Polyherbal Formulation, Immu-21 in Mice. *Indian J. Exp. Biol.* **2002,** *40* (3), 282–287.

Palmieri, A.; Scapoli, L.; Iapichino, A.; Mercolini, L.; et al., Berberine and *Tinospora cordifolia* Exert A Potential Anticancer Effect on Colon Cancer Cells by Acting on Specific Pathways. *Int. J. Immunopathol. Pharmacol.* **2019,** *33*, 2058738419855567.

Pandey, V. K.; Amin, P. J.; Shankar, B. S. G1–4A, a Polysaccharide from *Tinospora cordifolia* Induces Peroxynitrite Dependent Killer Dendritic Cell (KDC) Activity Against Tumor Cells. *Int. Immunopharmacol.* **2014,** *23* (2), 480–488.

Patil, S.; Ashi, H.; Hosmani, J.; Almalki, A. Y.; Alhazmi, Y. A.; Mushtaq, S.; Parveen, P. et al. *Tinospora cordifolia* (Thunb.) Miers (Giloy) Inhibits Oral Cancer Cells in a Dose-Dependent Manner by Inducing Apoptosis and Attenuating Epithelial-Mesenchymal Transition. *Saudi J. Biol. Sci.* **2021,** *28* (8), 4553–4559.

Polu, P. R.; Nayanbhirama, U.; Khan, S.; Maheswari, R. Assessment of Free Radical Scavenging and Anti-proliferative Activities of *Tinospora cordifolia* Miers (Willd). *BMC Complement. Altern. Med.* **2017,** *17* (1), art. no. 457.

Rao, S. K.; Rao P. S.; Rao, B. N. Preliminary Investigation of the Radio-Sensitizing Activity of Guduchi (*Tinospora cordifolia*) in Tumor-Bearing Mice. *Phytother. Res.* **2008,** *22* (11), 1482–1489.

Rao, S. K.; Rao, P. S. Alteration in the Radiosensitivity of Hela Cells by Dichloromethane Extract of Guduchi (*Tinospora cordifolia*). *Integr. Cancer Therap.* **2010,** *9* (4), 378–384.

Rashmi, K. C.; Atreya, H. S.; Harsha, R. M.; Salimath, B. P.; Aparna, H. S. A Pyrrole-Based Natural Small Molecule Mitigates HSP90 Expression In MDA-MB-231 Cells and Inhibits Tumor Angiogenesis in Mice by Inactivating HSF-1. *Cell Stress Chaperones* **2017,** *22* (5), 751–766.

Rashmi, K. C.; Harsha, R. M.; Paul M.; Girish K. S.; Salimath, B. P.; Aparna, H. S. A New Pyrrole Based Small Molecule from *Tinospora cordifolia* Induces Apoptosis in MDA-MB-231 Breast Cancer Cells via ROS Mediated Mitochondrial Damage and Restoration of P53 Activity. *Chemico-Biol. Interact.* **2019,** *299*, 20–130.

Sharma, N.; Kumar, A.; Sharma, P. R.; Qayum, A.; et al., A New Clerodane Furano Diterpene Glycoside from *Tinospora cordifolia* Triggers Autophagy and Apoptosis in HCT-116 Colon Cancer Cells. *J. Ethnopharmacol.* **2018**, *211*, 295–310.

Sharma, P.; Bharat P. D.; Dheeraj B., Dash, A. K.; Kumar D. The Chemical Constituents and Diverse Pharmacological Importance of *Tinospora cordifolia*. *Heliyon* **2019**, *5*, e02437.

Sharma, S.; Chauhan, R.; Dwivedi, J. Evaluation of Combined Herbal Extract of *Withania somnifera, Ocimum sanctum* and *Tinospora cordifolia* as Chemo-protective in Cancer. *Pharmacologyonline* **2011**, *2*, 619–625.

Singh, N., Singh S. M., Prakash, Singh G. Restoration of Thymic Homeostasis in a Tumor-Bearing Host by In Vivo Administration of Medicinal Herb *Tinospora cordifolia*. *Immunopharmacol. Immunotoxicol.* **2005a**, *27* (4), 585–599.

Singh, N., Singh S. M., Shrivastava P. Effect of *Tinospora cordifolia* on the Antitumor Activity of Tumor-Associated Macrophages-Derived Dendritic Cells. *Immunopharmacol. Immunotoxicol.* **2005b**, *27* (1), 1–14.

Singh, N.; Singh, S. M.; Shrivastava, P. Immunomodulatory and Antitumor Actions of Medicinal Plant *Tinospora cordifolia* Are Mediated Through Activation of Tumor Associated Macrophages. *Immunopharmacol. Immunotoxicol.* **2004**, *26*, 145.

Singh, N.; Singh, S. M.; Shrivastava, P. Immunomodulatory Effect of *Tinospora cordifolia* in Tumor Bearing Host. *Orient. Pharm. Exp. Med.* **2003**, *3*, 72.

Singh, S. M.; Singh, N.; Shrivastava, P. Effect of Alcoholic Extract of Ayurvedic Herb *Tinospora cordifolia* on the Proliferation and Myeloid Differentiation of Bone Marrow Precursor Cells in a Tumor-Bearing Host. *Fitoterapia.* **2006**, *77* (1), 1–11.

Subash-Babu, P.; Alshammari, G. M.; Ignacimuthu, S.; Alshatwi, A. A. Epoxy Clerodane Diterpene Inhibits MCF-7 Human Breast Cancer Cell Growth by Regulating the Expression of the Functional Apoptotic Genes Cdkn2A, Rb1, mdm2 and p53. *Biomed. Pharmacother.* **2017**, *87*, 388–396.

Verma, R.; Chaudhary, H. S.; Agrawal, R. C. Evaluation of Anticarcinogenic and Antimutagenic Effect of *Tinospora cordifolia* in Experimental Animals, *J. Chem. Pharm. Res.* **2011**, *3*, 877–881.

CHAPTER 11

Taxus brevifolia (Nutt.) Pilger.

K. VENKATA RATNAM[1], LEPAKSHI MD. BHAKSHU[2],
PULALA RAGHUVEER YADAV[3], ANU PANDITA[4], DEEPU PANDITA[5], and
K. M. SRINIVASA MURTHY[6]

[1]*Department of Botany, Rayalaseema University, Kurnool, Andhra Pradesh, India*

[2]*Department of Botany, PVKN, Government College (Autonomous), Chittoor, Andhra Pradesh, India*

[3]*Department of Biotechnology, Indian Institute of Technology, Hyderabad, Kandi, Telangana State, India*

[4]*Vatsalya Clinic, Krishna Nagar, New Delhi, India*

[5]*Government Department of School Education, Jammu, Jammu and Kashmir, India*

[6]*Department of Microbiology and Biotechnology, Jnanabharathi Campus Bangalore University, Bengaluru, India*

ABSTRACT

Taxus brevifolia (Nutt.) Pilger., known as pacific yew or mountain mahogany of the family Taxaceae, native of North America, distributed in coastal areas of Alaska to California. Local tribes of these areas used pacific yew to treat lung and skin diseases. Economically the plant has been used to make weapons, canoe paddles, and drum frames, etc. The major chemical constituents of the pacific yew are diterpene alkaloids called as taxanes. Some of the notable active principles of taxane family are paclitaxel, docetaxel, and 10-deacetylbaccatin III. Anticancer activity of taxol and its derivatives have been reported by several authors in different cancer cell lines like liver,

Potent Anticancer Medicinal Plants: Secondary Metabolite Profiling, Active Ingredients, and Pharmacological Outcomes. Deepu Pandita and Anu Pandita (Eds.)

prostate, lung, pancreatic, and breast. The components showed its mechanism of action in two ways, that is, mitotic action and apoptotic action. In mitotic action, the components target the formation microtubules, which play a crucial role in chromosome separation during cell division. In apoptotic action, the principles inactivate the proapoptotic genes such as Bcl-2 and p53 and activate the apoptosis-induced genes, that is, caspase-3 family.

11.1   INTRODUCTION

Taxus brevifolia (Nutt.) Pilger., known as pacific yew or mountain mahogany (family Taxaceae), native of North America and distributed in the coastal areas of Alaska to California (Little, 1979; Munz, 1973). The species belongs to Gymnosperm plant group. It is a slow growing coniferous tree species. Traditionally, the wood has been used to prepare weapons, poles, wooden boxes, drum frames, bows, and tool handles by the native Americans (Gunther, 1945; Hartzell, 1991). The chief phytochemical constituents of the pacific yew are diterpene alkaloids called as taxanes. Some of the notable active principles of taxane family are paclitaxel, docetaxel, and 10-deacetyl-baccatin III. These components showed their beneficial effects on cancer by stabilizing the microtubules present in the cells and by suppressing the mitotic cell division of cancer cells (Sinha, 2020). In the present chapter, we discussed anticancer activity of *T. brevifolia* extracts, nanoparticles, and active principles published by reviewing some of the noteworthy publication available online.

11.2   ANTICANCER ACTIVITY OF TAXOL/TAXANES

Taxol, a terpenoid substance, isolated from *Taxus* species, is widely used as a chemotherapeutic drug to treat various types of cancers. Paclitaxel generic name of taxol was reported to have strong tumor suppressing activity in distinct cancer cell lines such as ovarian, lung, breast, leukemia, and malignant melanoma cancers (Nwafor et al., 2001). A clinical experiment study results revealed that taxol exhibited noteworthy toxicity against leukocytes than thrombocytes and reticulocytes in a 22 days experimental schedule (McGuire et al., 1989). A sequence clinical therapy of taxol with cisplatin was assessed in untreated and pre-treated solid tumor patients, showed that sequence with cisplatin before taxol showed very less cytotoxic activity (Rowinsky et al., 1991). Anticancer and free radical production capacity of

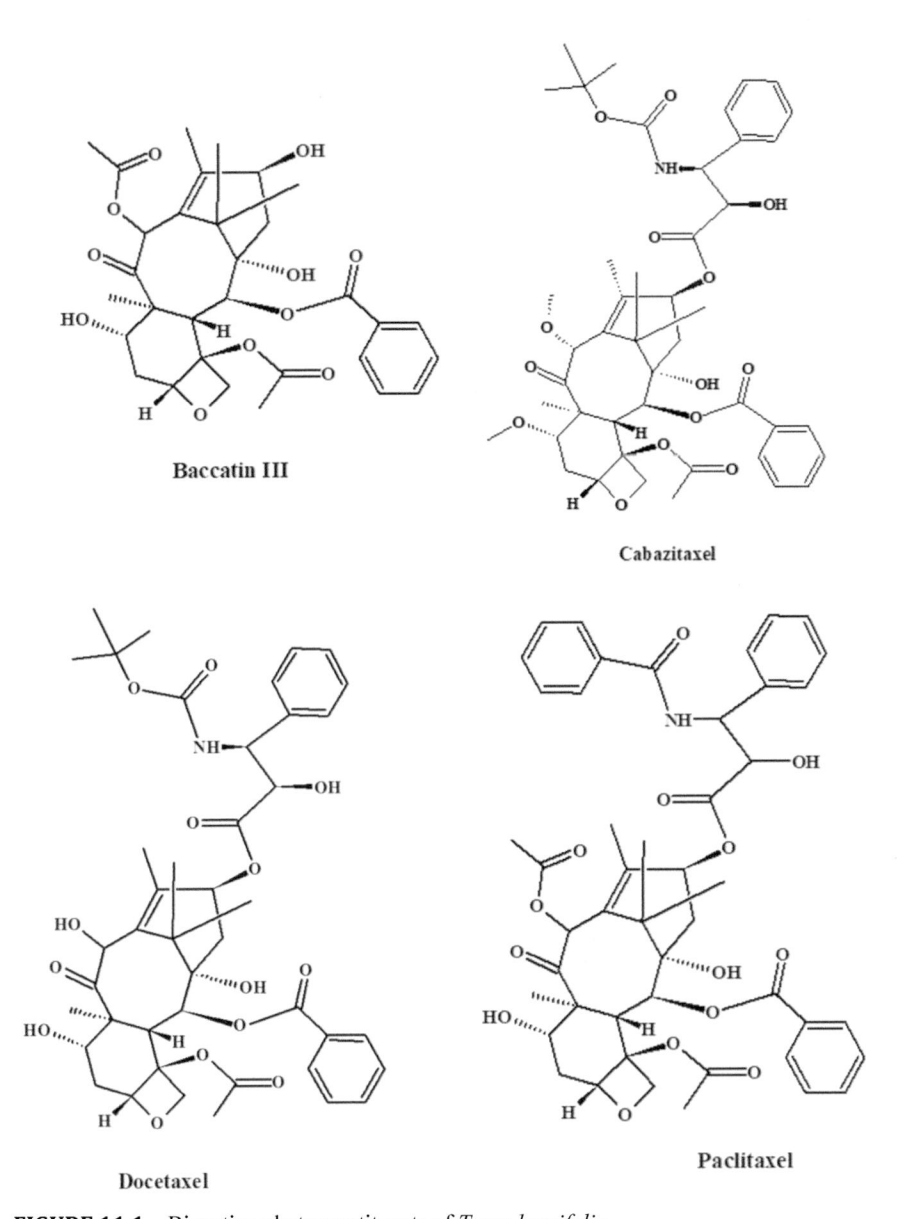

FIGURE 11.1 Bioactive phytoconstituents of *Taxus brevifolia.*

taxol was assessed against macrophages, neutrophils, and murine macrophages using a chemiluminescence indicator. Taxol modulated the phorbol myristate acetate-induced chemiluminescence in neutrophils and macrophages but not in murine macrophages (Czuba et al., 1998). While Manfredi et al. (1982) investigated the binding capacity of taxol to microtubules in murine macrophages. The drug showed maximal binding capacity at higher toxic doses. Cytotoxic efficiency of taxol along with cisplatin was studied in cancer cells, indicated that both the tested drugs significantly suppressed the cell proliferation by arresting the cell cycles events such as G1/S and G2 phases (Donaldson et al., 1994).

Antimitotic effect of paclitaxel was assessed in a murine xenograft human cancer model using high resolution *in vivo* microscopy method. The microscopic visuals revealed that paclitaxel altered the spindle assembly, mitotic cell division arrest, formation of multinucleate cells, and apoptosis (Orth et al., 2011). A comparative cytotoxic activity of taxol, cisplatin, and Taxotere was investigated against 12 human normal bone marrow cells, 25 tumor cells, and 35 primary cultures using *in vitro* studies. The results showed that taxol and docetaxel both are exhibited the potent antitumor activity than cisplatin (Braakhuis et al., 1994). Bissery et al. (1991) reported potent antitumor activity against B16 melanoma cells. It further cured colon adenocarcinoma and pancreatic ductal adenocarcinoma at earlier stage. The pre-treatment of different types of cancer cells with docetaxel for 24-h period time, showed 50% survival capacity of cervix, gastric, and pancreatic cell lines (Balcer-Kubiczek et al., 2006). Fabbri et al. (2006) conducted a cytometric analysis to understand the apoptotic mechanism of docetaxel against cancerous cells. The mechanism of action of docetaxel indicated that it induced cell shrinkage and mitochondrial membrane transition, reduced cell granularity, and DNA fragmentation.

The efficacy and toxicity of a new nano-formulation prepared using paclitaxel and abraxane were investigated in B16 melanoma xenografts growing in athymic mice, indicated that paclitaxel-based nano-formulation strongly inhibited the tumors in mice than abraxane (Feng et al., 2010). Ferlini et al. (2009) investigated the antitumor mechanism of paclitaxel in cancer cell lines, revealed that paclitaxel binds to the sites of tubulin and Bcl-2 sites, but functionally mimic it as Nurr77, a member pf nuclear subfamily, its expression and localization are closely connected with cell proliferation and programmed cell death. Derry et al. (1995) investigated the cytotoxic effects of taxol at different concentrations (10 nM to 1 µM) on the growth and condense dynamics of microtubules in bovine brain cells under *in vitro* conditions. At lower concentrations, taxol selectively suppressed the shortening the plus ends of microtubules and slightly increased the mass of microtubule protein. At intermediate concentration (100 nM to 1 µM), it inhibited the growth and condensation of microtubules at both ends and stopped the additional increase of microtubules mass. At higher concentrations (more than 1 µM), microtubular mass was sharply increased and microtubular dynamics were completely suppressed. A comparative anti-angiogenic effect of docetaxel and paclitaxel was investigated on human umbilical vein endothelial cells, demonstrated that endothelial cells are strongly inhibited with both the tested drugs (Grant et al., 2003).

The anticancer efficiency of paclitaxel at seven different concentrations was assessed against FM3A cell lines. It significantly suppressed the growth of cells and increased dead cell number. It also induced morphological changes like bleb formation and nuclear fragmentation (Ilgar and Arican, 2009). DNA fragmentation was confirmed with agarose gel electrophoresis. Antitumor activity of nanocrystals of paclitaxel was investigated in cancer cell lines and xenograft animal models. It demonstrated that nanocrystals of paclitaxel showed better therapeutic effects over taxol in both the studied models (Liu et al., 2010). Inoue et al. (2003) studied the docetaxel-enhanced antitumor effect of TNP-479 in 253J B-V growing human transitional cell carcinoma cell lines (*in vivo)* and HUVE cells (*in vitro*). Docetaxel increased anti-proliferation activity of TNP-479 in HUVEC cells, but no effect was observed on 253J B-V. The combination of docetaxel and TNP-479 was significantly suppressed the intra-tumor neovascularization. The cytotoxic activity of docetaxel-loaded lipid microbubbles was investigated on liver tumors of VX2 rabbits. It showed that higher tumor suppression and apoptotic index were observed in treated animals. Higher levels of caspase-3 enzymes levels indicated its antitumor activity (Kang et al., 2010).

11.2.1 LIVER CANCER

Anticancer activity and to enhance the bioavailability of paclitaxel (PTX), a biocompatible formulation prepared using dietary lipidic-based nanocarriers (NLCs) was investigated in the HepG2 cell line. The formulation along with PTX, significantly inhibited cell proliferation in a concentration-dependent manner and enhance ROS levels in HepG2 cell lines (Harshita et al., 2019). Wan et al. (2018) studied the anticancer properties of docetaxel and gold nanoparticles encapsulated apatite carrier against human liver cancer (HepG2) cells. Both docetaxel and nanoparticles showed noteworthy cell toxicity against the tested cancer cell lines. Docetaxel carboxymethyl coated cellulose nanoparticles strongly suppressed the tumor growth in mice. At the molecular level, it significantly altered the expression levels of collagen I and α- smooth muscle actin in hepatic satellite cells in a dose-dependent manner (Chang et al., 2018). Cabazitaxel is a semi-synthetic derivative of a natural component taxoid, significantly arrested the cell cycle progression in Huh-7 and SK-hep-1 cell lines. At the molecular level, it suppressed the expression levels of different types of cyclins such as Cdc2, Cdc25c, and pCdc2 at protein levels (Chen et al., 2018).

Anticancer effects of docetaxel against human hepatocellular carcinoma cells (SMMC-7721 HCC) were investigated. Colony-forming assay results revealed that docetaxel exhibited strong and concentration-dependent anti-proliferative activity. Flow cytometry and fluorescence microscopy results indicated that docetaxel arrested the cell cycle at the G2 phase of Mitotic division. Further, it strongly induced free radical generation and inhibited glutathione levels in cells (Geng et al., 2003). A trial of docetaxel-based chemotherapy was conducted in patients with advanced liver cancer, revealed that the drug was not safe for patients with advanced hepatocellular carcinoma (Hebbar et al., 2006).

11.2.2   LUNG CANCER

Nanoparticles encapsulated with PEG, which has detachable properties to transport siRNA along with paclitaxel against survivin gene used in lung cancer therapy. The nanoparticles capsulated with survivin siRNA effectively bring down the expression of mRNA of survivin and enhanced the uptake of nanoparticles by the cell. In the cell counting assay, nanoparticles expressed low toxicity and paclitaxel presented good antiproliferative activity in A549 cells (Jin et al., 2018). Min et al. (2018) investigated the anticancer efficiency of paclitaxel combined with caffeic acid using *in vivo* and *in vitro* methods against non-small-cell lung cancer h1299 cells. Paclitaxel along with caffeic acid exhibited noteworthy cytotoxic activity in h1299 cells by inducing apoptosis. At the molecular level, both significantly activated protein markers of the MAPK pathway. Anticancer activity of mixed micelles loaded with paclitaxel was assessed in A549, Caco-2 cells, and Lewis C57BL/6 mice. The results demonstrated that mixed micelles loaded with paclitaxel exhibited noteworthy cytotoxic effects against the tested cell lines models. In *in vivo* studies, paclitaxel-loaded mixed micelles showed better activity than paclitaxel alone treated Lewis C57BL/6 mice (Hou et al., 2017). Patel et al. (2016) assessed the tumor-disrupting activity of docetaxel-loaded PEGylated liposomes. Docetaxel-loaded PEGylated liposomes expressed high levels of physical stability and very feeble hemolysis. It further showed higher levels of bioavailability with significantly higher intra-tumoral uptake of the drug.

Antitumor activity of exisulind in combination with docetaxel was carried out using *in vivo* and *in vitro* experimental models. The *in vitro* study results revealed that both the tested competent increased the apoptosis in A549 human lung cancer cells. Whereas *in vivo* results indicated that both components moderately increased the survival capacity of nude rats with A549 orthotopic lung cancer, reduced tumor formation, and increased apoptosis in comparison with control rats (Chan et al., 2002). Anticancer effect of docetaxel was assessed in human lung cancer cell lines H1299 and A549 to understand the mechanism of action in *in vitro* conditions. The p53-null H1299 cells treated with docetaxel for 18-h period enhanced beta-tubulin gene expression levels and strongly induced the p53 RNA levels in A549 cells (Chang et al., 2006). The efficacy and toxicity of a new nano-formulation prepared using paclitaxel and abraxane were investigated against NCI H460 human lung cancer, indicated that paclitaxel-based nano-formulation showed significant anti-tumor activity than abraxane (Feng et al., 2010).

Antitumor effect of paclitaxel at two concentrations (low concentration 35 nM and higher concentration 70 µM) against human lung adenocarcinoma cells, revealed that paclitaxel at low concentration, it exhibited the potent cytotoxic effect by inducing DNA fragmentation, caspase-3, phosphatidylserine (PS) externalization, Bax translocation into mitochondria, and cell cycle arrest at G2/M phase. At higher concentrations, it induced vacuoles in cytoplasm and swellings in mitochondria and paraptosis like cell death (Guo et al., 2010). Antitumor activity of docetaxel and diindolylmethane was investigated using *in vitro* and *in vivo* models, demonstrated the increased apoptotic cell percentage, expression of Bax, and N-cadherin and cleaved poly (ADP-ribose) polymerase in A549 cells. *In vivo* studies indicated that the combination treatment significantly reduced the lung tumors in mice models (Ichite et al., 2009).

11.2.3 BREAST CANCER

Anticancer efficacies of nanoparticle-encapsulated versus free cabazitaxel were investigated in the xenograft mice model. The nanoparticle-encapsulated drug resulted in 75% reduction in tumors growth; whereas drug alone gave complete remission only 22% reductions in tumors growth (Fusser et al., 2019). Cytotoxic studies of lipospheres co-loaded with thymoquinone along with cabazitaxel were investigated against breast cancer cell lines like MCF-7 and MDA-MB-231. It revealed that in cell cycle analysis combination of cabazitaxel and thymoquinone co-loaded lipospheres significantly increased Sub G1 phase arrest and cell death in MCF-7 and MDA-MB-231 breast cancer cells (Kommineni et al., 2018). Anti-breast cancer efficiency of solid lipid nanoparticles loaded with cabazitaxel and hyaluronic acid was assessed in MCF-7 cells. Increased cytotoxicity was noted in nanoparticles-treated cells than with the cells treated with drug (Zhu and An, 2017). Anticancer effects of cabazitaxel-loaded micelles revealed that cabazitaxel-loaded micelles did not show much effect on the survival capacity of 4T1 cells, but remarkably reduced the cell migration activities (Zhong et al., 2017).

Antitumor efficiency of keratin nanoparticles loaded with paclitaxel was investigated against breast cancer cell lines such as MDA MB 231 and MCF-7 using 2D and 3D models. The nanoparticles significantly suppressed the proliferation of cancer cells in 2D models. The nanoparticles showed their anticancer mechanism by increasing the proapoptotic genes levels and proteins (Foglietta et al., 2018). Combined anticancer effect of sorafenib, paclitaxel, and radiation therapy was assessed in renal cell carcinoma and breast cancer xenograft models. It revealed that combination therapy exhibited more significant anticancer effects than cotreatment with paclitaxel or sorafenib and radiation (Choi et al., 2019). The anticancer activity of nanoparticles encapsulated with paclitaxel and gemcitabine, revealed that it enhanced anticancer activity than the paclitaxel and gemcitabine treated alone (Dong et al., 2018). Antiproliferative activity of paclitaxel on two cancer cell lines such as MCF-7 and Hela cell lines, revealed that paclitaxel strongly inhibited the growth of MCF-7 cells, and very less cytotoxic activity was observed against Hela cell lines (Ryang et al., 2019).

Docetaxel nanoformulation was prepared using folic acid, thiol, and chitosan to enhance cellular internalization and to improve oral absorption of docetaxel in *in vitro* and *in vivo* models. The formulation notably inhibited the cancerous cell proliferation and enhanced the successful transport of drug across the intestine in *in vivo* model and significantly improved the oral bioavailability of drug (Sajjad et al., 2019). Further, Fard et al. (2017) studied anticancer effects of combined docetaxel therapy and radiation therapy on MCF-7 cancer cell lines. Significant synergetic dose-dependent cytotoxicity was observed in neoadjuvant therapy when compared to monotherapy models. Antiproliferative efficiency of paclitaxel-loaded selenium nanoparticles was assessed against breast, cervical, and colon cancer cell lines. Nanoparticles showed noteworthy antiproliferative activity against the tested cancerous cell lines. It showed their mechanism of action by seizing the G2 phase of the M phase leading to apoptosis (Bidkar et al., 2017).

The cytotoxic effect of taxol was carried out on MCF-7 breast cancer and p53 null cell lines using *in vitro* studies. Taxol exhibited its anticancer activity mechanism by suppressing the cyclin inhibitor p21WAF1 in both cell lines (Blagosklonny et al., 1995). Further, they elucidated the anticancer mechanism of taxol, showed that Bel-2 phosphorylation will occur after the activation of Raf-1 protein. The cytotoxic activity of taxol was time and dose-dependent (Blagosklonny et al., 1996). Brown et al. (2004a) identified a new anticancer mechanism of action against docetaxel resistant two types of breast cancer cell lines, that is, MCF-7 and MDA-MB-231 using the cDNA microarray method. The reduction in p27 expression levels is associated with acquired docetaxel-resistant cancer cells. Chan et al. (2010) evaluated the anticancer efficiency of Bevacizumab with the combination of taxanes in metastatic breast cancers. The combined therapy exhibited more beneficial effects in a broad range of cancer patients than taxanes alone treatment.

A clinical study with taxane-based neoadjuvant chemotherapy was conducted in 45 breast cancer patients with stage II/IIIA. They suggested that patients with tumor size ≤4 cm after chemotherapy with taxanes are best candidates for breast conservation therapy (El-Sayed et al., 2010). A combined cytotoxic effect of docetaxel with oestrogen receptor (ICI 182,780) was assessed against two breast cancer cell lines, that is, MCF-7 ADRr and MDA-MB 231. Both the tested drugs showed synergetic cytotoxic effects on tested cell lines. ICI 182,780 increased the docetaxel activity. The mechanism involved is inhibition of P-glycoprotein activity, seizing G2 phase of M phase and inactivation of Bcl-2 protein and induction of apoptosis (Ferlini et al., 1998). Tumor suppression activity of taxol, vincristine, colchicine, and 2-Methoxyestradiol (2-ME) was investigated on C57BL/6 mice model. 2-ME and taxol exhibited moderately inhibited the expression levels of VEGF-induced neovascularization and basic fibroblast growth factor, but vincristine and colchicine showed no cytotoxic effect (Klauber et al., 1997).

Antiangiogenic and antitumor efficiency of paclitaxel, trastuzumab with/without combination was evaluated on mice treated with ErbB-2-overexpressing breast carcinoma cells. In combined therapy with paclitaxel/trastuzumab, exhibited significant reduction in tumor volume and micro-vessel density than control group. The antitumor activity of both the components may be mediated *via* the reduction of phosphorylated Akt (Klos et al., 2003). A comparative clinical study was conducted to evaluate anticancer activity of paclitaxel and with or without bevacizumab in metastatic breast cancer patients. Combination therapy with bevacizumab/paclitaxel extended the development of free survival rate but not overall survival rate (Miller et al., 2007).

11.2.4 PROSTATE CANCER

Anticancer effects of paclitaxel or noscapine alone and with the combination were investigated against prostate cancer cell (PCC) lines such as PC-3 and LNCaP. Both the cells treated with paclitaxel and noscapine significantly decreased the mRNA expression of B-cell CLL/Lymphoma and enhanced the expression levels of Bcl-2-associated X protein (Rabzia et al., 2017). The mechanism and synergetic effects of docetaxel combined with acankoreano-gein or impressic acid was studied by (Jiang et al., 2018). Anticancer potential of cabazitaxel was investigated in prostate cancer cell lines. It revealed that the component significantly suppressed the functionality of androgen receptor and its associated proteins HSP40, HSP90α, and HSP70/HSP90 (Rottach et al., 2019). Anticancer effects of hyaluronic acid encapsulated cationic liposomes of cabazitaxel was assessed against DU-145 and PC-3 cancer cell lines (Mishra et al., 2019). It revealed that liposomes exhibited potent cytotoxic activity with low IC_{50} values by inhibiting cell migration and inducing apoptosis. Further, the liposomes showed its synergetic cytotoxic effects on CD^{44+}-labeled DU-145 and PC-3 cancer cell lines.

Anticancer efficiency of nano-formulation prepared of cabazitaxel was investigated against metastatic prostate cancer. It showed that the formulation significantly reduced the pain of the bone tumor and improved bone structure (Gdowski et al., 2017). The combined anticancer efficiency of docetaxel and desmopressin was studied against the castrate-resistant prostate cancer cell (CRPC) lines-induced tumors in orthotopic nude mice models. The combination therapy significantly reduced the CRPC cell lines-induced tumor volume in mice model (Bass et al., 2019). Gallego-Yerga et al. (2017) investigated the cytotoxic effect of nanoparticles synthesized from docetaxel against human PCC lines PC3 and LnCap and glioblastoma rat C6 and human U87 cell lines. The nanoparticles encapsulated with docetaxel exhibited dose-dependent antitumor effects against the tested prostate and glioblastoma cell lines.

Combined cytotoxic effects of docetaxel- and angiostatin-mediated gene therapy were evaluated on PCC lines LNCaP, PC3, and DU145 and human endothelial cells. It revealed that human endothelial cells are more sensitive to docetaxel than prostate cancer cell PCC lines. Among the tested PCC lines, PC3 cells are more sensitive to docetaxel. Combined therapy exhibited significant antitumor effects in PC3-grafted athymic mice (Galaup et al., 2003). Anticancer efficiency of taxanes, that is, paclitaxel and docetaxel in combination with methylseleninic acid were evaluated on PCC lines in *in vitro* methods and *in vivo* method using DU145 xenograft mice model. It showed that combination therapy exerted good cytotoxic effects on tested PC-3 and DU145 cells. In *in vivo* study, the combination of paclitaxel/meth-ylseleninic acid was significantly inhibited tumors in xenograft mice (Hu et al.,2008).

The *in vitro* and *in vivo* experiments were conducted by Jiang and Huang (2010) to demonstrate the antitumor activity of taxol on 22Rv1 PCC lines and 22Rv1 xenografts mice models, revealed that *in vitro* experiment taxol upregulated PTEN, tumor suppressor gene, and inhibited transcripts of the androgen receptor. In *in vivo* study, it induced mitotic arrest and reduced the expression levels of the androgen receptor. The antimitotic effect of docetaxel and all-trans retinoic acid (ATRA) was assessed in drug and hormone-resis-tant prostate cancer cell lines, DU-145. The results demonstrate that both the tested components exhibited time and dose-dependent cytotoxic activity and strongly reduced the expression of lymphotoxin beta-receptor, surviving, and myeloid cell leukemia-1 at transcript levels (Kucukzeybek et al., 2008).

11.2.5 PANCREATIC CANCER

Anticancer efficiency of polymeric nanoparticles prepared using docetaxel along with IR in pancreatic cancer therapy was assessed (Park et al., 2018). The combination of nanoparticles along with IR showed a strong cytotoxic effect against the tested cell lines models. They concluded that this method is a novel and promising radio-sensitizing agent.

11.2.6 OVARIAN CANCER

Combined therapy of paclitaxel with cyclophosphamide and cisplatin was assessed in ovarian cancer patients, revealed positive clinical benefits to human ovarian cancer patients (Reed et al., 1995). Cytotoxic efficiency of taxol was investigated against ovarian cancer cells such as human A2780 and hamster CHO. It revealed that taxol was more toxic to mitotic phase cells but not interphase cells. Taxol exhibited a potent and noteworthy cytotoxic effect against human A2780 cells than CHO cells (Lopes et al., 1993). Antitumor effects of taxol against murine mammary and ovarian cancer cells induced tumors in mice were studied using *in vivo* experiment. The kinetic study results revealed that taxol induced antitumor effect in a concentration-dependent manner (Milas et al., 1995). Lorigan et al. (1996)

highlighted the introduction of paclitaxel as a potent chemotherapeutic drug to treat ovarian cancer. The anticancer effect of taxol was assessed in patients with three stages of ovarian cancer, that is, advanced, progressed, and drug refractory. Taxol significantly reduced cancer lesions drug refractory stage patients (McGuire et al., 1989). Nicoletti et al. (1993) reported the antitumor potential of taxol against ovarian cancer tumors growing in the peritoneal cavity of mice. Taxol exhibited notable enhancement in the survival time of mice with tumors than the reference drug cisplatin. A comparative cytotoxic study of natural and synthetic taxol was conducted against ovarian cancer cell lines using *in vitro* methods. Results revealed that taxol exhibited concentration-dependent cytotoxic activity against the tested cell lines. No significant difference was observed between synthetic and natural taxol (Modarresi-Darreh et al., 2018).

A combined clinical study was carried out with taxanes with or without platinum in 222 patients with sex cord-stromal ovarian tumors. Pre-treatment with taxanes showed significant antitumor activity in the tested patients. Taxanes in combination with platinum reduced the disease symptoms in patients (Brown et al., 2004b). du Bois et al. (2009) conducted a randomized clinical experiment with two combinations, that is, cisplatin/paclitaxel and carboplatin/paclitaxel in 798 ovarian cancer patients. Of the two combinations tested, carboplatin/paclitaxel combination affords better anticancer effects associated with the quality of life with better tolerability than cisplatin/paclitaxel combination. The efficacy and toxicity of a new nano-formulation prepared using paclitaxel and abraxane were investigated against 2008 human ovarian cancer, indicated that paclitaxel-based nano-formulation showed significant anti-tumor activity than abraxane (Feng et al., 2010).

11.2.7 CERVICAL CANCER

Microtubule suppression activity of taxol isolated from *T. brevifolia* was investigated against HeLa cell lines. No effect was observed at concentrations on DNA, RNA, and protein synthesis in cancer cell lines during the treatment period. The component significantly promoted the assembly of calf brain microtubules but decreased the lag time for microtubule assembly and the critical concentration of tubulin required for assembly (Schiff et al., 1979). Cytotoxic effects of paclitaxel assessed in HeLa cell demonstrated that the component induced breaks in DNA into nucleosome-sized fragments, which is a characteristic feature of apoptosis (Jordan et al., 1996). Further, taxol inhibited the proliferation of HeLa cells at 8 nM concentration. The mechanism of action showed that taxol altered the formation of metaphase plate and spindle fiber during metaphase cell division (Jordan et al., 1993).

11.2.8 LEUKEMIA

Anti-proliferative activity of taxol against leukemia cell lines, that is, HL-60 and KG-1 cells, demonstrate that taxol induced internucleosomal DNA fragmentation and apoptosis. It exhibited marked inhibition against HL-60 cells. It showed its anticancer mechanism by reducing the Bcl-2 and c-myc transcription factor levels (Bhall et al., 1993). A combined cytotoxic effect of docetaxel with oestrogen receptor (ICI 182,780) was assessed against two leukemia cell lines, that is, CEM VBLr and CEM. Both the tested drugs showed synergetic cytotoxic effects against the tested cell lines. ICI 182,780 increased the docetaxel activity. The mechanism involved is the inhibition of P-glycoprotein activity, seizing of the G2 phase of M phase, inactivation of Bcl-2 protein, and trigger apoptosis (Ferlini et al., 1998).

11.2.9 GASTRIC CANCER

Docetaxel and radiotherapy (hypersensitivity)-based cytotoxic effects against gastric cancer cell lines, revealed that hypersensitivity was not correlated with ATM kinase-dependent early G2 phase checkpoint arrest (Balcer-Kubiczek et al., 2008). Chang et al. (2005) investigated the combined anticancer effect of carboplatin and paclitaxel in patients with advanced gastric cancer, treated with 5-fluorouracil and platinum. The results revealed that the combination therapy showed beneficial effects to reduce the gastric cancer in patients.

11.2.10 EPITHELIAL CANCER

Cytotoxic activity of docetaxel was assessed in human epidermoid KB, breast HCC1937, and colon HT-29 cancer cells. The pre-treatment of cells with docetaxel for 48 h resulted in increased levels of RAS protein, the phosphory-lated isoforms of Erk-1/2 and Raf-1. In another study, they used R115777, a farnesyl transferase inhibitor that suppresses RAS, along with docetaxel, resulted in strong synergetic inhibition of the growth of the tested cancer cell lines (Caraglia et al., 2005). Epithelial cancer (A549) cells treated with pacli-taxel demonstrated that paclitaxel at low concentrations (3–6 nM) significantly inhibited its proliferation. It upregulated both p53 and p21 genes, which leads to cell cycles arrest at G1 and G2 phases (Giannakakou et al., 2001).

11.2.11   COLON/ COLORECTAL CANCER

The antimitotic activity of toxoids (paclitaxel and docetaxel) and nocodazole was evaluated using RT-PCR, immunoblot, and flow cytometry analysis assays against HT29-D4 cell line. Both the components exhibited 30–50% inhibition in c-myc induction in HT29-D4 cell cultures. Flow cytometry analysis revealed that c-myc inhibition is not linked to the mitotic arrest (el Khyari et al., 1997). Colon cancer (HCT 116) cells treated with paclitaxel demonstrated that paclitaxel at low concentrations (3–6 nM) exhibited cytotoxic effects with no significant activity (Giannakakou et al., 2001). The anti-angiogenetic activity of non-toxic concentration of docetaxel was investigated on LS174T Cells. It significantly suppressed the expression levels of transcripts such as basic fibroblast growth factor, matrix metalloproteinase-2, matrix metalloproteinase-9, and VEGF at both gene and protein levels (Guo et al., 2003).

11.2.12   HEAD AND NECK SQUAMOUS CELL CARCINOMA

Cytotoxic effects of bevacizumab combined with paclitaxel against head and neck squamous cell carcinoma cell lines and tumor xenografts mice models, resulted in strong inhibition of tumors in xenografts mice models when compared to individual components (Fujita et al., 2007).

11.2.13   FIBRO SARCOMA

A comparative anti-tumor/anti-angiogenic effect of docetaxel and paclitaxel was investigated on fibro sarcoma cells, demonstrated that docetaxel exhibited noteworthy anti-angiogenic activity in fibro sarcoma cells (Grant et al., 2003).

11.2.14   THYROID CANCER

A combined treatment taxanes with dehydroxymethylepoxyquinomicin (DHMEQ) on anaplastic thyroid cancer cells, demonstrate that it strongly induced programmed cell death, and expression of poly (ADP-ribose) polymerase, caspase 3, and survival. In the xenograft animal model, the combined therapy significantly exhibited the greater antitumor activity (Meng et al., 2008).

11.3   ANTICANCER ACTIVITY OF BREVIFOLIOL

Brevifoliol is another diterpenoid cyclodecane (Figure 11.2), isolated from needles of Himalayan yew tree reported to have potent anti-proliferative activity against all cancer cell lines than paclitaxel and it exhibited note-worthy cytotoxic activity against colon cancer cell lines than taxol. It showed its mechanism of action by inducing apoptosis through the process of apoptosomes formation in $CaCo_2$ cell lines (Chattopadhyay et al., 2008; Kaur et al., 2014).

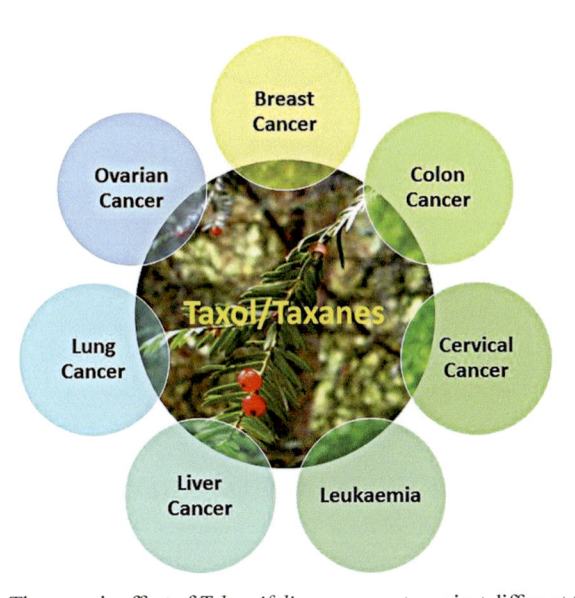

FIGURE 11.2   Therapeutic effect of *T. brevifolia* components against different types of cancers.

11.4   ANTICANCER ACTIVITY OF BACCATIN III

Tumor suppression activity of baccatin III, a precursor used to synthesize taxol, in combination with the antitumor drug gemcitabine was investigated. The combination of gemcitabine with baccatin III was resulted in a signifi-cant reduction in tumor growth. Baccatin III alone enhanced the immune adjuvant activity (Kim et al., 2011). A comparative antitumor effect of standard baccatin III and biochemically produced baccatin III was against four human cancer cell lines such as cervical, lung, skin, and liver cancers. Enzymatically synthesized baccatin III showed its potent cytotoxic effects against cervical cancer cell lines followed by lung, skin, and liver cancer cell lines. Further, it suppressed the free radical formation in the tested cell lines (Sah et al., 2020).

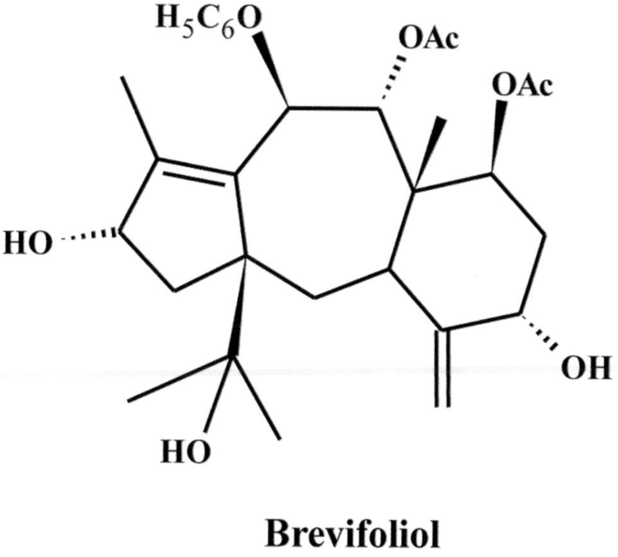

Brevifoliol

FIGURE 11.3   Structure of brevifoliol.

11.5   ANTICANCER ACTIVITY OF *TAXUS BREVIFOLIA* EXTRACTS/ NANOPARTICLES

Silver nanoparticles synthesized from *T. brevifolia* leaf aqueous extract were subjected to evaluate anticancer activity against MCF-7 cell lines using *in vitro* methods. It revealed that the nanoparticles showed strong antimitotic activity on MCF-7 cell lines than standard drug taxol (Sarli et al., 2020). A comparative cytotoxic/anticancer activity of taxol derived from *T. brevifolia* and *Aspergillus terreus* was investigated in Ehrlich ascites carcinoma mice. In results, taxol derived from both sources exhibited noteworthy cytotoxic activity in mice models and further it enhanced the antioxidant potential of mice. Histopathological studies showed that toxol significantly reduced Ehrlich ascites carcinoma-induced oxidative stress (Zein et al., 2019). Anti-proliferative efficiency of taxol and baccatin III isolated from *Fusarium solani* was investigated against different types of cancer cell lines like Ovcar3, HeLa, Jurkat, HepG2, and T47D. Both baccatin and taxol were derived from the fungal source and significantly suppressed the growth of tested cancerous cell lines *in vitro* conditions (Chakravarthi et al., 2013).

KEYWORDS

- ***Taxus breviflia***
- **taxol**
- **anticancer activity**
- **paclitaxel**
- **docetaxel**
- **and 10-deacetylbaccatin III**

REFERENCES

Balcer-Kubiczek, E. K.; Attarpour, M.; Jiang, J.; Kennedy, A. S.; Suntharalingam, M. Cytotoxicity of Docetaxel (Taxotere) Used as a Single Agent and in Combination with Radiation in Human Gastric, Cervical and Pancreatic Cancer Cells. *Chemotherapy* **2006,** *52* (5), 231–240.

Balcer-Kubiczek, E. K.; Attarpour, M.; Wang, J. Z.; Regine, W. F. The Effect of Docetaxel (Taxotere) on Human Gastric Cancer Cells Exhibiting Low-Dose Radiation Hypersensitivity. *Clin. Med. Oncol.* **2008,** *2*, 301–311.

Bass, R.; Roberto, D.; Wang, D. Z.; Cantu, F. P.; Mohamadi, R. M.; Kelley, S. O.; Klotz, L.; Venkateswaran, V. Combining Desmopressin and Docetaxel for the Treatment of Castrationresistant Prostate Cancer in an Orthotopic Model. *Anticancer Res.* **2019,** *39* (1), 113–118.

Bhall, K.; Ibvado, A. M.; Tourkina, E.; Tang, C.; Mahoney, M. E.; Huang, Y. Taxol Induces Internucleosomal DNA Fragmentation Associated with Programmed Cell Death in Human Myeloid Leukemia Cells. *Leukem21ia* **1993,** *7*, 563–568.

Bidkar, A. P.; Sanpui, P.; Ghosh, S. S. Efficient Induction of Apoptosis in Cancer Cells by Paclitaxel-Loaded Selenium Nanoparticles. *Nanomedicine (Lond)* **2017,** *12* (21), 2641–2651.

Bissery, M. C.; Guénard, D.; Guéritte-Voegelein, F. Experimental Antitumor Activity of Taxotere (RP 56976, NSC 628503), a Taxol Analogue. *Cancer Res.* **1991,** *51*, 4845–4852.

Blagosklonny, M. V.; Schulte, T.; Nguyen, P.; Trepel, J.; Neckers, L. M. Taxol-Induced Apoptosis and Phosphorylation of Bcl-2 Protein Involves c-Raf-1 and Represents a Novel c-Raf-1 Signal Transduction Pathway. *Cancer Res.* **1996,** *56* (8), 1851–1854.

Blagosklonny, M. V.; Schulte, T. W.; Nguyen, P.; Mimnaugh, E. G.; Trepel, J.; Neckers, L. Taxol Induction of p21WAF1 and p53 Requires c-raf-1. *Cancer Res.* **1995,** *55* (20), 4623–4626.

Braakhuis, B. J.; Hill, B. T.; Dietel, M. In Vitro Antiproliferative Activity of Docetaxel (Taxotere), Paclitaxel (Taxol) and Cisplatin Against Human Tumour and Normal Bone Marrow Cells. *Anticancer Res.* **1994,** *14,* 205–208.

Brown, I.; Shalli, K.; McDonald, S. L.; Moir, S. E.; Hutcheon, A. W.; Heys, S. D.; Schofield, A. C. Reduced Expression of p27 Is a Novel Mechanism of Docetaxel Resistance in Breast Cancer Cells. Breast *Cancer Res.* **2004a,** *6* (5), R601–R607.

Brown, J.; Shvartsman, H. S.; Deavers, M. T.; Burke, T. W.; Munsell, M. F.; Gershenson, D. M. The Activity of Taxanes in the Treatment of Sex Cord-Stromal Ovarian Tumors. *J. Clin. Oncol.* **2004b,** *22* (17), 3517–3523.

Caraglia, M.; Giuberti, G.; Marra, M.; Di Gennaro, E.; Facchini, G.; Caponigro, F.; Iaffaioli, R.; Budillon, A.; Abbruzzese, A. Docetaxel Induces p53-Dependent Apoptosis and Synergizes with Farnesyl Transferase Inhibitor r115777 in Human Epithelial Cancer Cells. *Front. Biosci.* **2005,** *10,* 2566–2575.

Chakravarthi, B. V. S. K.; Ramanathan Sujay, Gini C Kuriakose, Anjali A Karande, Chelliah Jayabaskaran. Inhibition of Cancer Cell Proliferation and Apoptosis-Inducing Activity of Fungal Taxol and Its Precursor Baccatin III Purified from Endophytic *Fusarium solani. Cancer Cell Int.* **2013,** *13,* 105.

Chan, A.; Miles, D. W.; Pivot, X. Bevacizumab in Combination with Taxanes for the First-Line Treatment of Metastatic Breast Cancer. *Ann. Oncol.* **2010,** *21* (12), 2305–2315.

Chan, D. C.; Earle, K. A.; Zhao, T. L.; Helfrich, B.; Zeng, C.; Baron, A.; Whitehead, C. M.; Piazza, G.; Pamukcu, R.; Thompson, W. J.; Alila, H.; Nelson, P.; Bunn, P. A. Jr. Exisulind in Combination with Docetaxel Inhibits Growth and Metastasis of Human Lung Cancer and Prolongs Survival in Athymic Nude Rats with Orthotopic Lung Tumors. *Clin. Cancer Res.* **2002,** *8* (3), 904–912.

Chang, C. C.; Yang, Y.; Gao, D. Y.; Cheng, H. T.; Hoang, B.; Chao, P. H.; Chen, L. H.; Bteich, J.; Chiang, T.; Liu, J. Y.; Li, S. D.; Chen, Y. Docetaxel-Carboxymethyl Cellulose Nanoparticles Ameliorate CCl4 -Induced Hepatic Fibrosis in Mice. *J. Drug Target.* **2018,** *26* (5–6), 516–524.

Chang, H. M.; Kim, T. W.; Ryu, B. Y.; Choi, S. J.; Park, Y. H.; Lee, J. S.; Kim, W. K.; Kang, Y. K. Phase II Study of Paclitaxel and Carboplatin in Advanced Gastric Cancer Previously Treated with 5-Fluorouracil and Platinum. *Jpn J. Clin. Oncol.* **2005,** *35* (5), 251–255.

Chang, J. T.; Chang, G. C.; Ko, J. L.; Liao, H. Y.; Liu, H. J.; Chen, C. C.; Su, J. M.; Lee, H.; Sheu, G. T. Induction of Tubulin by Docetaxel Is Associated with p53 Status in Human Non-Small Cell Lung Cancer Cell Lines. *Int. J. Cancer* **2006,** *118* (2), 317–325.

Chattopadhyay, S. K.; Tripathi, S.; Darokar, M. P.; Faridi, U.; Sisodia, B.; Negi, S. Syntheses and Cytotoxicities of the Analogues of the Taxoid Brevifoliol. *Eur. J. Med. Chem.* **2008,** *43* (7), 1499–1505.

Chen, R.; Cheng, Q.; Owusu-Ansah, K. G.; Chen, J.; Song, G.; Xie, H.; Zhou, L.; Xu, X.; Jiang, D.; Zheng, S. Cabazitaxel, a Novel Chemotherapeutic Alternative for Drug-Resistant Hepatocellular Carcinoma. *Am. J. Cancer Res.* **2018,** *8* (7), 1297–1206.

Choi, K. H.; Jeon, J. Y.; Lee, Y. E.; Kim, S. W.; Kim, S. Y.; Yun, Y. J.; Park, K. C. Synergistic Activity of Paclitaxel, Sorafenib, and Radiation Therapy in Advanced Renal Cell Carcinoma and Breast Cancer. *Transl. Oncol.* **2019,** *12* (2), 381–388.

Czuba, Z. P.; Król, W.; Hasiński, P.; Nowowiejska, A. The Effects of Taxol (Paclitaxel) on Chemiluminescence of Neutrophils, Macrophages and J.774.2 Cell Line. *Acta Biochim. Pol.* **1998,** *45* (1), 103–106.

Derry, W. B.; Wilson, L.; Jordan, M. A. Substoichiometric Binding of Taxol Suppresses Microtubule Dynamics. *Biochemistry* **1995,** *34* (7), 2203–2211.

Donaldson, K. L.; Goolsby, G. L.; Wahl, A. F. Cytotoxicity of the Anticancer Agent's Cisplatin and Taxol During Cell Proliferation and the Cell Cycle. *Int. J. Cancer* **1994,** *57* (6), 847–855.

Dong, S.; Guo, Y.; Duan, Y.; Li, Z.; Wang, C.; Niu, L.; Wang, N.; Ma, M.; Shi, Y.; Zhang, M. Co-Delivery of Paclitaxel and Gemcitabine by Methoxy Poly (Ethylene Glycol)-Poly (Lactide-Coglycolide)-Polypeptide Nanoparticles for Effective Breast Cancer Therapy. *Anti-Cancer Drugs* **2018,** *29* (7), 637–45.

du Bois, A.; Lück, H. J.; Meier, W.; Adams, H. P.; Möbus, V.; Costa, S.; Bauknecht, T.; Richter, B.; Warm, M.; Schröder, W.; Olbricht, S.; Nitz, U.; Jackisch, C.; Emons, G.; Wagner, U.; Kuhn, W.; Pfisterer, J. Arbeit gemeinschaft Gynäkologische Onkologie Ovarian Cancer Study Group. A Randomized Clinical Trial of Cisplatin/Paclitaxel Versus Carboplatin/Paclitaxel as First-Line Treatment of Ovarian Cancer. *J. Natl. Cancer Inst.* **2009,** *5* (17), 1320–1329.

el Khyari, S.; Bourgarel, V.; Barra, Y.; Braguer, D.; Briand, C. Pretreatment by Tubulin Agents Decreases C-MYC Induction in Human Colon Carcinoma Cell Line HT29-D4. *Biochem. Biophys. Res. Commun.* **1997,** *231* (3), 751–754.

El-Sayed, M. I.; Maximous, D. W.; Aboziada, M. A.; Abdel-Wanis, M. E.; Mikhail, N. N. Feasibility of Breast Conservation After Neoadjuvant Taxene Based Chemotherapy in Locally Advanced Breast Cancer: A Prospective Phase I Trial. *Ann. Surg. Innov. Res.* **2010,** *31*, 4:5. DOI: 10.1186/1750-1164-4-5.

Fabbri, F.; Carloni, S.; Brigliadori, G. et al. Sequential Events of Apoptosis Involving Docetaxel, a Microtubule-Interfering Agent: A Cytometric Study. *BMC Cell Biol.* **2006,** *7*, 6. https://doi.org/10.1186/1471-2121-7-6.

Fard, A. E.; Tavakoli, M. B.; Salehi, H.; Emami, H. Synergetic Effects of Docetaxel and Ionizing Radiation Reduced Cell Viability on MCF-7 Breast Cancer Cell. *Appl. Cancer Res.* **2017,** *37*, 29.

Feng, Z.; Zhao, G.; Yu, L.; Gough, D.; Howell, S. B. Preclinical Efficacy Studies of a Novel Nanoparticle-Based Formulation of Paclitaxel That Out-Performs Abraxane. *Cancer Chemother. Pharmacol.* **2010,** *65* (5), 923–930.

Ferlini, C.; Cicchillitti, L.; Raspaglio, G.; Bartollino, S.; Cimitan, S.; Bertucci, C.; Mozzetti, S.; Gallo, D.; Persico, M.; Fattorusso, C.; Campiani, G.; Scambia, G. Paclitaxel Directly Binds to Bcl-2 and Functionally Mimics Activity of Nur77. *Cancer Res.* **2009,** *69* (17), 6906–6914.

Ferlini, C.; Di Stefano, M.; Marone, M. et al. The Synergistic Anti-Tumour Activity of ICI 182,780 in Combination with Docetaxel Is Mediated by P-Glycoprotein Inhibition. *Endocr. Relat. Canc.* **1998,** *5*, 315–24.

Foglietta, F.; Spagnoli, G. C.; Muraro, M. G.; Ballestri, M.; Guerrini, A.; Ferroni, C.; Aluigi, A.; Sotgiu, G.; Varchi, G. Anticancer Activity of Paclitaxel-Loaded Keratin Nanoparticles in Two-Dimensional and Perfused Three-Dimensional Breast Cancer Models. *Int. J. Nanomed.* **2018,** *13*, 4847–4867.

Fujita, K.; Sano, D.; Kimura, M.; Yamashita, Y.; Kawakami, M.; Ishiguro, Y.; Nishimura, G.; Matsuda, H.; Tsukuda, M. Anti-Tumor Effects of Bevacizumab in Combination with Paclitaxel on Head and Neck Squamous Cell Carcinoma. *Oncol. Rep.* **2007,** *18* (1), 47–51.

Fusser, M.; Øverbye, A.; Pandya, A. D.; Mørch, Ý.; Borgos, S. E.; Kildal, W.; Snipstad, S.; Sulheim, E.; Fleten, K. G.; Askautrud, H. A.; Engebraaten, O.; Flatmark, K.; Iversen, T. G.; Sandvig, K.; Skotland, T.; Mælandsmo, G. M. Cabazitaxel-Loaded Poly (2-Ethylbutyl Cyanoacrylate) Nanoparticles Improve Treatment Efficacy in a Patient Derived Breast Cancer Xenograft. *J. Control. Release* **2019,** *293*, 183–192.

Galaup, A.; Opolon, P.; Bouquet, C.; Li, H.; Opolon, D.; Bissery, M. C.; Tursz, T.; Perricaudet, M.; Griscelli, F. Combined Effects of Docetaxel and Angiostatin Gene Therapy in Prostate Tumor Model. *Mol. Ther.* **2003,** *7* (6), 731–740.

Gdowski, A. S.; Ranjan, A.; Sarker, M. R.; Vishwanatha, J. K. Bone-Targeted Cabazitaxel Nanoparticles for Metastatic Prostate Cancer Skeletal Lesions and Pain. *Nanomedicine (Lond)* **2017,** *12* (17), 2083–2095.

Geng, C. X.; Zeng, Z. C.; Wang, J. Y. Docetaxel Inhibits SMMC-7721 Human Hepatocellular Carcinoma Cells Growth and Induces Apoptosis. *World J. Gastroenterol.* **2003,** *9* (4), 696–700.

Giannakakou, P.; Robey, R.; Fojo, T.; Blagosklonny, M. V. Low Concentrations of Paclitaxel Induce Cell Type-Dependent p53, p21 and G1/G2 Arrest Instead of Mitotic Arrest: Molecular Determinants of Paclitaxel-Induced Cytotoxicity. *Oncogene* **2001,** *20* (29), 3806–3813.

Grant, D. S.; Williams, T. L.; Zahaczewsky, M.; Dicker, A. P. Comparison of Antiangiogenic Activities Using Paclitaxel (Taxol) and Docetaxel (Taxotere). *Int. J. Cancer* **2003,** *104* (1), 121–129.

Gunther, E. Ethnobotany of Western Washington. *Seattle: Univ. Washington Pub. Anthropol.* **1945,** *10* (1), 1–62.

Guo, W. J.; Chen, T. S.; Wang, X. P.; Chen, R. Taxol Induces Concentration-Dependent Apoptotic and Paraptosis-Like Cell Death in Human Lung Adenocarcinoma (ASTC-a-1) Cells. *J X-Ray Sci. Technol.* **2010,** *18* (3), 293–308.

Guo, X. L.; Lin, G. J.; Zhao, H.; Gao, Y.; Qian, L. P.; Xu, S. R.; Fu, L. N.; Xu, Q.; Wang, J. J. Inhibitory Effects of Docetaxel on Expression of VEGF, bFGF and MMPs of LS174T Cell. *World J. Gastroenterol.* **2003,** *9* (9), 1995–1998.

Harshita, B. M. A.; Rizwanullah, M.; Beg, S.; Pottoo, F. H.; Siddiqui, S.; Ahmad, F. J. Paclitaxel-loaded Nanolipidic Carriers with Improved Oral Bioavailability and Anticancer Activity Against Human Liver Carcinoma. *AAPS Pharm. Sci.Tech.* **2019,** *20* (2), 87. DOI: 10.1208/s12249–019–1304–4. PMID: 30675689.

Hartzell Jr., Hal. *The Yew Tree: A Thousand Whispers*; Hulogosi: Eugene, OR, 1991.

Hebbar, M.; Ernst, O.; Cattan, S.; Dominguez, S.; Oprea, C.; Mathurin, P.; Triboulet, J. P.; Paris, J. C.; Pruvot, F. R. Phase II Trial of Docetaxel Therapy in Patients with Advanced Hepatocellular Carcinoma. *Oncology* **2006,** *70* (2), 154–158.

Hou, J.; Sun, E.; Zhang, Z. H.; Wang, J.; Yang, L.; Cui, L.; Ke, Z. C.; Tan, X. B.; Jia, X. B.; Lv, H. Improved oral Absorption and Anti-Lung Cancer Activity of Paclitaxel-Loaded Mixed Micelles. *Drug Deliv.* **2017,** *24* (1), 261–269.

Hu, H.; Li, G. X.; Wang, L.; Watts, J.; Combs, G. F. Jr.; Lü, J. Methylseleninic Acid Enhances Taxane Drug Efficacy Against Human Prostate Cancer and Down-Regulates Antiapoptotic Proteins Bcl-XL and Survivin. *Clin. Cancer Res.* **2008,** *14* (4), 1150–1158.

Ichite, N.; Chougule, M. B.; Jackson, T. et al. Enhancement of Docetaxel Anticancer Activity by a Novel Diindolylmethane Compound in Human Non-Small Cell Lung Cancer. *Clin. Cancer Res.* **2009,** *15*, 543–552.

Ilgar, N. N.; Arıcan, G. O. Induction of Apoptosis and Cell Proliferation Inhibition by Paclitaxel in FM3A Cell Cultures. *Afr. J. Biotech* **2009,** *8*, 547–555.

Inoue, K.; Chikazawa, M.; Fukata, S.; Yoshikawa, C.; Shuin, T. Docetaxel Enhances the Therapeutic Effect of the Angiogenesis Inhibitor TNP-470 (AGM-1470) in Metastatic Human Transitional Cell Carcinoma. *Clin. Cancer Res.* **2003,** *9* (2), 886–899.

Jiang, J.; Huang, H. Targeting the Androgen Receptor by Taxol in Castration-Resistant Prostate Cancer. *Mol. Cell Pharmacol.* **2010,** *2* (1), 1–5.

Jiang, S.; Zhang, K.; He, Y.; Xu, X.; Li, D.; Cheng, S.; Zheng, X. Synergistic Effects and Mechanisms of Impressic Acid or Acankoreanogein in Combination with Docetaxel on Prostate Cancer. *RSC Adv.* **2018**, *8*, 2768–2776.

Jin, M.; Jin, G.; Kang, L.; Chen, L.; Gao, Z.; Huang, W. Smart Polymeric Nanoparticles with pH-Responsive and PEGdetachable Properties for Co-Delivering Paclitaxel and Survivin siRNA to Enhance Antitumor Outcomes. *Int. J. Nanomed.* **2018**, *13*, 2405–2426.

Jordan, M. A.; Toso, R. J.; Thrower, D.; Wilson, L. Mechanism of Mitotic Block and Inhibition of Cell Proliferation by Taxol at Low Concentrations. *Proc. Natl. Acad. Sci. USA* **1993**, *90*, 9552–9556.

Jordan, M. A.; Wendell, K.; Gardiner, S.; Derry, W. B.; Copp, H.; Wilson, L. Mitotic Block Induced in HeLa Cells by Low Concentrations of Paclitaxel (Taxol) Results in Abnormal Mitotic Exit and Apoptotic Cell Death. *Cancer Res* **1996**, *56* (4), 816–825.

Kang, J.; Wu, X.; Wang, Z.; Ran, H.; Xu, C.; Wu, J.; Wang, Z.; Zhang, Y. Antitumor Effect of Docetaxel-Loaded Lipid Microbubbles Combined with Ultrasound-Targeted Microbubble Activation on VX2 Rabbit Liver Tumours. *J. Ultrasound Med.* **2010**, *29* (1), 61–70.

Kaur, R.; Chattopadhyay, S. K.; Chatterjee, A.; Prakash, O.; Khan, F.; Suri, N. Synthesis and In Vitro Anticancer Activity of Brevifoliol Derivatives Substantiated by In Silico Approach. *Med. Chem. Res.* **2014**, *23* (9), 4138–4148.

Kim, K-H.; Park, C-S.; Park, D-E.; Im, A-A.; Lee, C-K. Immune Adjuvant and Antitumor Activity of Baccatin III, a Synthetic Precursor of Taxol. *J. Immunol.* **2011**, *186* (1 Suppl), 155.15.

Klauber, N.; Parangi, S.; Flynn, E.; Hamel, E.; D'Amato, R. J. Inhibition of Angiogenesis and Breast Cancer in Mice by the Microtubule Inhibitors 2-Methoxyestradiol and Taxol. *Cancer Res.* **1997**, *57* (1), 81–86.

Klos, K. S.; Zhou, X.; Lee, S.; Zhang, L.; Yang, W.; Nagata, Y.; Yu, D. Combined Trastuzumab and Paclitaxel Treatment Better Inhibits ErbB-2-Mediated Angiogenesis in Breast Carcinoma Through a More Effective Inhibition of Akt Than Either Treatment Alone. *Cancer* **2003**, *98* (7), 1377–1385.

Kommineni, N.; Saka, R.; Bulbake, U.; Khan, W. Cabazitaxel and Thymoquinone Co-Loaded Lipospheres as a Synergistic Combination for Breast Cancer. *Chem. Phys. Lipids* **2018**, S0009–3084.

Kucukzeybek, Y.; Gul, M. K.; Cengiz, E.; Erten, C.; Karaca, B.; Gorumlu, G.; Atmaca, H.; Uzunoglu, S.; Karabulut, B.; Sanli, U. A.; Uslu, R. Enhancement of Docetaxel-Induced Cytotoxicity and Apoptosis by All-Trans Retinoic Acid (ATRA) Through Downregulation of Survivin (BIRC5), MCL-1 and LT Beta-R in Hormone- and Drug Resistant Prostate Cancer Cell Line, DU-145. *J. Exp. Clin. Cancer Res.* **2008**, *27* (1), 37.

Little Elbert, L. Jr. **1979**. Checklist of United States Trees (Native and Naturalized). Agric. Handb. 541; U. S. Department of Agriculture, Forest Service: Washington, DC, 375 p. [2952].

Liu, Y.; Huang, L.; Liu, F. Paclitaxel Nanocrystals for Overcoming Multidrug Resistance in Cancer. *Mol. Pharm.* **2010**, *7* (3), 863–869.

Lopes, N. M.; Adams, E. G.; Pitts, T. W.; Bhuyan, B. K. Cell Kill Kinetics and Cell Cycle Effects of Taxol on Human and Hamster Ovarian Cell Lines. *Cancer Chemother. Pharmacol.* **1993**, *32* (3), 235–242.

Lorigan, P. C.; Crosby, T.; Coleman, R. E. Current Drug Treatment Guidelines for Epithelial Ovarian Cancer. *Drugs* **1996**, *51*, 571–584.

Manfredi, J. J.; Parness, J.; Horwitz, S. B. Taxol Binds to Cellular Microtubules. *J. Cell. Biol.* **1982**, *94* (3), 688–696.

McGuire, W. P.; Rowinsky, E. K.; Rosenshein, N. B.; Grumbine, F. C.; Ettinger, D. S.; Armstrong, D. K.; Donehower, R. C. Taxol: A Unique Antineoplastic Agent with Significant Activity in Advanced Ovarian Epithelial Neoplasms. *Ann. Intern. Med.* **1989**, *111*, 273–279.

Meng, Z.; Mitsutake, N.; Nakashima, M.; Starenki, D.; Matsuse, M.; Takakura, S.; Namba, H.; Saenko, V.; Umezawa, K.; Ohtsuru, A.; Yamashita, S. Dehydroxymethylepoxyquinomicin, a Novel Nuclear Factor-kappaB Inhibitor; Enhances Antitumor Activity of Taxanes in Anaplastic Thyroid Cancer Cells. *Endocrinology* **2008**, *149* (11), 5357–5365.

Milas, L.; Hunter, N. R.; Kurdoglu, B. et al. Kinetics of Mitotic Arrest and Apoptosis in Murine Mammary and Ovarian Tumors Treated with Taxol. *Cancer Chemother. Pharmacol* **1995**, *35*, 297–303.

Miller, K.; Wang, M.; Gralow, J.; Dickler, M.; Cobleigh, M.; Perez, E. A.; Shenkier, T.; Cella, D.; Davidson, N. E. Paclitaxel Plus Bevacizumab Versus Paclitaxel Alone for Metastatic Breast Cancer. *N. Engl J Med.* **2007**, *357* (26), 2666–2676.

Min, J.; Shen, H.; Xi, W.; Wang, Q.; Yin, L.; Zhang, Y.; Yu, Y.; Yang, Q.; Wang, Z. N. Synergistic Anticancer Activity of Combined Use of Caffeic Acid with Paclitaxel Enhances Apoptosis of Non-Small-Cell Lung Cancer H1299 Cells In Vivo and In Vitro. *Cell. Physiol. Biochem.* **2018**, *48* (4), 1433–1442.

Modarresi-Darreh, B.; Kazem Kamali, Seyed Mehdi Kalantar, Hamid Dehghanizadeh, Behrouz Aflatoonian. Comparison of Synthetic and Natural Taxol Extracted from Taxus Plant (Taxus baccata) on Growth of Ovarian Cancer Cells Under In Vitro Condition. *Eurasia J. Biosci.* **2018**, *12*, 413–418.

Munz, Philip A. *A California Flora and Supplement*; University of California Press: Berkeley, CA, 1973; 1905 p. [6155].

Nicoletti, M. I.; Lucchini, V.; Massazza,G.; Abbott, B. J.; D'Incalci, M.; Giavazzi, R. Antitumor Activity of Taxol (NSC-125973) in Human Ovarian Carcinomas Growing in the Peritoneal Cavity of Nude Mice. *Ann. Oncol.* **1993**, *4*, 151–155.

Nwafor, S. V.; Akah, P. A.; Okol, C. O. Potentials of Plant Products as Anticancer Agents. *J. Nat. Remed.* **2001**, *1/2*, 75–88.

Orth, J. D.; Kohler, R. H.; Foijer, F.; Sorger, P. K.; Weissleder, R.; Mitchison, T. J. Analysis of Mitosis and Antimitotic Drug Responses in Tumors by In Vivo Microscopy and Single-Cell Pharmacodynamics. *Cancer Res.* **2011**, *71* (13), 4608–4616.

Park, J.; Park, S. S.; Lee, K. J.; Ju, J. E.; Shin, S. H.; Ko, E. J.; Lee, S. W.; Seo, M. H.; Lee, J. S.; Song, S. Y.; Jeong, S. Y.; Choi, E. K. Docetaxel-Polymeric Nanoparticle Enhances Radiotherapeutic Efficacy in Human Pancreatic Cancer. *Transl. Cancer Res.* **2018**, *7* (1), 60–67.

Patel, K.; Doddapaneni, R.; Chowdhury, N.; Boakye, C. H.; Behl, G.; Singh, M. Tumor Stromal Disrupting Agent Enhances the Anticancer Efficacy of Docetaxel Loaded PEGylated Liposomes in Lung Cancer. *Nanomedicine (Lond)* **2016**, *11* (11), 1377–1392.

Rabzia, A.; Khazaei, M.; Rashidi, Z.; Khazaei, M. R. Synergistic Anticancer Effect of Paclitaxel and Noscapine on Human Prostate Cancer Cell Lines. *Iran J. Pharma. Res.* **2017**, *16* (4), 1432–1442.

Reed, E.; Kohn, E. C.; Sarosy, G.; Dabholkar, M.; Davis, P.; Jacob, J.; Maher, M. Paclitaxel, Cisplatin, and Cyclophosphamide in Human Ovarian Cancer: Molecular Rationale and Early Clinical Results. *Semin. Oncol.* **1995**, *22* (3 Suppl 6), 90–96.

Rottach, A. M.; Ahrend, H.; Martin, B.; Walther, R.; Zimmermann, U.; Burchardt, M.; Stope M. B. Cabazitaxel Inhibits Prostate Cancer Cell Growth by Inhibition of Androgen Receptor and Heat Shock Protein Expression. *World J. Urol.* **2019**. DOI: 10.1007/s00345–018–2615-x.

Rowinsky, E. K.; Gilbert, M. R.; McGuire, W. P.; Noe, D. A.; Grochow, L. B.; Forastiere, A. A.; Ettinger, D. S.; Lubejko, B. G.; Clark, B.; Sartorius, S. E. et al. Sequences of Taxol and Cisplatin: A Phase I and Pharmacologic Study. *J. Clin. Oncol.* **1991,** *9* (9), 1692–1703.

Ryang, J.; Yan, Y.; Song, Y. Anti-HIV, Antitumor and Immunomodulatory Activities of Paclitaxel from Fermentation Broth Using Molecular Imprinting Technique. *AMB Expr.* **2019,** *9*, 194. https://doi.org/10.1186/s13568–019–0915–1.

Sah, B.; Kumari, M.; Subban, K. Evaluation of the Anticancer Activity of Enzymatically Synthesized Baccatin III: An Intermediate Precursor of Taxol®. *3 Biotech* **2020,** *10*, 465. https://doi.org/10.1007/s13205–020–02457–1.

Sajjad, M.; Khan, M. I.; Naveed, S.; Ijaz, S.; Qureshi, O. S.; Raza, S. A.; Shahnaz, G.; Sohail, M. F. Folate-Functionalized Thiomeric Nanoparticles for Enhanced Docetaxel Cytotoxicity and Improved Oral Bioavailability. *AAPS Pharm. Sci. Tech.* **2019,** *20* (2), 81.

Sarli, S.; Kalani, M. R.; Moradi, A. A Potent and Safer Anticancer and Antibacterial *Taxus*-Based Green Synthesized Silver Nanoparticle. *Int. J. Nanomed.* **2020,** *15*, 3791–3801. https://doi.org/10.2147/IJN.S251174.

Schiff, P.; Fant, J.; Horwitz, S. Promotion of Microtubule Assembly In Vitro by Taxol. *Nature* **1979,** *277*, 665–667.

Sinha, D. A Review on Taxanes: An Important Group of Anticancer Compounds Obtained from Taxus Species. *IJPSR* **2020,** *11* (5), 1969–1985.

Wan, J.; Ma, X.; Xu, D.; Yang, B.; Yang, S.; Han, S. Docetaxel-Decorated Anticancer Drug and Gold Nanoparticles Encapsulated Apatite Carrier for the Treatment of Liver Cancer. *J. Photochem. Photobiol. B* **2018,** *185*. 10.1016/j.jphotobiol.2018.05.021.

Zein, N.; Aziz, S. W.; El-Sayed, A. S.; Sitohy, B. Comparative Cytotoxic and Anticancer Effect of Taxol Derived from Aspergillus terreus and Taxus brevifolia. *Biosci. Res.* **2019,** *16* (2), 1500–1509.

Zhong, T.; He, B.; Cao, H. Q.; Tan, T.; Hu, H. Y.; Li, Y. P.; Zhang, Z. W. Treating Breast Cancer Metastasis with Cabazitaxel Loaded Polymeric Micelles. *Acta Pharmacol. Sinca* **2017,** *38* (6), 924–930.

Zhu, C. J.; An, C. G. Enhanced Antitumor Activity of Cabazitaxel Targeting CD44+ Receptor in Breast Cancer Cell Line via Surface Functionalized Lipid Nanocarriers. *Trop. J. Pharma. Res.* **2017,** *16* (6), 1383–1390.

CHAPTER 12

Glycyrrhiza glabra

VIJAY KUMAR VEENA[1], B. UMA REDDY[2],
ADHIKESAVAN HARIKRISHNAN[3], and RAMASAMY SHANMUGAVALLI[3]

[1]*Department of Biotechnology, School of Applied Sciences, REVA University, Bangalore, Karnataka, India*

[2]*Department of Studies in Botany, Vijayanagara Sri Krishnadevaraya University, Ballari, Karnataka, India*

[3]*Department of Chemistry, School of Arts and Science, Vinayaka Mission Research Foundation-AV Campus, Chennai, Tamil Nadu, India*

ABSTRACT

Glycyrrhiza glabra, a medicinal botanical belonging to the family *Fabaceae* (*Leguminosae*) that has been traditionally used for various diseases and disorders worldwide. The medicinal metabolites attributed to cure or treat the diseases are mostly belonging to the class of triterpenoids, flavonoids, and saponins. *G. glabra* and its bioactive molecules, including glycyrrhizin, glycyrrhetinic acid, isoliquiritigenin, Lic-A, B, and E, liquirtigin, isoangustone-A, licoricidin, glycerol, licocoumarone, and β-hydroxy-DHP are reported to exhibit the anticancer properties. Mode of anticancer action of these molecules are different depending on the type of cancer and characteristics of these molecules. The major mode of anticancer mechanisms reported to induce cell cycle arrest and antimetastatic potential produce the reactive oxygen species (ROS), inhibit the protein kinases and transcription factors, and activation of programmed cell death or autophagy through downregulating the various oncogenes and oncoproteins. The anti-inflammatory and

Potent Anticancer Medicinal Plants: Secondary Metabolite Profiling, Active Ingredients, and Pharmacological Outcomes. Deepu Pandita and Anu Pandita (Eds.)

antioxidative potentials of the metabolites have an indirect anticancer mechanism. However, their interaction with the oncoproteins and their regulation seems to be crucial for the bioactive molecules in *G. glabra* to be cancer chemopreventive and to have anticancer properties.

12.1 INTRODUCTION

Glycyrrhiza glabra is a perennial plant that grows up to a height of 1 m with pinnate leaves of 7–15 cm in length with pale to purple flowers in hermaphrodite arrangement type of inflorescence with oblong legume fruit with several seeds. The *Glycyrrhiza* genus is reported to have more than 30 species and some important ones are *G. glabra, G. uralensis, G. inflata, G. aspera, G. korshinskyi*, and so on. Among them, *G. glabra* is well reported with medicinal properties and it can fix nitrogen fixation in humid soils (Fiore et al., 2005). The root part of the *G. glabra* is commonly utilized for major medicinal and food usage, whereas the leaves are used as agrochemical waste (Hayashi et al., 2009; Siracusa et al., 2011).

G. glabra, a medicinal plant from the family *Fabaceae* (*Leguminosae*) is traditionally used for food additive and feeding purposes. The word *Glycyrrhiza* means sweet root in Greek and commonly referred as liquorice, licorice, glycyrrhiza, sweet-wood, etc. *G. glabra* is native to Mediterranean regions and is also present in India, China, and Russia. *G. glabra* extracts have been well-documented for various medicinal values. Hence, the extract of plants is being utilized in functional food preparations, food supplements in food industries, and pharma industries (Hayashi et al., 2016; Herrera et al., 2009; Armanini et al., 2002). In traditional medicine of different geographical areas of the world, the *G. glabra* extracts have been documented to treat the disorders, such as cough, bronchitis, arthritis, and gastrointestinal issues. Also, it is commonly used to treat peptic ulcers, gastritis, tremors, and respiratory disorders in folk medicines.

Tea prepared from *G. glabra* roots is efficient in quenching the thirst, the dried root powder is used as a tooth cleanser and also as food additive (Armanini et al., 2002: Mukhopadhyay and Panja, 2008; Rizzato et al., 2017). In the cosmetics industries, the *G glabra* extracts are used as pigmentation reducing agents in tropical uses. The root fibers are used as insulation, wallboards, and boxboard materials in fire extinguishers, and beer industries (Asl and Hosseinzadeh, 2008; Fiore et al., 2005).

12.2   PHYTOCHEMICAL COMPOSITION OF *GLYCYRRHIZA GLABRA*

Phytochemical analysis of *G. glabra* has revealed that it is rich in proteins, amino acids, simple and complex sugars with essential minerals, and phytochemicals, including magnesium, calcium, iron, sodium, potassium, phosphorous, silicon, selenium, manganese, copper, zinc, pectin, resin, phytosterols, and gum (s) (Wang et al., 2015a, b). Metabolites, such as estrogens, phytosterols, glycosides, tannins, coumarins, and vitamins, such as B-complex, E, and C are reported. The major classes of phytomolecules present in *G. glabra* include triterpenes, saponins, and flavonoids (Rizzato et al., 2017). Glucuronides and glycyrrhizin, which are saponins present as major constituents in *G. glabra*, are reported to be responsible for the sweet taste (Yu et al., 2015). Oral intake of glycyrrhizin has been reported to break down it into the 18-glycyrrhetic acid 3-O-monoglucuronide, and glycyrrhetic acid by bacteria present in the human gut microbiome (Albermann et al., 2010). The yellowish color of *G. glabra* is due to the presence of flavonoids, such as flavones, flavanones, flavanonols, isoflavans, chalcones, isoflavenes, and isoflavanones.

The majority of flavonoids reports include glycosides-based liquiritigenin and isoliquiritigenin that includes liquiritins, isoliquiritin, liquiritin-apioside, and licuraside (Rizzato et al., 2017). More recently, new flavonoids, such as glucoliquiritin apioside, shin-flavanone, shinpterocarpin, prenyl-licoflavanone A, and 1-methoxyphaseloin have been reported from the dried roots (Rizzato et al., 2017) of *G. glabra*. Glabridin is reported as a major isoflavanone in the dried root (Simmler et al., 2013). Small quantities of phenolics, such as isoprenoid-substituted flavonoids, chromene, coumarins, comestans, dihydroxyphenylalanine, dihyrostilbenes, and benzofurans are also reported. Some of the volatile compounds present in the root include geraniol, hexanol, pentanol, terpinene 4-ol, and α-terpineol which are responsible for the characteristics and smell of the plant. Essential oils are also reported from *G. glabra* that include propionic acid, furfuraldehyde, 2, 3-butanediol, benzoic acid, furfuryl-formate, maltol, 1-methyl-2-formylpyrrole, and trimethyl-pyrazine (Chouitah et al., 2011).

12.3   PHARMACOLOGICAL PROPERTIES OF *GLYCYRRHIZA GLABRA*

The use of *G. glabra* dates back to a long period of traditional and folk medicine in various parts of the globe. The first reports are linked be in ancient

Egypt, China, and India. However, Theophrastu and Pedanius Dioscorides have documented the *G. glabra* as medicinal herb that describes about the medicinal effects (Armanini et al., 2002). Hence, it is the most important and popular component in herbal and traditional medicine practiced worldwide.

Glabridin and *G. glabra* extracts have been reported to have neuroprotective properties in orally administrated in vivo mice models by improving the learning and memory through interaction with the adrenergic and dopaminergic neurons (Hasanein, 2011; Dhingra and Sharma, 2006; Parle et al., 2004). Similar reports have shown that *G glabra* extract and its purified metabolites have been reported for their sedative (Jin et al., 2013; Cho et al., 2010), antidepressant (Dhingra and Sharma, 2006), and estrogenic activities (Sharma et al., 2012; Su et al., 2015; Tamir et al., 2001; Tung et al., 2014).

Glycyrrhizin of *G. glabra* has been reported to have antiviral activities against the viruses, such as HIV (Sasaki et al., 2002), DHV (Soufy et al., 2012), EV71 (Kuo et al., 2009), Hepatitis-C (Yasui et al., 2011), and influenza virus (Michaelis et al., 2011). Glycyrrhetic acid, glycyrrhizin, glabridin, and *G. glabra* extracts have been reported to possess hepatoprotective properties (Huo et al., 2011; Hajiaghamohammadi et al., 2012; Yin et al., 2017; Liu et al., 2017) and anti-inflammatory activities (Xiao et al., 2010; Rackova et al., 2007; Wang et al., 2017a, b) in vivo and in vitro studies.

12.4 ANTICANCER METABOLITES FROM *GLYCYRRHIZA GLABRA*

Several reports have shown the anticancer potential of *G. glabra* extracts and purified metabolites. The 18-β-glycyrrhetinic and glycyrrhizic acid have shown anticancer properties by inducing the mitochondrial-mediated cell death in the tumor cells, such as cervical and uterus cancer cells (Lee et al., 2008). Glycyrrhizin acid and glycyrrhetinic acid have been reported to have anticancer activities against gastric cancer. However, glycyrrhizin exhibited the suppression of thromboxane A-2 in lung cancerous cells with less toxic effects (Deng et al., 2017). The 18-β-glycyrrhetinic acid is reported for anticancer properties against the ovarian, gastric, liver cancer, and leukemic cells through the induction of ROS, RNS, and collapse of the membrane potential of mitochondria (Hasan et al., 2016).

Derivatives of glycyrrhetic acids have been shown for the inhibition of breast cancerous cells (MCF-7, MDA-MB-231 cells) (Li et al., 2016). Another study has shown that these molecules have actively inhibited the leukemic cells (HL-60 cells) through the induction of apoptotic death through intrinsic or mitochondrial-mediated and extrinsic or ligand receptor-mediated pathway (Huang et al., 2016). Lic-E has very efficient than Lic-A and isoliquiritigenin in the anticancer properties (Xiao et al., 2011; Yu et al., 2017; Park et al., 2015).

12.5 ANTICANCER MECHANISM AND MOLECULAR TARGETS

Phytochemical analysis of *G. glabra* especially commonly used root part consisted of alkaloids, oils, polyamines, phenolics, triterpenes, flavones, flavans, chalcones, flavonoids, and isoflavanoids (Kusano et al., 1991; Vaya et al., 1997; Hatano et al., 2000). Antioxidant and anti-inflammatory properties of *G. glabra* extracts and its components play a crucial role in various diseases including cancer. The anticancer molecules reported in *G. glabra* include glycyrrhizin, glycyrrhetinic acid, isoliquiritigenin. Lic-A, B and E, Liquirtigin, iso-angustone A, licoricidin, glycryol, licocoumarones, and β-hydroxy-DHP (Table 12.1)

Glycyrrhizin and glycyrrhetinic acid of *G. glabra* are reported to control the proliferation and metastasis of skin and melanoma cancer in mice models (Agarwal et al., 1991; Aydemir et al., 2011; Kobayashi et al., 2002). Similarly, the anticancer potential of prostate cancer (Thirugnanam et al., 2008), gastric cancers (Lin et al., 2014), leukemia (Hibasami et al., 2006), endometrial cancer (Niwa et al., 2007), and breast cancers (Rossi et al., 2003).

Glabridin (GBD) have been reported to interfere in the migrative, invasive, and epithelial mesenchymal transition (EMT) in lung (A549), breast (MDA-MB-231), and human endothelial (HUVEC) cells by downregulation of FAK, Src, Akt, myosin, and Rho activities (Tsai et al., 2011; Hsu et al., 2011). Similarly, in Huh-7 and Sk Hep 1 cancer cells, GBD decreased the MMP, phosphorylated ERKs, NF-κB, and upregulated the MMP-1 inhibitors (Hsieh et al., 2014).

Glycyrrhetinic acid has been shown to have anticancer activities in the lung (A549, NCI-H460) cancer cells by inhibiting thromboxane synthase activity without any toxicity to normal lung cells such as NCH-123 cells and 16HBE-T bronchial epithelial cells (Huang et al., 2014a, b, c). Further, it is known to induce G1/M cell cycle arrest and programmed cell death in the NCI-H460 HSCLC cells with poly-ADP ribosyl polymerase (PARP) cleavage and activation of caspases-3 with downregulation of antiapoptotic Bcl-2 proteins and cyclin-D1-Cyclin-E proteins (Song et al., 2014). Also, the glycyrrhetinic acid has been shown to reduce the phosphorylated forms of protein kinase C (PKC αβ), extracellular activated proteins kinases (ERKs), and C-Jun NH-2 terminal kinases that eventually induced cell death in the

TABLE 12.1 Various Anticancer Molecules Reported in *G. glabra* and Their Mode of Anticancer Action.

Anticancer molecules from *G. glabra*	Mechanistic activity	Molecular targets	References
Mixed Extracts	Cell cycle arrest, apoptosis, autophagy, anti-metastasis	Downregulation of mTOR, upregulation of p53 and Bax	Agarwal et al. (1991); Aydemir et al. (2011); Kobayashi et al. (2002)
Glycyrrhizin	Cell cycle arrest, apoptosis, autophagy, anti-metastasis, antiangiogenetic, and sensitization	Downregulating the NF-κB, Selectin, TNF-α, and HMGB1	Thirugnanam et al. (2008); Lin et al. (2014); Hibasami et al. (2006); Niwa et al. (2007); Rossi et al. (2003)
Glycyrrhetinic acid	Cell cycle arrests, apoptotic cell death, autophagy, anti-metastasis, differentiation, and antiangiogenetic	Downregulation of NF-κB, PKC and GSH	Cherng et al. (2011); Chueh et al. (2012); Song et al. (2014); Huang et al. (2014a, b, c)
Isoliquiritigenin	Cell cycle arrest, apoptosis, autophagy, anti-metastasis, differentiation, antiangiogenetic, and sensitization	Downregulation of NF-κB and JNK	Cuendet et al. (2010); Ye et al. (2009); Vandooren et al. (2013); Ma et al. (2013); Lorusso and Marech (2013); Zhao et al. (2014); Takahashi et al. (2004); Hsu et al. (2009); Bode and Dong (2015)
Lic-A	Cell cycle arrests, apoptotic cell death, autophagy, anti-metastasis, differentiation, and antiangiogenetic sensitization	Blockade of ERKs, NF- κB and Ki-67; Downregulating the Bcl 2 and Akt	Kwon et al. (2008); Jiang et al. (2014); Yuan et al. (2013); Yo et al. (2009); Park et al. (2008); Kim et al. (2010); Yao et al. (2014)
Lic-B and E	Cell cycle arrest, apoptotic cell death, autophagy, anti-metastasis, and antiangiogenetic	Downregulation of Bcl-2, surviving, DNA topoisomerase, Ki67 and upregulation of Bax	Kwon et al. (2013); Kim et al. (2012) and Lee et al. (2013a, b)
Liquirtigin	Apoptosis, autophagy, anti-metastasis, differentiation, and antiangiogenetic	Downregulation of Akt, activation of p53, and production of ROS	Liu et al. (2011); Liu et al. (2012); Wang et al. (2012); Zhou et al. (2010)

TABLE 12.1 *(Continued)*

Anticancer molecules from *G. glabra*	Mechanistic activity	Molecular targets	References
Isoangustone-A	Cell cycle arrests, apoptotic cell death, autophagy, anti-metastasis, and antiangiogenetic	Downregulation of mTOR, JNK, and activation of death receptors (DR)	Huang et al. (2014a, b, c); Seon et al. (2010, 2012); Lee et al. (2013a, b); Liang et al. (2002); Song et al. (2013)
Glabridin	Anti-metastasis, antiangiogenetic	Downregulation of Rho-A, FAK, Akt, and Src	Tsai et al. (2011); Hsu et al. (2011); Hsieh et al. (2014)
Licoricidin	Anti-metastasis	Downregulation of uPA, MMP-9, and ICAM	Kobayashi et al. (1995)
Licocoumarone	Apoptosis	Increase in DNA damage	Shin et al. (2011); Watanabe et al. (2002)
Glycryol	Cell cycle arrests, apoptosis, and autophagy	Activation of caspases-8, 9 and FAS	Shin et al. (2011); Tang et al. (2015)
β-hydroxy-DHP	Apoptosis	Downregulation of Bcl-2 protein	Tang et al. (2015)

tumor cells. However, the molecular and cellular targets of *G. glabra* extracts and their components are not yet identified.

G. glabra is proven to be cancer chemopreventive in the case of UV-irradiated skin tumor mice model, where it delayed the incidence of tumor and reduced the tumor proliferation and size. In this study, the cells had decreased thymidine dimers in UV-exposed skin cells, downregulated the gene expression of proliferative cell nuclear antigen (PCNA), and exhibited few DNA tail in terminal deoxy-nucleotidyl transferase-mediated d-UTP nick labeling (TUNEL) assay of treated cells. The treated animals exhibited increased p53 and p21/Cip1 levels in the epidermis (Cherng et al., 2011). Further, the inhibition of NF-κB activity and cyclooxygenase-2 (COX-2) activities with decreased prostaglandin E2 (PGE2) and nitric oxide (NO) levels has also been observed. This evidence suggested that glycyrrhetinic acid of *G. glabra* extract exhibited anticancer effects by blocking the translocation of NF-κB by interrupting the degrading IκB (Cherng et al., 2011). In another study, glycyrrhetinic acid-supplemented Wistar rats have been shown to suppress the development of 1,2-dimethylhydrazine (DMH)-induced pre-cancerous lesions. This study also suggested the reduction in the mast cells by reducing the gene expression of Ki67, NF-κB/p65, COX 2, iNOS, and vascular endothelial growth factor (VEGF) with increased gene expression of p53, connexin 43, caspases 9, and activated caspases 3 to activate the apoptosis (Khan et al., 2013). In the case of WEHI-3 leukemic cells, glycyrrhetinic acid has been shown for the induction of G0/ G1 cell cycle arrest and apoptotic cell death through DNA damage death-receptor, mitochondrial-mediated, and endoplasmic reticulum (ER) stress-induced signaling pathway (Chueh et al., 2012).

Isoliquiritigenin [(E-1-(2,4-dihydroxyphenyl)-3-(4-hydroxyphenyl)prop-2-en-1-one] (ILQ) chalconoids have been identified from *G. glabra* for its anticancer activities. Dietary supplements of ILQ have been reported to delay the DMBA-induced breast cancer in female Sprague-Dawly rats without any toxic effects through aromatase inhibition (Cuendet et al., 2010; Ye et al., 2009). But ILQ is also known to inhibit the matrix metalloproteases (MMP), COX 2, and induced apoptosis (Vandooren et al., 2013). ILQ is also known to downregulate the metastatic gene expressions, such as hypoxia-inducible factor (HIF)-1 (Ma et al., 2013), VEGF (Cao, 2014), and MMP-2/-9 protein (Kessenbrock et al., 2010) levels in breast cancerous cells. The mentioned anticancer activity could be because of the regulation of gene expression of PI3K, NF-κB, and p38 signaling pathways (Lorusso and Marech, 2013). ILQ is shown to inhibit the phorbol myristate acetate (PMA)-induced gene expression of MMPs in HuVeCs through inhibition of JNKs and p38 signaling pathway (Kang et al., 2010). Oral supplementation of ILQ has reduced the colon cancer proliferation, incidence and size of the tumor in azoxymethane-induced carcinogenesis in colitis-associated tumor of mice model (Zhao et al., 2014). The same study has explained the downregulation of M2 macrophage polarization in colon carcinogenesis through the downregulation of PGE2 and IL- 6 inflammatory signaling pathway (Zhao et al., 2014). A majority of the reports suggested that ILQ has induced the apoptotic cell death in cancer cells through the upregulation of p53 and p21 levels (Zhou et al., 2014; Takahashi et al., 2004). In the case of HeLa cervical cancer cells, ILQ has been shown to induce G2/M cell cycle arrest and induced apoptosis (Hsu et al., 2009). This mode of anticancer activities was observed through the activation of ataxia telangiectasia mutated (ATM), increased phosphorylated p53, and cell cycle regulators, including cyclin B, A, cell division cycle protein (Cdc2) cdc25C, and checkpoint kinases (Chk2) even tyrosine kinase inhibitor resistance NSCLC cells (Hsu et al., 2009; Bode and Dong, 2015).

Isoangustone-A (IAA), a new flavonoid-based compound has been reported from the roots of *G. glabra*. IAA has been reported to have anticancer activities mainly by interfering with the cell cycle and apoptosis. IAA is reported to induce caspase-dependent and mitochondrial-mediated apoptotic cell death in SW480 colorectal cancerous cells (Huang et al., 2014a, b, c). IAA can also induce similar mechanism in hormone-resistant DU145 prostate cancerous cells and xenograft cancer BALB/c mice models through the induction of G1 cell cycle arrest via downregulation of mTOR, Cdk-2, Cdk-4, cyclin-A, and D1 levels (Seon et al., 2010, 2012; Lee et al., 2013a, b; Liang et al., 2002). IAA is also known to suppress the growth of SKMEL 28 cells by direct binding to PI3K, MKK 4, and MKK 7 even in xenograft cancer (Song et al., 2013).

Licochalcone-A (Lic-A), a major phenolic compound reported in *G. glabra* is shown to exhibit antiproliferative and anti-inflammatory potentials, but it also has strong antioxidant properties (Xiao et al., 2011). Lic-A has been reported to induce apoptosis in breast cancer (MCF-7) and human promyelo-cytic leukemic (HL-60) cells through downregulation of Bcl-2 proteins (Rafi et al., 2000). In combination with geldanamycin, it enhances the apoptosis in ovarian cancer cells (OVCAR-3 and SKOV-3) (Kim et al., 2013; Lee et al., 2013a, b). Lic-A also suppresses the proliferation of colon cancer (CT26) implants in BAL-B/c by inducing apoptosis (Kim et al., 2010). Lic-A is also reported to induce apoptotic cell death in the prostate cancerous (DU145) cells (Park et al., 2014) while in human prostate cancerous cells (LNCaP), it induced autophagy (Yo et al., 2009). An interesting result, Lic-A reduced the viable cells in oral (KB) cancer cells with less effect on primary normal keratinocytic cells (Kim et al., 2014). Further, it induced apoptosis through gene expression of factor-associated suicide ligand (FasL) and eventually the activation of caspases in KB cells (Kim et al., 2014). Others reported that Lic-A inhibited gastric cancer cells (MKN28, AGS, and MKN45) by inducing the G2 /M cell cycle arrest and apoptosis with increasing Rb levels and decreasing cyclin-A/B and MDM-2 levels with increased cleavage of PARP and activation of caspases-3 (Xiao et al., 2011; Fu et al., 2004). Few reports suggest that Lic-A exhibits the inhibition of platelet-derived growth factor (PDGF) induction of rat vascular smooth muscle cells (r-VSMC) proliferations to reduce cyclin-A, D1, CDK 2/4, ERKs, and phosphorylated Rb (Park et al., 2008).

The anticancer effect of Lic-A is partially because of its antioxidant and anti-inflammatory potential. A study of C57BL mice with DSS-induced colon cancer showed that Lic-A has reduced tumor formation and decreased the expression of PCNA, β-catenin, iNOS, COX-2, and proinflammatory levels (Kim et al., 2010). Another study suggested that when Lic-A has been combined with cisplatin the treated mice of CT-26 have shown to inhibit the cisplatin-induced liver and kidney damage (Lee et al., 2004).

Lic-A also has anti-inflammatory potential that inhibited the production of NO and COX 2 in LPS-inflamed macrophage (Raw-264.7) cells (Kwon et al., 2008). Lic-A also inhibited the human bladder cancer by generating ROS and induction of apoptosis (Jiang et al, 2014; Yuan et al., 2013). Several reports have suggested that Lic-A has been reported to suppress the NF-κB and also mitogen-activated protein kinases (MAPKs) signaling pathway (Furusawa et al., 2009). Other reports have suggested that Lic-A inhibited the migrative and invasive potentials of lung (A549 and H460) cancerous cells with the suppression of phosphorylated Akt (Huang et al., 2014a, b, c). Other mechanisms reported by Lic-A is JNKs that is observed in other diseases, such as diabetes or Parkinsons' diseases (Yao et al., 2014).

Licochalcone-E (Lic-E) from *G. glabra* appears to induce the cell cycle arrest and apoptotic cell death similar to the inactivation of NF-κB and downregulating the antiapoptotic Bcl 2 proteins in breast cancerous cells in mice model studies (Kwon et al., 2013). These studies also suggested that Lic-E has an antiangiogenesis activity that reduced the migrative and invasive potential of breast cancer cells. Like Lic-A, Lic-E and Lic-B have anti-inflammatory and antioxidant potential as similar mechanism (Kim et al., 2012; Lee et al., 2013a, b).

Liquiritigenin (LG), a flavonoid that possess anticancer activities in liver cancer AMMM-7721 and HeLa cell lines through upregulation of Bax and activation of caspase through inhibiting the Bcl-2 and survinin by the induction of ROS in the cancer cells. In the lung cancer A549 cells and Hela xenograft mice studies, the LG inhibited the migrations and attachment of cells by inactivating the MMP 2, reducing the phosphorylated Akt and activation of ERKs (Liu et al., 2011; Liu et al., 2012; Wang et al., 2012; Zhou et al., 2010). There are few reports on licocoumarone (LCM) and β-hydroxy-DHP that are known to have antiproliferative potential through induction of apoptosis (Shin et al., 2011; Watanabe et al., 2002). Glycryol has been shown to induce apoptosis and cell cycle arrest by activating the Fas and caspases in Jurkat cell lines. In the case of AGS and HCT-116, glycryol has induced autophagy through the activation of JNK (Shin et al., 2011; Tang et al., 2015).

12.6 CONCLUSIONS

G. glabra and its molecules are reported to be used in the tradition-based medicinal system and folk medicinal systems for long years. A literature survey has suggested that the anticancer properties of phytochemicals present in *G. glabra* had been reported to possess various modes of actions due to various interactions of oncoproteins. Major responsible bioactive anticancer molecules reported in *G. glabra* include glycyrrhizin, glycyrrhetinic acid, isoliquiritigenin, Lic-A, B and E, Liquirtigin, isoangustone-A, licoricidin, glycryol, licocoumarone, and β-hydroxy-DHP. The major mode of anticancer action includes inducing the cancerous cell cycle arrest, inducting the apoptotic cell deaths, and antioxidant activity. Major molecular targets to exhibit the anticancer potential from *G. glabra* have been identified that include the protein markers related to cell cycle arrests, apoptosis-inducing proteins, MMPs, COXs, GSK- β, PI3Ks, NF- κB, and MAPK that have been evidenced by in vitro, in vivo, and in silico studies. These survey results in *G. glabra*, suggest that its extract and components have no obvious toxicity but can be chemopreventive and has anticancer potential.

ACKNOWLEDGMENT

We acknowledge Dr. Honoureen Beatrice Gamble, Department of Chemistry, School of Arts and Science, Vinayaka Mission Research Foundation- AV campus, Chennai.

KEYWORDS

- *Glycyrrhiza glabra*
- glycyrrhizin
- 18β- glycyrrhetinic acid
- glabrin
- isoliquiritigenin
- isoangustone-A
- Lic-A
- anticancer mechanism

REFERENCES

Agarwal, R.; Wang, Z. Y.; Mukhtar, H. Inhibition of Mouse Skin Tumorinitiating Activity of DMBA by Chronic Oral Feeding of Glycyrrhizin in Drinking Water. *Nutr. Cancer.* **1991,** *15* (3–4), 187–93.

Albermann, M. E.; Musshoff, F.; Hagemeier, L.; Madea, B. Determination of Glycyrrhetic Acid After Consumption of Liquorice and Application to a Fatality. *Forensic Sci. Int.* **2010,** *197* (1), 35–39.

Armanini, D.; Fiore, C.; Mattarello, M. J.; Bielenberg, J.; Palermo, M. History of the Endocrine Effects of Licorice. *Exp. Clin. Endocrinol. Diab.* **2002,** *110* (06), 257–261.

Asl, M. N.; Hosseinzadeh, H. Review of Pharmacological Effects of Glycyrrhiza sp. and Its Bioactive Compounds. *Phytother. Res.* **2008,** *22* (6), 709–724.

Aydemir, E. A.; Oz, E. S.; Gokturk, R. S.; Ozkan, G.; Fiskin, K. Glycyrrhiza Flavescens Subsp. Antalyensis Exerts Antiproliferative Effects on Melanoma Cells via Altering TNF-Alpha and IFN-Alpha Levels. *Food Chem Toxicol.* **2011,** *49* (4), 820–828.

Bode, A. M.; Dong, Z. Chemopreventive Effects of Licorice and Its Components. *Curr. Pharmacol. Rep.* **2015,** *1,* 60–71.

Cao, Y. VEGF-Targeted Cancer Therapeutics-Paradoxical Effects in Endocrine Organs. *Nat. Rev. Endocrinol.* **2014,** *10* (9), 530–539.

Cherng, J. M.; Tsai, K. D.; Yu, Y. W.; Lin, J. C. Molecular Mechanisms Underlying Chemopreventive Activities of Glycyrrhizic Acid Against UVB-Radiation-Induced Carcinogenesis in SKH-1 Hairless Mouse Epidermis. *Radiat. Res.* **2011,** *176* (2), 177–186.

Cho, S.-M.; Shimizu, M.; Lee, C. J.; Han, D. S.; Jung, C. K.; Jo, J. H.; Kim, Y. M. Hypnotic Effects and Binding Studies for GABAA and 5- HT2C Receptors of Traditional Medicinal Plants Used in Asia for Insomnia. *J. Ethnopharmacol.* **2010,** *132* (1), 225–223.

Chouitah, O.; Meddah, B.; Aoues, A.; Sonnet, P. Chemical Composition and Antimicrobial Activities of the Essential Oil from *Glycyrrhiza glabra* Leaves. *J. Essential Oil-Bear. Plants* **2011,** *14* (3), 284–288.

Chueh, F. S.; Hsiao, Y. T.; Chang, S. J.; Wu, P. P.; Yang, J. S.; Lin, J. J. et al. Glycyrrhizic Acid Induces Apoptosis in WEHI-3 Mouse Leukemia Cells Through the Caspase- and Mitochondria-Dependent Pathways. *Oncol. Rep.* **2012**, *28* (6), 2069–2076.

Cuendet, M.; Guo, J.; Luo, Y.; Chen, S.; Oteham, C. P.; Moon, R. C. et al. Cancer Chemopreventive Activity and Metabolism of Isoliquiritigenin, a Compound Found in Licorice. *Cancer Prev. Res. (Phila).* **2010**, *3* (2), 221–232.

Deng, Q. P.; Wang, M. J.; Zeng, X.; Chen, G. G.; Huang, R. Y. Effects of Glycyrrhizin in a Mouse Model of Lung Adenocarcinoma. Cell. Physiol. Biochem. **2017**, *41* (4), 1383–1392.

Dhingra, D.; Sharma, A. Antidepressant-Like Activity of Glycyrrhiza glabra L. in Mouse Models of Immobility Tests. *Progress Neuro-Psychopharmacol. Biol. Psych.* **2006**, 30 (3), 449–454.

Fiore, C.; Eisenhut, M.; Ragazzi, E.; Zanchin, G.; Armanini, D. A History of the Therapeutic Use of Liquorice in Europe. *J. Ethnopharmacol.* **2005**, *99* (3), 317–324.

Fu, Y.; Hsieh, T. C.; Guo, J.; Kunicki, J.; Lee, M. Y.; Darzynkiewicz, Z. et al. Licochalcone-A, a Novel Flavonoid Isolated from Licorice Root (Glycyrrhiza glabra), Causes G2 and late-G1 Arrests in Androgen Independent PC-3 Prostate Cancer Cells. *Biochem. Biophys. Res. Commun.* **2004**, *322* (1), 263–270.

Furusawa, J.; Funakoshi-Tago, M.; Tago, K.; Mashino, T.; Inoue, H.; Sonoda, Y. et al. Licochalcone A significantly suppresses LPS 70 Signaling Pathway Through the Inhibition of NF-kappaB p65 Phosphorylation at Serine 276. *Cell Signal.* **2009**, *21* (5), 778–785.

Hajiaghamohammadi, A. A.; Ziaee, A.; Samimi, R. The Efficacy of Licorice Root Extract in Decreasing Transaminase Activities in Non-Alcoholic Fatty Liver Disease: A Randomized Controlled Clinical Trial. *Phytother. Res.* **2012**, *26* (9), 1381–1384.

Hasan, S. K.; Siddiqi, A.; Nafees, S.; Ali, N.; Rashid, S.; Ali, R. et al. Chemopreventive Effect of 18β-Glycyrrhetinic Acid via Modulation of Inflammatory Markers and Induction of Apoptosis in Human Hepatoma Cell Line (HepG2). *Mol. Cell. Biochem.* **2016**, *416* (1–2), 169–177.

Hasanein, P. Glabridin as a Major Active Isoflavan from *Glycyrrhiza glabra* (Licorice) Reverses Learning and Memory Deficits in Diabetic Rats. *Acta Physiologica Hungarica* **2011**, *98* (2), 221–230.

Hatano, T.; Shintani, Y.; Aga, Y.; Shiota, S.; Tsuchiya, T.; Yoshida, T. Phenolic Constituents of Licorice. VIII. Structures of Glicophenone and Glicoisoflavanone, and Effects of Licorice Phenolics on Methicillin-Resistant *Staphylococcus aureus*. *Chem. Pharm. Bull.* **2000**, *48* (9), 1286–1292.

Hayashi, H.; Tamura, S.; Chiba, R.; Fujii, I.; Yoshikawa, N.; Fattokhov, I.; Saidov, M. Field Survey of Glycyrrhiza Plants in Central Asia (4). Characterization of *G. glabra* and *G. bucharica* Collected in Tajikistan. *Biol. Pharma. Bull.* **2016**, *39* (11), 1781–1786.

Herrera, M.; Herrera, A.; Ariño, A. Estimation of Dietary Intake of Ochratoxin A from Liquorice Confectionery. *Food Chem. Toxicol.* **2009**, *47* (8), 2002–2002.

Hibasami, H.; Iwase, H.; Yoshioka, K.; Takahashi, H. Glycyrrhetic Acid (a Metabolic Substance and Aglycon of Glycyrrhizin) Induces Apoptosis in Human Hepatoma, Promyelotic Leukemia and Stomach Cancer Cells. *Int. J. Mol. Med.* **2006**, *17* (2), 215–219.

Hsieh, M. J.; Lin, C. W.; Yang, S. F.; Chen, M. K.; Chiou, H. L. Glabridin Inhibits Migration and Invasion by Transcriptional Inhibition of Matrix Metalloproteinase 9 Through Modulation of NF-kappaB and AP-1 Activity in Human Liver Cancer Cells. *Br. J. Pharmacol.* **2014**, *171*, 3037–3050.

Hsu, Y. L.; Chia, C. C.; Chen, P. J.; Huang, S. E.; Huang, S. C.; Kuo, P. L. Shallot and Licorice Constituent Isoliquiritigenin Arrests Cell Cycle Progression and Induces Apoptosis Through the Induction of ATM/ p53 and Initiation of the Mitochondrial System in Human Cervical Carcinoma HeLa Cells. *Mol. Nutr. Food Res*. **2009,** *53* (7), 826–835.

Hsu, Y. L.; Wu, L. Y.; Hou, M. F.; Tsai, E. M.; Lee, J. N.; Liang, H. L.; Jong, Y. J.; Hung, C. H.; Kuo, P. L. Glabridin, an Isoflavan from Licorice Root, Inhibits Migration, Invasion and Angiogenesis of MDA-MB-231 Human Breast Adenocarcinoma Cells by Inhibiting Focal Adhesion Kinase/Rho Signaling Pathway. *Mol. Nutr. Food Res*. **2011,** *55*, 318–327.

Huang, H. C.; Tsai, L. L.; Tsai, J. P.; Hsieh, S. C.; Yang, S. F.; Hsueh, J. T. et al. Licochalcone A Inhibits the Migration and Invasion of Human Lung Cancer Cells via Inactivation of the Akt Signaling Pathway with Downregulation of MMP-1/-3 Expression. *Tumour Biol*. **2014a,** PubMed PMID: 25149157.

Huang, R. Y.; Chu, Y. L.; Huang, Q. C.; Chen, X. M.; Jiang, Z. B.; Zhang, X. et al. 18beta-Glycyrrhetinic Acid Suppresses Cell Proliferation Through Inhibiting Thromboxane Synthase in Non-Small Cell Lung Cancer. *PLoS ONE* **2014b,** *9* (4), e93690.

Huang, W.; Tang, S.; Qiao, X.; Ma, W.; Ji, S.; Wang, K. et al. Isoangustone A Induces Apoptosis in SW480 Human Colorectal Adenocarcinoma Cells by Disrupting Mitochondrial Functions. *Fitoterapia* **2014c,** *94*, 36–47.

Huang, Y. C.; Kuo, C. L.; Lu, K. W.; Lin, J. J.; Yang, J. L.; Wu, R. S. et al. 18α-Glycyrrhetinic Acid Induces Apoptosis of HL-60 Human Leukemia Cells Through Caspases- and Mitochondria-Dependent Signaling Pathways. *Molecules* **2016,** *21* (7).

Huo, H. Z.; Wang, B.; Liang, Y. K.; Bao, Y. Y.; Gu, Y. Hepatoprotective and Antioxidant Effects of Licorice Extract Against CCl (4)- Induced Oxidative Damage in Rats. *Int. J. Mol. Sci*. **2011,** *12* (10), 6529–6565.

Jiang, J.; Yuan, X.; Zhao, H.; Yan, X.; Sun, X.; Zheng, Q. Licochalcone A Inhibiting Proliferation of Bladder Cancer T24 Cells by Inducing Reactive Oxygen Species Production. *Biomed Mater Eng*. **2014,** *24* (1), 1019–1025.

Jin, Z.; Kim, S.; Cho, S.; Kim, I. H.; Han, D.; Jin, Y. H. Potentiating Effect of Glabridin on GABAA Receptor-Mediated Responses in Dorsal Raphe Neurons. *Planta Medica* **2013,** *79* (15), 1408–1412.

Kang, S. W.; Choi, J. S.; Choi, Y. J.; Bae, J. Y.; Li, J.; Kim, D. S. et al. Licorice Isoliquiritigenin Dampens Angiogenic Activity via Inhibition of MAPK-Responsive Signaling Pathways Leading to Induction of Matrix Metalloproteinases. *J. Nutr. Biochem*. **2010,** *21* (1), 55–65.

Kessenbrock, K.; Plaks, V.; Werb, Z. Matrix Metalloproteinases: Regulators of the Tumor Microenvironment. *Cell*. **2010,** *141* (1), 52–67.

Khan, R.; Khan, A. Q.; Lateef, A.; Rehman, M. U.; Tahir, M.; Ali, F. et al. Glycyrrhizic Acid Suppresses the Development of Precancerous Lesions via Regulating the Hyperproliferation, Inflammation, Angiogenesis and Apoptosis in the Colon of Wistar Rats. *PLoS ONE* **2013,** *8* (2), e56020.

Kim, J. K.; Shin, E. K.; Park, J. H.; Kim, Y. H. Antitumor and Antimetastatic Effects of Licochalcone A in Mouse Models. *J. Mol. Med. (Berl)*. **2010,** *88* (8), 829–838.

Kim, J. S.; Park, M. R.; Lee, S. Y.; Kim, K.; Moon, S. M.; Kim, C. S. et al. Licochalcone A Induces Apoptosis in KB Human Oral Cancer Cells via a Caspase-Dependent FasL Signaling Pathway. *Oncol Rep*. **2014,** *31* (2), 755–762.

Kim, S. S.; Lim, J.; Bang, Y.; Gal, J.; Lee, S. U.; Cho, Y. C. et al. Licochalcone E Activates Nrf2/Antioxidant Response Element Signaling Pathway in Both Neuronal and Microglial Cells: Therapeutic Relevance to Neurodegenerative Disease. *J. Nutr. Biochem*. **2012,** *23* (10), 1314–1323.

Kim, Y. H.; Shin, E. K.; Kim, D. H.; Lee, H. H.; Park, J. H.; Kim, J. K. Antiangiogenic Effect of Licochalcone A. *Biochem Pharmacol.* **2010,** *80* (8), 1152–1159.

Kim, Y. J.; Jung, E. B.; Myung, S. C.; Kim, W.; Lee, C. S. Licochalcone A Enhances Geldanamycin-Induced Apoptosis Through Reactive Oxygen Species-Mediated Caspase Activation. *Pharmacology* **2013,** *92* (1– 2), 49–59.

Kobayashi, M.; Fujita, K.; Katakura, T.; Utsunomiya, T.; Pollard, R. B.; Suzuki, F. Inhibitory Effect of Glycyrrhizin on Experimental Pulmonary Metastasis in Mice Inoculated with B16 Melanoma. *Anticancer Res.* **2002,** *22* (6C), 4053–4058.

Kobayashi, S.; Miyamoto, T.; Kimura, I.; Kimura, M. Inhibitory Effect of Isoliquiritin, a Compound in Licorice Root, on Angiogenesis In Vivo and Tube Formation In Vitro. *Biol Pharm Bull.* **1995,** *18*, 1382–1386.

Kuo, K. K.; Chang, J. S.; Wang, K. C.; Chiang, L. C. Water Extract of Glycyrrhiza Uralensis Inhibited Enterovirus 71 in a Human Foreskin Fibroblast Cell Line. *Am. J. Chin. Med.* **2009,** *37* (2), 383–394.

Kusano, A.; Nikaido, T.; Kuge, T.; Ohmoto, T.; Delle Monache, G.; Botta, B. et al. Inhibition of Adenosine 3′,5′-Cyclic Monophosphate Phosphodiesterase by Flavonoids from Licorice Roots and 4- Arylcoumarins. *Chem. Pharm. Bull.* **1991,** *39* (4), 930–933.

Kwon, H. S.; Park, J. H.; Kim, D. H.; Kim, Y. H.; Shin, H. K.; Kim, J. K. Licochalcone A Isolated from Licorice Suppresses Lipopolysaccharide-Stimulated Inflammatory Reactions in RAW264.7 Cells and Endotoxin Shock in Mice. *J. Mol. Med. (Berl).* **2008,** *86* (11), 1287–1295.

Kwon, S. J.; Park, S. Y.; Kwon, G. T.; Lee, K. W.; Kang, Y. H.; Choi, M. S. et al. Licochalcone E Present in Licorice Suppresses Lung Metastasis in the 4T1 Mammary Orthotopic Cancer Model. *Cancer Prev. Res. (Phila).* **2013,** *6* (6), 603–613.

Lee, C. K.; Son, S. H.; Park, K. K.; Park, J. H.; Lim, S. S.; Kim, S. H. et al. Licochalcone A inhibits the Growth of Colon Carcinoma and Attenuates Cisplatin-Induced Toxicity Without a Loss of Chemotherapeutic Efficacy in Mice. *Basic Clin. Pharmacol. Toxicol.* **2008,** *103* (1), 48–54.

Lee, C. S.; Kwak, S. W.; Kim, Y. J.; Lee, S. A.; Park, E. S.; Myung, S. C. et al. Guanylate Cyclase Activator YC-1 Potentiates Apoptotic Effect of Licochalcone A on Human Epithelial Ovarian Carcinoma Cells via Activation of Death Receptor and Mitochondrial Pathways. *Eur. J. Pharmacol.* **2012,** *683* (1–3), 54–62.

Lee, E.; Son, J. E.; Byun, S.; Lee, S. J.; Kim, Y. A.; Liu, K. et al. CDK2 and mTOR Are Direct Molecular Targets of Isoangustone A in the Suppression of Human Prostate Cancer Cell Growth. *Toxicol Appl Pharmacol.* **2013a,** *272* (1), 12–20.

Lee, H. N.; Cho, H. J.; Lim Do, Y.; Kang, Y. H.; Lee, K. W.; Park, J. H. Mechanisms by Which Licochalcone E Exhibits Potent Anti-inflammatory Properties: Studies with Phorbol Ester-Treated Mouse Skin and Lipopolysaccharide-Stimulated Murine Macrophages. *Int. J. Mol. Sci.* **2013b,** *14* (6), 10926–1043.

Lee, C. S.; Kim, Y. J.; Lee, M. S.; Han, E. S.; Lee, S. J. 18β- Glycyrrhetinic Acid Induces Apoptotic Cell Death in Siha Cells and Exhibits a Synergistic Effect Against Antibiotic Anti-Cancer Drug Toxicity. *Life Sci.* **2008,** *83* (13–14), 481–489.

Lee, J. Y.; Lee, J. H.; Park, J. H.; Kim, S. Y.; Choi, J. Y.; Lee, S. H. et al. Liquiritigenin, a Licorice Flavonoid, Helps Mice Resist Disseminated Candidiasis Due to Candida Albicans by Th1 Immune Response, Whereas Liquiritin, Its Glycoside Form, Does Not. *Int. Immunopharmacol.* **2009,** *9* (5), 632–638.

Li, Y.; Feng, L.; Song, Z. F.; Li, H. B.; Huai, Q. Y. Synthesis and Anticancer Activities of Glycyrrhetinic Acid Derivatives. *Molecules* **2016,** *21* (2).

Liang, J.; Zubovitz, J.; Petrocelli, T.; Kotchetkov, R.; Connor, M. K.; Han, K. et al. PKB/Akt Phosphorylates p27, Impairs Nuclear Import of p27 and Opposes p27-Mediated G1 Arrest. *Nat. Med.* **2002,** *8* (10), 1153–1160.

Lin, D.; Zhong, W.; Li, J.; Zhang, B.; Song, G.; Hu, T. Involvement of BID Translocation in Glycyrrhetinic Acid and 11-Deoxy Glycyrrhetinic Acid-Induced Attenuation of Gastric Cancer Growth. *Nutr. Cancer* **2014,** *66* (3), 463–473.

Liu, C.; Wang, Y.; Xie, S.; Zhou, Y.; Ren, X.; Li, X.; Cai, Y. Liquiritigenin Induces Mitochondria-Mediated Apoptosis via Cytochrome C Release and Caspases Activation in HeLa Cells. *Phytother. Res.* **2011,** *25,* 277–283.

Liu, Y.; Xie, S.; Wang, Y.; Luo, K.; Wang, Y.; Cai, Y. Liquiritigenin Inhibits Tumor Growth and Vascularization in a Mouse Model of HeLa Cells. *Molecules* **2012,** *17,* 7206–7216.

Liu, K.; Pi, F.; Zhang, H.; Ji, J.; Xia, S.; Cui, F. et al. Metabolomics Analysis to Evaluate the Anti-Inflammatory Effects of Polyphenols: Glabridin Reversed Metabolism Change Caused by LPS in RAW 264.7 Cells. *J. Agric. Food Chem.* **2017,** *65* (29), 6070–6079.

Lorusso, V.; Marech, I. Novel Plant-Derived Target Drugs: A Step Forward from Licorice? *Expert Opin Ther Targets* **2013,** *17* (4), 3.

Ma, J.; Zhang, Q.; Chen, S.; Fang, B.; Yang, Q.; Chen, C. et al. Mitochondrial Dysfunction Promotes Breast Cancer Cell Migration and Invasion Through HIF1alpha Accumulation via Increased Production of Reactive Oxygen Species. *PLoS ONE.* **2013,** *8* (7), e69485.

Michaelis, M.; Geiler, J.; Naczk, P.; Sithisarn, P.; Leutz, A.; Doerr, H. W.; Cinatl, J. Glycyrrhizin Exerts Antioxidative Effects in H5N1 Influenza a Virus-Infected Cells and Inhibits Virus Replication and Pro- Inflammatory Gene Expression. *PLoS One* **2011,** *6* (5), e19705.

Mukhopadhyay, M.; Panja, P. A Novel Process for Extraction of Natural Sweetener from Licorice (*Glycyrrhiza glabra*) Roots. *Sep. Purif. Technol.* **2008,** *63* (3), 539–545.

Niwa, K.; Lian, Z.; Onogi, K.; Yun, W.; Tang, L.; Mori, H. et al. Preventive Effects of Glycyrrhizin on Estrogen-Related Endometrial Carcinogenesis in Mice. *Oncol. Rep.* **2007,** *17* (3), 617–622.

Park, J. H.; Lim, H. J.; Lee, K. S.; Lee, S.; Kwak, H. J.; Cha, J. H. et al. Antiproliferative Effect of Licochalcone A on Vascular Smooth Muscle Cells. *Biol. Pharm. Bull.* **2008,** *31* (11), 1996–2000.

Park, S. Y.; Kim, E. J.; Choi, H. J.; Seon, M. R.; Lim, S. S.; Kang, Y. H. et al. Anti-Carcinogenic Effects of Non-Polar Components Containing Licochalcone A in Roasted Licorice Root. *Nutr. Res. Pract.* **2014,** *8* (3), 257–266.

Park, M. R.; Kim, S. G.; Cho, I. A.; Oh, D.; Kang, K. R.; Lee, S. Y. et al. Licochalcone-A Induces Intrinsic and Extrinsic Apoptosis via ERK1/2 and p38 Phosphorylation-Mediated TRAIL Expression in Head and Neck Squamous Carcinoma FaDu Cells. *Food Chem. Toxicol.* **2015,** *77,* 34–43.

Parle, M.; Dhingra, D.; Kulkarni, S. K. Memory-Strengthening Activity of *Glycyrrhiza glabra* in Exteroceptive and Interoceptive Behavioral Models. *J. Med. Food* **2004,** *7* (4), 462–466.

Rackova, L.; Jancinova, V.; Petrikova, M.; Drabikova, K.; Nosal, R.; Stefek, M. et al. Mechanism of Anti-Inflammatory Action of Liquorice Extract and Glycyrrhizin. *Nat. Product Res.* **2007,** *21* (14), 1234–1241.

Rafi, M. M.; Rosen, R. T.; Vassil, A.; Ho, C. T.; Zhang, H.; Ghai, G. et al. Modulation of BCL-2 and Cytotoxicity by Licochalcone-A, a Novel Estrogenic Flavonoid. *Anticancer Res.* **2000,** *20* (4), 2653–2658.

Rizzato, G.; Scalabrin, E.; Radaelli, M.; Capodaglio, G.; Piccolo, O. A New Exploration of Licorice Metabolome. *Food Chem.* **2017,** *221,* 959–968.

Rossi, T.; Castelli, M.; Zandomeneghi, G.; Ruberto, A.; Benassi, L.; Magnoni, C. et al. Selectivity of Action of Glycyrrhizin Derivatives on the Growth of MCF-7 and HEP-2 Cells. *Anticancer Res.* **2003,** *23* (5A), 3813–3818.

Sasaki, H.; Takei, M.; Kobayashi, M.; Pollard, R. B.; Suzuki, F. Effect of Glycyrrhizin, an Active Component of Licorice Roots, on HIV Replication in Cultures of Peripheral Blood Mononuclear Cells from HIV-Seropositive Patients. *Pathobiology* **2002,** *70* (4), 229–236.

Seon, M. R.; Lim, S. S.; Choi, H. J.; Park, S. Y.; Cho, H. J.; Kim, J. K. et al. Isoangustone A Present in Hexane/Ethanol Extract of Glycyrrhiza Uralensis Induces Apoptosis in DU145 Human Prostate Cancer Cells via the Activation of DR4 and Intrinsic Apoptosis Pathway. *Mol. Nutr. Food Res.* **2010,** *54* (9), 1329–1339.

Seon, M. R.; Park, S. Y.; Kwon, S. J.; Lim, S. S.; Choi, H. J.; Park, H. et al. Hexane/Ethanol Extract of Glycyrrhiza Uralensis and Its Active Compound Isoangustone A Induce G1 Cycle Arrest in DU145 Human Prostate and 4T1 Murine Mammary Cancer Cells. *J Nutr Biochem.* **2012,** *23* (1), 85–92.

Sharma, G.; Kar, S.; Palit, S.; Das, P. K. 18β-Glycyrrhetinic Acid Induces Apoptosis Through Modulation of Akt/FOXO3a/Bim Pathway in Human Breast Cancer MCF-7 Cells. *J. Cell. Physiol.* **2012,** *227* (5), 1923–1931.

Sharma, V.; Katiyar, A.; Agrawal, R. C. *Glycyrrhiza glabra*: Chemistry and Pharmacological Activity. In *Sweeteners: Pharmacology, Biotechnology, and Applications*; Merillon, J.-M., Ramawat, K. G., Eds.; Springer International Publishing, Cham, 2016; pp. 1–14.

Shin, E. M.; Kim, S.; Merfort, I.; Kim, Y. S. Glycyrol Induces Apoptosis in Human Jurkat T Cell Lymphocytes via the Fas-FasL/Caspase-8 Pathway. *Planta Med* **2011,** *77,* 242–247.

Simmler, C.; Pauli, G. F.; Chen, S.-N. Phytochemistry and Biological Properties of Glabridin. *Fitoterapia* **2013,** *90,* 160–184.

Siracusa, L.; Saija, A.; Cristani, M.; Cimino, F.; D'Arrigo, M.; Trombetta, D. et al. Phytocomplexes from Liquorice (*Glycyrrhiza glabra* L.) Leaves—Chemical Characterization and Evaluation of Their Antioxidant, Anti-Genotoxic and Anti-Inflammatory Activity. *Fitoterapia* **2011,** *82* (4), 546–556.

Song, J.; Ko, H. S.; Sohn, E. J.; Kim, B.; Kim, J. H.; Kim, H. J. et al. Inhibition of Protein Kinase C Alpha/BetaII and Activation of C-Jun NH2- Terminal Kinase Mediate Glycyrrhetinic Acid Induced Apoptosis in Non-Small Cell Lung Cancer NCI-H460 Cells. *Bioorg. Med. Chem. Lett.* **2014,** *24* (4), 1188–1191.

Song, N. R.; Lee, E. J.; Byun, S.; Kim, J. E.; Mottamal, M.; Park, J. H. et al. Isoangustone, A.; A Novel Licorice Compound, Inhibits Cell Proliferation by Targeting PI3-K, MKK4 and MKK7 in Human Melanoma. *Cancer Prev. Res. (Phila).* Oct 8. PubMed PMID: 24104352. Epub 2013/10/10.

Soufy, H.; Yassein, S.; Ahmed, A. R.; Khodier, M. H.; Kutkat, M. A.; Nasr, S. M.; Okda, F. A. Antiviral and Immune Stimulant Activities of Glycyrrhizin Against Duck Hepatitis Virus. *Afr. J. Tradit. Complement. Altern. Med.* **2012,** *9* (3), 389–395.

Su Wei Poh, M.; Voon Chen Yong, P.; Viseswaran, N.; Chia, Y. Y. Estrogenicity of Glabridin in Ishikawa Cells. *PLoS One* **2015,** *10* (3), e0121382.

Takahashi, T.; Takasuka, N.; Iigo, M.; Baba, M.; Nishino, H.; Tsuda, H. et al. Isoliquiritigenin, a Flavonoid from Licorice, Reduces Prostaglandin E2 and Nitric Oxide, Causes Apoptosis, and Suppresses Aberrant Crypt Foci Development. *Cancer Sci.* **2004,** *95* (5), 448–453.

Tamir, S.; Eizenberg, M.; Somjen, D.; Izrael, S.; Vaya, J. Estrogen-Like Activity of Glabrene and Other Constituents Isolated from Licorice Root. *J. Steroid Biochem. Mol. Biol.* **2001,** *78* (3), 291–298.

Tamir, S.; Eizenberg, M.; Somjen, D.; Stern, N.; Shelach, R.; Kaye, A.; Vaya, J. Estrogenic and Antiproliferative Properties of Glabridin from Licorice in Human Breast Cancer Cells. *Cancer Res.* **2000,** *60* (20), 5704–5709.

Tang, Z.; Li, T.; Tong, Y.; Chen, X. et al. A Systematic Review of Anticancer Properties of Compounds Isolated from Licorice (Gancao). *Plant Med.* **2015,** *81,* 1670–1687.

Thirugnanam, S.; Xu, L.; Ramaswamy, K.; Gnanasekar, M. Glycyrrhizin Induces Apoptosis in Prostate Cancer Cell Lines DU145 and LNCaP. *Oncol Rep.* **2008,** *20* (6), 138.

Tsai, Y. M.; Yang, C. J.; Hsu, Y. L.; Wu, L. Y.; Tsai, Y. C.; Hung, J. Y.; Lien, C. T.; Huang, M. S.; Kuo. P. L. Glabridin Inhibits Migration, Invasion, and Angiogenesis of Human Non-Small Cell Lung Cancer A549 Cells by Inhibiting the FAK/Rho Signaling Pathway. *Integr. Cancer Ther.* **2011,** *10,* 341–349.

Tung, N. H.; Shoyama, Y.; Wada, M.; Tanaka, H. Improved In Vitro Fertilization Ability of Mouse Sperm Caused by the Addition of Licorice Extract to the Preincubation Medium. *Open Reprod. Sci. J.* **2014,** *6,* 1–7.

Vandooren, J.; Van den Steen, P. E.; Opdenakker, G. Biochemistry and Molecular Biology of Gelatinase B or Matrix Metalloproteinase-9 (MMP-9), the Next Decade. *Crit. Rev. Biochem. Mol. Biol.* **2013,** *48* (3), 222–272.

Vaya, J.; Belinky, P. A.; Aviram, M. Antioxidant Constituents from Licorice Roots: Isolation, Structure Elucidation and Antioxidative Capacity Toward LDL Oxidation. *Free Radic. Biol. Med.* **1997,** *23* (2), 302–313.

Wang, Y.; Xie, S.; Liu, C.; Wu, Y.; Liu, Y.; Cai, Y. Inhibitory Effect of Liquiritigenin on Migration via Downregulation proMMP-2 and PI3 K/Akt Signaling Pathway in Human Lung Adenocarcinoma A549 Cells. *Nutr Cancer* **2012,** *64,* 627–634.

Wang, L.; Yang, R.; Yuan, B.; Liu, Y.; Liu, C. The Antiviral and Antimicrobial Activities of Licorice, a Widely-Used Chinese Herb. *Acta Pharmaceutica Sinica* B **2015a,** *5* (4), 310–315.

Wang, Q.; Qian, Y.; Wang, Q.; Yang, Y.-F.; Ji, S.; Song, W. et al. Metabolites Identification of Bioactive Licorice Compounds in Rats. *J. Pharma. Biomed. Analys.* **2015b,** *115,* 515–522.

Wang, S.; Shen, Y.; Qiu, R.; Chen, Z.; Chen, Z.; Chen, W. 18β- Glycyrrhetinic Acid Exhibits Potent Antitumor Effects Against Colorectal 2338 Pastorino et al. Cancer via Inhibition of Cell Proliferation and Migration. *Int. J. Oncol.* **2017a,** *51* (2), 615–624.

Wang, X. R.; Hao, H. G.; Chu, L. Glycyrrhizin Inhibits LPS-Induced Inflammatory Mediator Production in Endometrial Epithelial Cells. *Microb. Pathogenesis* **2017b,** *109,* 110–113.

Wang, X.; Zhang, H.; Chen, L.; Shan, L.; Fan, G.; Gao, X. Liquorice, a Unique "Guide Drug" of Traditional Chinese Medicine: A Review of Its Role in Drug Interactions. *J. Ethnopharmacol.* **2013,** *150* (3), 781–790.

Watanabe, M.; Hayakawa, S.; Isemura, M.; Kumazawa, S.; Nakayama, T.; Mori, C.; Kawakami, T. Identification of Licocoumarone as an Apoptosis-Inducing Component in Licorice. *Biol. Pharm. Bull.* **2002,** *25,* 1388–1390.

Xiao, X. Y.; Hao, M.; Yang, X. Y.; Ba, Q.; Li, M.; Ni, S. J. et al. Licochalcone A Inhibits Growth of Gastric Cancer Cells by Arresting Cell Cycle Progression and Inducing Apoptosis. *Cancer Lett.* **2011,** *302* (1), 69–75.

Xiao, Y.; Xu, J.; Mao, C.; Jin, M.; Wu, Q.; Zou, J. et al. 18β- Glycyrrhetinic Acid Ameliorates Acute Propionibacterium Acnes-Induced Liver Injury Through Inhibition of Macrophage Inflammatory Protein-1α. *J. Biol. Chem.* **2010,** *285* (2), 1128–1137.

Yao, K.; Chen, H.; Lee, M. H.; Li, H.; Ma, W.; Peng, C. et al. Licochalcone, A.; A Natural Inhibitor of C-Jun N-Terminal Kinase 1. *Cancer Prev. Res. (Phila).* **2014,** *7* (1), 139–149.

Yasui, S.; Fujiwara, K.; Tawada, A.; Fukuda, Y.; Nakano, M.; Yokosuka, O. Efficacy of Intravenous Glycyrrhizin in the Early Stage of Acute Onset Autoimmune Hepatitis. *Digest. Dis. Sci.* **2011,** *56* (12), 3638–3647.

Ye, L.; Gho, W. M.; Chan, F. L.; Chen, S.; Leung, L. K. Dietary Administration of the Licorice Flavonoid Isoliquiritigenin Deters the Growth of MCF-7 Cells Overexpressing Aromatase. *Int. J. Cancer* **2009,** *124* (5), 1028–1036.

Yin, X.; Gong, X.; Zhang, L.; Jiang, R.; Kuang, G.; Wang, B. et al. Glycyrrhetinic Acid Attenuates Lipopolysaccharide-Induced Fulminant Hepatic Failure in D-Galactosamine-Sensitized Mice by Up-Regulating Expression of Interleukin-1 Receptor-Associated Kinase-M. *Toxicol. Appl. Pharmacol.* **2017,** *320*, 8–16.

Yo, Y. T.; Shieh, G. S.; Hsu, K. F.; Wu, C. L.; Shiau, A. L. Licorice and Licochalcone-A Induce Autophagy in LNCaP Prostate Cancer Cells by Suppression of Bcl-2 Expression and the mTOR Pathway. *J. Agric. Food Chem.* **2009,** *57* (18), 8266–8273.

Yu, J. Y.; Ha, J. Y.; Kim, K. M.; Jung, Y. S.; Jung, J. C.; Oh, S. Anti-Inflammatory Activities of Licorice Extract and Its Active Compounds, Glycyrrhizic Acid, Liquiritin and Liquiritigenin, in BV2 Cells and Mice Liver. *Molecules* **2015,** *20* (7), 13041–13054.

Yu, S. J.; Cho, I. A.; Kang, K. R.; Jung, Y. R.; Cho, S. S.; Yoon, G. et al. Licochalcone-E Induces Caspase-Dependent Death of Human Pharyngeal Squamous Carcinoma Cells Through the Extrinsic and Intrinsic Apoptotic Signaling Pathways. *Oncol. Lett.* **2017,** *13* (5), 3662–3636.

Yuan, X.; Li, D.; Zhao, H.; Jiang, J.; Wang, P.; Ma, X. et al. Licochalcone A-Induced Human Bladder Cancer T24 Cells Apoptosis Triggered by Mitochondria Dysfunction and Endoplasmic Reticulum Stress. *Biomed. Res. Int.* **2013,** *2013*, 474272.

Zhao, H.; Zhang, X.; Chen, X.; Li, Y.; Ke, Z.; Tang, T. et al. Isoliquiritigenin, a Flavonoid from Licorice, Blocks M2 Macrophage Polarization in Colitis-Associated Tumorigenesis Through Downregulating PGE2 and IL-6. *Toxicol. Appl. Pharmacol.* **2014,** *279* (3), 311–321.

Zhou, M.; Higo, H.; Cai, Y. Inhibition of Hepatoma 22 Tumor by Liquiritigenin. *Phytother. Res.* **2010,** *24*, 827–833.

Zhou, Y.; Ho, W. S. Combination of Liquiritin, Isoliquiritin and Isoliquirigenin Induce Apoptotic Cell Death Through Upregulating p53 and p21 in the A549 Non-Small Cell Lung Cancer Cells. *Oncol Rep.* **2014,** *31* (1), 298–304.

CHAPTER 13

Ocimum sanctum

SHARMISTHA BANERJEE[1], RAJESH SINGH TOMAR[2], and
SHUCHI KAUSHIK[3]

*[1]Biomedical Engineering and Bioinformatics, University Teaching
Department, Chhattisgarh Swami Vivekanand Technical University,
Newai, Bhilai, Chhattisgarh, India*

*[2]Amity Institute of Biotechnology, Amity University Madhya Pradesh,
Maharajpura Dang, Madhya Pradesh, India*

[3]State Forensic Science Laboratory, Sagar, Madhya Pradesh, India

ABSTRACT

Ocimum sanctum (*O. sanctum*) is a tropical annual herb belonging to the
family Lamiaceae (Labiatae) and is popularly called Tulasi or holy basil, and
is cultivated across the whole of India. It ascends to a height of 1800 m in
the Himalayas, usually grown in courtyards and gardens. It is a Perennial
herb with multiple branching and poses a unique aroma. It is considered an
excellent drug with various essential oils and secondary metabolites and can
be used for curing dysentery, diarrhea, malaria, bronchitis, bronchial asthma,
arthritis, skin diseases, eye diseases, and chronic fever. *O. sanctum* also poses
antifungal, anticancerous, antifertility, antimicrobial, immunomodulatory,
anti-ulcer, anti-inflammatory, hepatoprotective, antihypertensive, antispasmodic, radioprotective, antiemetic, cardio protective, analgesic, antidiabetic,
diaphoretic and adaptogenic activity. Extracts from *Ocimum* sp. and phytochemicals or bioactive components from *O. sanctum*, such as rosmarinic acid,
eugenol, apigenin, β–sitosterol, carnosic acid, luteolin, oleic acid, linoleic
acid, isorientin, ocimarin, orientin, aesculectin, aesculin, chlorogenic acid,

Potent Anticancer Medicinal Plants: Secondary Metabolite Profiling, Active Ingredients, and
Pharmacological Outcomes. Deepu Pandita and Anu Pandita (Eds.)

galuteolin, citronellal, gallic acid, sabinene, camphene, dimethyl benzene, vitamin C, ethyl benzene, and calcium are known to have several pharmacological properties. Few among them have been reported to prevent oral, lung, and liver cancers and chemical-induced skin cancer by enhancing the antioxidant potential, altering genetic expressions, inducing apoptosis, inhibiting metastasis and angiogenesis by deactivating matrix metalloproteinases. The anticancer potential of *O. sanctum* extract has been tested against various cancer cells both in vitro and in vivo, for example, human fibrosarcoma cells, Swiss albino mice having Ehrlich ascites carcinoma (EAC), hamster buccal pouch carcinogenesis, papillomas. This article is an attempt to study the properties of secondary metabolites and bioactive components present in Tulsi and its ability to act as a potent anticancer drug along with its mechanism.

13.1 INTRODUCTION

Tulsi also popularly known as holy basil, *is* extensively used in the Indian medicine system. Another name for it is *Vishnupriya* which symbolizes the one who is having the power to amuse Lord Vishnu. It is easily available and cultivated across the whole country and its scientific name is *O. sanctum*. Tulsi belongs to the family Lamiaceae/Labiatae, and in Sanskrit, it is named as surasa. It has been used for more than 1000 years in Ayurveda for its multiple healing ability and therapeutic potential (Kavyashree et al., 2019).

Tulsi is also referred to in the Charaka Samhita, the conventional ayurvedic text (NIIR, 2004). Due to its astringent taste and strong smell, it is considered a life saver in Ayurveda and is also known for increasing the life span of individuals (Puri and Rasayana, 2002). Tulsi, the queen of herbs, mythically unbeatable is one of the sacred and most cherished herbs in the universe because of its immense medical and healing ability. Tulsi is mentioned in conventional Ayurveda and the ancient Unani system of herbal medicine and holistic health because of its immense healing potential and a remedy for several ailments (Warrier, 1995). Fresh extract or decoction of Tulsi leaves has been used since time immemorial for treating various disorders (Siva et al., 2016).

Tulsi is present throughout the country and ascends to a height of 1800 m (30–60 cm) in the Himalayas, usually grown in courtyards and gardens. It is a perennial herb with multiple branching and a peculiar aroma. Stems and branches are usually purple in color, sub-quadrangular, sometimes woody toward the bottom, clothed with soft spreading hairs. Leaves are 2.5

by 1.6–3.2 cm in size, Elliptic –oblong, obtuse or acute, entire or serrate, pubescent on both sides, minutely gland—dotted, with obtuse base or acute, petioles are 1.3–2.5 cm in length, hairy and slender (Kavyashree et al., 2019).

Inflorescence: Verticillaster. Flowers are racemes, 15–20 cm long in close whorls, bract is nearly 3 mm long, width is mostly equal to the length, usually ovate, long slender acuminate, ciliate; pedicels are longer than the calyx, slender, pubescent. Fruits are basically shaped like nutlets, 1.25 mm long, majorly ellipsoidal, nearly smooth, yellow in color with black markings (Sharma et al., 2005; Basu, 1987; Kavyashree et al., 2019).

13.1.1 VERNACULAR NAMES

Vernacular names are names of the drug by which they are commonly known in several regions of the world, and therefore helps in easy identification of the drug globally. Vernacular names of Tulsi in different languages are presented in Table 13.1.

TABLE 13.1 Vernacular Names of Tulsi.

Languages	Vernacular names
Hindi	Kalatulasi, Tulasi
Kannada	Vishu tulasi, Kari tulasi, Sri tulasi, Tulashi-gida
English	Holy basil
Malayalam	Tulasi, Trttava, karuttarttavu, Niella tirtua, Krishna tulasi, Shiva tulasi
Telugu	Tulasi, Gaggera-chettu
Tamil	Tulasi, Karuttulasi
Bengali	Tulasi, Krishna tulasi
Gujarati	Tulasi, Talasi
Punjab	Bantulsi, Tulasi
Marathi	Tulasa, Tulasi
Konkani	Tulsi

Source: Adapted from Siva et al. (2016).

Three varieties of *Ocimum* sp. are as follows:

1. Light or Rama Tulsi (*Ocimum sanctum*)
2. Dark or Shyama Tulsi (*Ocimum sanctum*)
3. Vana Tulsi (*Ocimum gratissimum*)

Scientific classification:

- Kingdom: Plantae
- Division: Magnoliophyta
- Class: Magnoliopsida
- Order: Lamiales
- Family: Lamiaceae
- Genus: *Ocimum*
- Species: *O. Tenuiflorum*
- Binomial name: *Ocimum tenuiflorum* or *Ocimum sanctum* L. (Siva et al., 2016)

13.2 PROPERTIES

Decoction or extracts of Tulsi are used in the treatment of headaches, common cold, heart disease, stomach disorders, inflammation, malaria, and several forms of poisoning. In conventional times, *O. sanctum* L. is used in several ways, as a dried powder, fresh leaf, or herbal tea. Numerous researchers have used these extracts and have shown their antioxidant, anti-inflammatory, anti-stress, and immunomodulatory activities (Singh et al., 1996; Mauli et al., 1997; Mediratta et al., 2002; Sen et al., 1992). In addition to this, it has also been known to pose anticarcinogenic and radioprotective properties (Siva et al., 2016).

Several parts of *O. sanctum* L., such as flowers, stem, leaves, seeds, root are reported to exhibit medicinal properties and have been used by several conventional medicinal practitioners as analgesic, anticancer, expectorant, antiemetic, antiasthmatic, antidiabetic, antistress, diaphoretic, hepatoprotective, antifertility, hypotensive, hypolipidemic drugs. It is also known to be used in treating bronchitis, fever, convulsions, arthritis, etc. (Siva et al., 2016).

Tulsi is also known to have anti-fatigue characteristic, adaptogenic property, antimicrobial, anticonvulsant, antifertility, antidiabetic, radioprotective, anti-inflammatory cardioprotective, immunomodulatory, hepatoprotective, anticarcinogenic, mosquito repellent, and analgesic activities (Venkatachalam and Muthusamy, 2018).

The current article has tried to include information from several works of literature regarding Tulsi, including synonyms, classification, chemical composition, bioactive compounds, and anticancer properties.

13.2.1 CHEMICAL COMPOSITION

The chemical composition of *O. sanctum* L. bears high complexity as it contains several nutrients and bioactive compounds. The composition and concentration of these components are dependent on varying cultivation regions, processing, storage, and harvesting environment. The reason and mechanism behind this are yet unexplored completely (Pattanayak et al., 2010; Siva et al., 2016).

13.3 BIOACTIVE COMPONENTS

The pharmacological and nutritional activities of Tulsi in its natural form are because of the synergistic action of several phytometabolites present in it. *O. sanctum* contains eugenol, volatile oil, carvacrol, urosolic acid, limatrol, linalool, sesquiterpine, methyl eugenol, estragol and caryophyllene. Sugars present in it contain polysaccharides and xylose (Pattanayak et al., 2010; Kelm et al., 2000). Phytochemical analysis of leaves and stems of *O. sanctum* has shown compounds, such as flavonoids, saponins, tannins, and triterpenoids (Pattanayak et al., 2010; Jaggi et al., 2003). The herb also contains Vitamin A, C, and minerals, such as iron, zinc, calcium, chlorophyll, and several other phytonutrients (Anbarasu and Vijayalakshmi, 2007). Different phytochemicals and chemical compounds present in several parts of Tulsi are depicted in Table 13.2.

TABLE 13.2 Different Phytochemicals and Bioactive Components Present in Several Parts of *Ocimum sanctum.*

S. No.	Plant part	Phytochemicals	Bioactive component	References
1.	Leaves	Flavonoids, saponins, alkaloids, tannins, anthocyanins, sterols, phenols, terpenoids	Eugenal, eugenol, carvacol, limatrol, urosolic acid. linalool methyl carvicol, caryophyllene, anthocyans	Pattanayak et al. (2010); Kelm et al. (2000); Jaggi et al. (2003)
2.	Stem	Phenols, flavonoids, tannins, triterpenoids, saponins	Crisimartin, apigenin, Romarinic acid, isothymonin, isothymocin	Pattanayak et al. (2010); Kelm et al. (2000)
3.	Seeds	Sitosterol, Fatty acids	Xylose and polysaccharides	Pattanayak et al. (2010); Kelm et al. (2000)

Source: Adapted from Siva et al. (2016). https://creativecommons.org/licenses/by-nc/3.0/

13.3.1 ANTICARCINOGENIC PROPERTY

Tulsi is a rich source of phytoconstituents and possesses an outstanding role in medicine. These phytochemicals are not essential for its survival but are of substantial importance to the human community to carry out various protective functions in the human body.

Plant contains polyphenolic compounds such as flavonoids, tannins, curcuminoids, gallocatechin; stilbenes like resveratrol; anthocyanidin like delphinidin; they have a broader range of pharmacological activities and are considered to have chemopreventive and remedial properties against malignant growth (Singh et al., 2012). Additionally, phenolic phytochemicals regulate oxidation stress-related sicknesses because of their immediate association with quenching the free radicals (Boots et al., 2008). Anticancer activity of phytopolyphenols includes preparation of endogenous copper potentially chromatin-associated copper and ensuing pro-oxidant activity (Hadi et al., 2007). It has additionally been proposed that the cytotoxic movement of plant phenolics displays pro-oxidant action (Galati and O'Brien, 2004, Summers and Felton, 1994) and cytotoxic properties (Yamanaka et al., 1997; Sugihara et al., 1999).

The test studies completed on biological models utilizing *O. sanctum* (OS) extract on fibrosarcoma cells in culture have shown that *O. sanctum* displays anticancer action (Karthikeyan et al., 1999). Younger leaves of the plant display additional resistance, and furthermore are known to have anticancer activity in experimental organisms (Mondal et al., 2009). *O. sanctum* has additionally been shown to display restoring properties against septic and hostile hypersensitive impacts (Godhwani et al., 1988). Tulsi has numerous gainful properties with insignificant harmfulness and is an ideal antistress/adaptogenic specialist for the advancement of well-being and the counteraction and treatment of illness. Methanolic concentrate of *Ocimum* varieties has been displayed to have malignant growth preventive potential through the decrease of the overabundance measure of nitric oxide (Kim et al., 1998). Tulsi also diminishes the frequency of benzo(a) pyrine-actuated neoplasia and 3-methyl di-methyl amino azobenzene initiated hematomas in trial creatures (Aruna and Sivaramarishnan, 1990). Ethanolic l extract of Tulsi leaves administered through a topical route resulted in a significant decrease in the upsides of growth occurrence (Paplliomas) in the skin of abino mice (Prashar et al., 1994).

Tricyclic sesquiterpenoids 2-(hydroxymethyl)- 5,5,9-trimethylcyclo [7.2.0.03,6]undecan-2-ol extracted from the oil of OS leaves have been found to display antiproliferative activity against MCF-7 cell line (IC50 30 ± 0.5 μM) utilizing doxorubicin as standard (IC50 9.7 μg/mL) (Singh and Chaudhuri, 2013; Singh et al., 2014). In continuation of antiproliferative screening against MCF-7 cell line, 5β-hydroxycaryophyllene, 4,5-epoxy-caryophyllene, and sesquiterpenes β-caryophyllene displayed IC50 esteems 4.8, 7.0, and 73.0 μg/mL, individually. Mixtures of apigenin, rosmarinic acid, orientin, luteolin, vicenin-2, oleanolic acid, and ursolic acid are reviewed and assayed for their anticancer activity (Nagaprashantha et al., 2011). Flavonoids and terpenoids are the significant class of mixtures answerable for the anticarcinogenic movement of *O. sanctum*.

Eugenol (1-hydroxy-2-methoxy-4-allylbenzene), the dynamic compound present in Tulsi has been observed to be a great extent answerable for the remedial possibilities of Tulsi in the treatment of different chronic sicknesses including cancer (Prakash and Gupta, 2005). It was tracked down that the ethanolic concentrate of *Ocimum* leaves hindered the expansion and angiogenesis-related protein through the downregulation of Bcl-2 and vascular endothelial development factor (VEGF) articulation and overarticulation of capase-3 during N-methyl-N"- nitro-N-nitrosoguanidine incited gastric malignancy bearing rates, and with the decrease in cancer cell size, an expansion in life expectancy of mice bearing Sarcoma-180 and Lewis-lung carcinoma in animal models was additionally noted (Prashar et al., 1998; Manikandan et al., 2007a, b; Nakamura et al., 2004). It is also observed that the dynamic phytoconstituents, in particular, oleanlic acid and urosolic acid in Tulsi can display anticancer properties (Singh et al., 2010). The alcoholic concentrate has also been known to enhance the activity of cytochrome b5, cytochrome p450, glutathione S-transferase, and aryl hydrocarbon hydroxylase that plays a significant part in the detoxification of cancer-causing agents and mutagens (Govind and Madhuri, 2006). The test study directed on animal models has shown that *O. sanctum* has the ability to diminish the occurrence of benzo(a)pyrine instigated neoplasia of forestomach of mice and 3-methyl-4-dimethylaminoazobenzene prompted hepatomas in rodents (Aruna and Sivaramarishnan, 1992).

The ethanolic *O. sanctum* leaf concentrate represses 7, 12-dimethylbenz[a] anthracene (DMBA)- incited oxidative stress and genotoxicity by modulating the xenobiotic metabolizing enzymes, thus decreasing the degree of protein and lipid oxidation and up-managing cancer prevention agents (Manikandan et al., 2007a, b). Manikandan et al. (2008) contemplated the combinatorial chemopreventive viability of *O. sanctum* (OS) and *Azadirachta indica* (AI) against N-methyl-N'- nitro-N-nitrosoguanidine (MNNG)-incited gastric carcinogenesis, in light of changes in oxidant-cancer prevention agent status, cell multiplication, angiogenesis, and apoptosis in a rodent forestomach carcinogenesis model and found that OS and AI mix might be interceded by their cancer prevention agent, antiproliferative, antiangiogenic, and apoptosis activating properties.

OS extract diminished the potency of ornithine decarboxylase, a protein participating in the regulation of cell expansion and increase of malignant growth. There was additionally a contemporary decrease in the Phase I catalysts and lipid peroxidation inferring that *O. sanctum* interferes in the movement of cancer-causing agents and that this causes a decline in the development of extreme cancer-causing moiety (Jemal et al., 2007). The anticancer potential of seed oil of Tulsi was inspected against 20-methylcholanthrene-induced fibrosarcoma cancers in the thigh of Swiss albino mice. An increase of 100 µL/kg body weight (most extreme endured portion) of the oil essentially diminished 20-methaylcholathrene-initiated cancer predominance and growth volume. The mice supplemented with the *O. sanctum* seed oil showed an enhanced survival rate and the incidence of tumor was also delayed in this case (Aggarwal et al. 2006).

13.3.2 ANTIRADIATION ACTIVITY

O. sanctum can shield the DNA of the body from risky radiation (Singh, 2005; Panda and Kar, 1998; Devi and Gonasoundari, 1995). It is important to refer that the flavonoids specifically vicenin and orientin extracted from Tulsi leaves displayed comparatively better radioprotective impact in comparison to commercially available chemical radioprotectors. They have shown critical protection to the human lymphocytes against the clastogenic impact of radiation at low, non-poisonous concentrations (Devi et al., 2000). The blend of OS leaf extract with WR-2721 (an engineered radioprotector) enhances higher bone marrow cell protection and show a decrease in the harmfulness of WR-2721 at higher portions, recommended that the blend would have promising radioprotection in humans (Gonasoundari et al., 1998).

Bhartiya et al. (2006) explored radiodefensive effect of aqueous extract of *O. sanctum* L. (40 mg/kg, for 15 days) in mice exposed to high dosages (3.7 MBq) of oral 131 iodine by examining the organ loads, lipid peroxidation, and cell reinforcement protection compound in different target organs, such as stomach, kidney, liver, and salivary organs at 24 h after administration.

The consequences of the investigations showed that the pretreatment with *O. sanctum* L. in radioiodine-exposed group displayed a critical decrease in lipid peroxidation in both salivary glands, kidney, and in the liver; diminished glutathione (GSH) levels displayed a huge decrease after radiation exposure while pre-therapy with *O. sanctum* L. has shown less exhaustion in GSH level even after exposure to 131 iodine. Nonetheless, no such changes were seen in the stomach. The outcomes demonstrate the chance of utilizing aqueous extract of *O. sanctum* L. for improving 131 iodine-activated harm to the salivary glands.

Subramanian et al. (2005) have seen that two polysaccharides isolated from *O. sanctum* L. have the capacity to forestall oxidative harm to liposomal lipids and plasmid DNA prompted by different oxidants, such as γ-radiation, 2,2-azobis (2-amidino-propane) dihydrochloride (AAPH) and iron. Vrinda and Uma (2001) detailed that two water-dissolvable flavonoids, Vicenin (Vc) and Orientin (Ot), isolated from the leaves of *O. sanctum* L. give huge protection against lethality, radiation, and chromosomal distortion in vivo. The effect of aqueous concentrate of the leaves of *O. sanctum* L. against lethality, radiation, and chromosomal damage was investigated by radiation-instigated lipid peroxidation in liver and the outcomes have displayed that aqueous concentrate itself increased the GSH and compounds fundamentally above the normal level, though radiation essentially decreased all the qualities and altogether increased the lipid peroxidation rate, arriving at a most extreme worth at 2 h after exposure (3.5 times of control) (Devi and Gonasoundari, 1999) In another examination, the liquid concentrate of OS has been found to decrease the lipid peroxidation and to speed up recuperation to normal levels in test organisms and *Ocimum* flavonoids delivered promising anti-radiation effect (Devi et al., 1998) Ganasoundari et al. (1998) examined the radioprotective effect of the leaf extract of *O. sanctum* L. (OE) in combination with WR-2721 (WR) on mouse bone marrow and noticed a huge abatement in unusual cells just as various kinds of distortions. The counter radiation impact of Tulsi is especially pertinent to individually presented to abundance radiation like working with radiodetermination and treatment (e.g., atomic medication, angiography, activity under X-beam control), getting radiography for malignomas, working in nuclear reactors and different units with exposure to radiation, routinely presented to high height sunlight-based radiation (e.g., aircraft staff), persistently presented to TV and PC screens. Subsequently, Tulsi can securely be utilized in anticipation of sick impacts of radiation in people presented to different radiations (Singh et al., 2010).

13.3.3 ANTIOXIDANT ACTIVITY

The antioxidant action of Tulsi has been accounted for by numerous researchers (Govind, 2009). The antioxidant activity of flavonoids and their connection to membrane protection was studied (Saija et al., 1995) Antioxidant potential of flavonoids (vicenin and orientin) in vivo was expressed in a critical decrease in the radiation-prompted lipid peroxidation in mouse liver (Gonasoundari et al., 1998). OS extract has the critical capacity to scavenge profoundly responsive-free radicals (Kelm et al., 2000). The phenolic compounds viz., cirsimaritin, apigenin, irsilineol, rosmarinic acid, isothymusin, and calculable amounts of eugenol (a significant part of the unstable oil) from OS concentrate of stem and new leaves had great antioxidant activity (Nair and Gunasegaran, 1982). The antioxidant potential of essential oils acquired by steam hydrorefining from *O. sanctum* L. was assessed utilizing high performance liquid chromatography (HPLC)-based hypoxanthine/xanthine oxidase and DPPH (1,1-Diphenyl-2-picrylhydrazyl) assays. It was also observed that in hypoxanthine/xanthine oxidase assay, effective cancer prevention activity was shown by *O. sanctum* L. (Trevisan et al., 2006). In another examination, the aqueous concentrate of *O. sanctum* L. was identified to altogether increase the action of antioxidants (Gupta et al., 2006). Oral intake also gives critical aortic and liver tissue security from hypercholestrolemia-prompted peroxidative damage (Yanpallewar et al., 2004).

Godhwani et al. (1988) examined the immunoregulatory profile of aqueous extract and methanolic concentrate of *O. sanctum* L. passes on to antigenic test of *Salmonella typhosa* and sheep erythrocytes by evaluating agglutinating antibodies utilizing the Widal agglutination and sheep erythrocyte agglutination tests and E-rosette development in albino rodents. The results of the investigation demonstrate an immunostimulation of humoral immunogenic reaction as addressed by an increase in counteracting agent titer in both the Widal and sheep erythrocyte agglutination tests just as by cell immunologic reaction addressed by E-rosette development and lymphocytosis.

Banerjee et al. (1996), observed the anticarcinogenic potential of *O. sanctum* against many cancer-causing agents. Decoction of fresh Tulsi leaves has anticancer potential. Alcoholic extracts of *O. sanctum* act on the activities of cytochrome b5, aryl hydrocarbon hydroxylase, and cytochrome P-450 in the liver and glutathione-S-transferase (GST) and a reduced glutathione level in lungs and liver have been observed. All these cofactors and enzymes are reported to exhibit an essential role in the detoxification of mutagens and carcinogens. Tulsi leaves when fed to experimental rats for 10 weeks remarkedly led to the reduction of hematoma and squamous cell carcinoma incidences (Aruna and Sivaramakrishnan, 1992; Shiva et al., 2016).

The anticancer activity of *O. sanctum* has been reported against human fibrosarcoma cells culture, wherein the alcoholic extract induced cytotoxicity @ 50 mg/mL and above, morphologically, the cells displayed condensed nucleus and shriveled cytoplasm. DNA was also observed to be disintegrated upon visualization in agarose gel electrophoresis (Pandey and Madhuri, 2006; Kathiresan et al., 1999).

13.4 CONCLUSIONS

O. sanctum (Os) has been examined for the bioassay-directed isolation of compounds and also for looking for novel particles from various concentrates. A few synthetic classes of mixtures including flavonoids, phenolics, phenylpropanoids, terpenoids, neolignans, unsaturated fat subordinates, coumarins, and fixed and essential oil have been accounted for from this spice. The essential oil present in Tulsi is a decent source of normal eugenol and is well investigated in analytical, compound, and organic aspects because of its high business significance in drug, beauty care products, and food industries.

Fixed oil present in Tulsi seeds is rich in ω-3 unsaturated fats and is a new research interest, because of its wide scope of pharmacological properties particularly in cardioprotection. Flavonoids are a significant class of mixtures extracted from *Ocimum* and have been reported as the fundamental dynamic components. The water-soluble flavonoids, that is, vicenin and orientin have been very much investigated as far as their radiation defensive impacts at both low and high dosages. The hydrophilic nature of both the flavonoids makes them valuable for their antioxidant impact in detoxification just as radiation defender in malignant growth treatment. The conventional significance of OS as immunomodulator spice was additionally upheld by the isolation of antistress particles, ocimumoside A and ocimumoside B, which recommend the use of Tulsi in treating neurological disorders. Further examinations are expected to investigate the atomic and cellular mechanism of antistress action of ocimumoside A and ocimumoside B. Until this point, the exploration work completed on OS is predominantly centered on the natural potential of its various concentrates and essential oil. The literature showed that ursolic acid, eugenol, orientin, rosmarinic acid, ocimumoside A and ocimumoside B, and vicenin are the principle dynamic phytocomponents of Tulsi and proposing a chance of discovering novel bioactive particles.

However, there are a few points that are expected to be investigated and examined further, such as work on the enormous plant material to extract adequate quantity of major and minor chemical constituents to investigate their pharmacological potential and component for restorative potential; investigate preclinically known compounds for clinical practices, particularly in anti-stress and radiation security; limited work has been done for the isolation and biological examinations on the root concentrates of *O. sanctum*. More than 60 compounds have been isolated, but only a few have been investigated for their pharmacological actions and preclinical examinations. In general, there is a demand for additional examination of the compound parts of OS to get new constituents with enhanced pharmacological potential. The long history of conventional uses, wide range of pharmacological properties, and toxicological studies recommended OS as a protected significant spice for clinical studies. The current knowledge of the active constituents alongside their pharmacological properties will be useful in future examinations on *O. Sanctum* plant as well as in looking for new leads for drug revelation.

KEYWORDS

- **secondary metabolites**
- **apoptosis**
- **angiogenesis**
- **matrix metalloproteinases**
- **metastasis**
- **carcinoma**
- **papilloma**

REFERENCES

Aggarwal, B. B.; Ichikawa, H.; Garodia, P.; Weerasinghe, P.; Sethi, G.; Bhatt, I. D. From Traditional Ayurvedic Medicine to Modern Medicine: Identification of Therapeutic Targets for Suppression of Inflammation and Cancer. *Expert Opin. Ther. Targets* **2006**, *10*, 87–118.

Anbarasu, K.; Vijayalakshmi, G. Improved Shelf Life of Protein-Rich Tofu Using *Ocimum sanctum* (Tulsi) Extracts to Benefit Indian Rural Population. *J. Food Sci.* **2007**, *72*, 305.

Aruna, K.; Sivaramakrishnan, V. M. Anticarcinogenic Effects of Some Indian Plant Products. *Food Chem. Toxicol.* **1992**, *30*, 953–956.

Aruna, K.; Sivaramarishnan, V. M. Plant Products as Protective Agents Against Cancer. *Ind. J. Exp. Biol.* **1990**, *28* (11), 1008–1011.

Banerjee, S.; Prashar, R. A.; Kumar, A.; Rao, A. R. Modulatory Influence of Alcoholic Extract of *Ocimum sanctum* on Carcinogen Metabolizing Enzyme Activities and Reduced Glutathione Level in Mouse. *Nutr. Cancer* **1996**, *25*, 205–217.

Basu, K. *Indian Medicinal Plants*, 2nd ed., Vol. 3; International Book Distributors, Dehradun, Reprint, **1987**; pp 1965–1996.

Bhartiya, U. S.; Raut, Y. S.; Joseph, L. J. Protective Effect of *Ocimum sanctum* L After High-Dose 131Iodine Exposure in Mice: An In Vivo Study. *Ind. J. Exp. Biol.* **2006**, *44*, 647–652.

Boots, A. W.; Haenen, G. R. M. M.; Bast, A. Health Effects of Quercetin: From Antioxidant to Nutraceutical. *Eur. J. Pharmacol.* **2008**, *585*, 325–337.

Devi, P. U.; Bisht, K. S.; Vinitha, M. A Comparative Study of Radioprotection by Ocimum Flavonoids and Synthetic Aminothiol Protectors in the Mouse. *Br. J. Radiol.* **1998**, *71*, 782–784.

Devi, P. U.; Ganasoundari, A. Modulation of Glutathione and Antioxidant Enzymes by *Ocimum sanctum* and Its Role in Protection Against Radiation Injury. *Ind. J. Exp. Biol.* **1999**, *37*, 262–268.

Devi, P. U.; Gonasoundari, A.; Vrinda, B.; Srinivasan, K. K.; Unnikrishanan, M. K. Radiation Protection by the *Ocimum sanctum* Flavonoids Orientin and Vicenin: Mechanism of Action. *Radiat. Res.* **2000**, *154* (4), 455–460.

Devi, P. U.; Gonasoundari, A. Radioprotective Effect of Leaf Extract of Indian Medicinal Plant *Ocimum sanctum*. *Ind. J. Exp. Biol.* **1995,** *33*, 205.

Galati, G.; O'Brien, P. J. Potential Toxicity of Flavonoids and Other Dietary Phenolics: Significance for Their Chemopreventive and Anticancer Properties. *Free Radical Biol. Med.* **2004**, *37*, 287–303.

Ganasoundari, A.; Devi, P. U.; Rao, B. S. Enhancement of Bone Marrow Radioprotection and Reduction of WR-2721 Toxicity by *Ocimum sanctum*. *Mutat. Res.* **1998**, *397*, 303–312.

Ganesan, V.; Amutha, M. Phytochemical Evaluation and Anticancer Activity of *Ocimum sanctum* L.—A Review. *J. Pharm. Res.* **2018**, *12* (7), 917–923.

Godhwani, S.; Godhwani, J. L., Vyas, D. S. *Ocimum sanctum*-A Preliminary Study Evaluating Its Immunoregulatory Profile in Albino Rats. *J. Ethnopharm.* **1988**, *24*, 193–198.

Govind, P.; Madhuri, S. Medicinal Plants: Better Remedy for Neoplasm. *Ind. Drug* **2006**, *43* (11), 869–874.

Govind, P. An Overview on Certain Anticancer Natural Products. *J. Pharm. Res.* **2009**, *2* (12), 1799–1803.

Gupta, S.; Mediratta, P. K.; Singh, S.; Sharma, K. K.; Shukla, R. Antidiabetic, Antihypercholesterolaemic and Antioxidant Effect of *Ocimum sanctum* (Linn) Seed Oil. *Ind. J. Exp. Biol.* **2006**, *44*, 300–304.

Hadi, S. M.; Showket, H. B.; Asfar, S. A.; Sarmad, H.; Uzma, S.; Ullah, M. F. Oxidative Breakage of Cellular DNA by Plant Polyphenols: A Putative Mechanism for Anticancer Properties. *Semin. Cancer Biol.* **2007**, *17*, 370–376.

Jaggi, R. K.; Madaan, R.; Singh, B. Anticonvulsant Potential of Holy Basil, *Ocimum sanctum* Linn. and Its Cultures. *Ind. J. Exp. Biol.* **2003**, *41*, 1329–1333.

Jemal, A.; Siegel, R.; Ward, E.; Murray, T.; Xu, J.; Thun, M. J. Cancer Statistics. *CA Cancer J. Clin.* **2007**, *57*, 43–66.

Karthikeyan, K.; Gunasekaran, P.; Ramamurthy, N.; Govindasamy, S. Anticancer Activity of *Ocimum sanctum*. *Pharma. Bio.* **1999**, *37 (4)*, 285–290.

Kavyashree, M. R.; Harini, A.; Prakash, L.; Pradeep, H. A Review on Tulasi (*Ocimum sanctum* Linn.). *J. Drug Del. Therap.* **2019**, *9* (2-s), 562–569.

Kelm, M. A.; Nair, M. G.; Strasburg, G. M.; DeWitt, D. L. Antioxidant and Cyclooxygenase Inhibitory Phenolic Compounds from *Ocimum sanctum* Linn. *Phytomedicine* **2000**, *7*, 7–13.

Kim, O. K.; Murakami, A.; Nakamura, Y.; Ohigashi, H. Screening of Edible Japanese Plants for Nitric Oxide Generation Inhibitory Activities in RAW 246.7 Cells. *Cancer Lett.* **1998**, *125* (1–2), 199–207.

Manikandan, P.; Letchoumy, P. V.; Prathiba, D.; Nagini, S. Proliferation, Angiogenesis and Apoptosis-Associated Proteins Are Molecular Targets for Chemoprevention of MNNG-Induced Gastric Carcinogenesis by Ethanolic *Ocimum sanctum* Leaf Extract. *Singapore Med. J.* **2007a**, *48*, 645–651.

Manikandan, P.; Murugan, R. S.; Abbas, H.; Abraham, S. K.; Nagini, S. *Ocimum sanctum* Linn. (Holy Basil) Ethanolic Leaf Extract Protects Against 7, 12-dimethylbenz (a) Anthracene-Induced Genotoxicity, Oxidative Stress, and Imbalance in Xenobiotic-Metabolizing Enzymes. *J. Med. Food* **2007b**, *10*, 495–502.

Manikandan, P.; Vidjaya, L. P.; Prathiba, D.; Nagini, S. Combinatorial Chemopreventive Effect of Azadirachta indica and *Ocimum sanctum* on Oxidant-Antioxidant Status, Cell Proliferation, Apoptosis and Angiogenesis in a Rat Forestomach Carcinogenesis Model. *Singapore Med. J.* **2008**, *49* (10), 814.

Mauli, G.; Maulik, N.; Bhandari, V.; Kagan, V. E.; Pakrashi, S.; Das, D. K. Evaluation of antioxidants effectiveness of few herbal plants. *Free Radic Res* **1997**, *27*, 221–228.

Mediratta, P. K.; Sharma, K. K.; Singh, S. Evaluation of Immmunomodulatory Potential of *Ocimum sanctum* Seed Oil and Its Possible Mechanism of Action. *J. Ethnopharmacol.* **2002**, *80*, 15–20.

Mondal, S.; Mirdha, B. R.; Mahapatra, S. C. The Science Behind Sacredness of Tulsi (*Ocimum sanctum* Linn.). *Indian J. Physiol. Pharmacol.* **2009**, *53* (4), 291–306.

Nagaprashantha, L. D.; Vatsyayan, R.; Singhal, J.; Fast, S.; Roby, R.; Awasthi, S.; Singhal, S. S. Anti-Cancer Effects of Novel Flavonoid Vicenin-2 as a Single Agent and in Synergistic Combination with Docetaxel in Prostate Cancer. *Biochem. Pharmacol.* **2011**, *82*, 1100–1109.

Nair, A. G. R.; Gunasegaran, R. Chemical Investigation of Certain South Indian Plants. *Indian J. Chem.* **1982**, *21B*, 979–980.

Nakamura, C. V.; Ishida, K.; Faccin, L. C.; Filho, B. P. D.; Cortez, D. A. G. In vitro activity of essential oil from *Ocimum gratissimum* L. Against Four Candida Species. *Res. Microbiol.* **2004**, *155*, 579–586.

NIIR Board. National Institute of Industrial Research (India) *Compendium of Medicinal Plants*; National Institute of Industrial Research, 2004; p 320.

Panda, S.; Kar, A. *Ocimum sanctum* Leaf Extract in the Regulation of Thyroid Function in the Male Mouse. *Pharmacol. Res.* **1998**, *38* (2), 107–110.

Pandey, G.; Madhuri, S. Medicinal Plants: Better Remedy for Neoplasm. *Ind. Drug.* **2006**, *43* (11), 869–874.

Pattanayak, P.; Debajyoti, D.; Sangram, K. P. *Ocimum sanctum* Linn. A Reservoir Plant for Therapeutic Applications. *Phcog. Rev.* **2010**, *4*, 95–105.

Prakash, J.; Gupta, S. K. Chemopreventive Activity of *Ocimum sanctum* Seed Oil. *J. Ethnopharmacol.* **2000**, *72*, 29–34.

Prakash, P.; Gupta, N. Therapeutic Uses of *Ocimum sanctum* Linn (Tulsi) with a Note on Eugenol and Its Pharmacological Actions: A Short Review. *Indian J. Physiol. Pharmacol.* **2005**, *49* (2), 125–131.

Prashar, R.; Kumar, A.; Banerjee, S.; Rao, A. R. Chemopreventive Action by an Extract from *Ocimum sanctum* on Mouse Skin Papillomagenesis and Its Enhancement of Skin Glutathione S-Transferase Activity and Acid Soluble Sulfydril Level. *Antican. Drugs* **1994**, *5* (5), 567–572.

Prashar, R.; Kumar, A.; Hewer, A.; Cole, K. J.; Davis, W.; Phillips, D. H. Inhibition by an Extract of *Ocimum sanctum* of DNA-Binding Activity of 7, 12-Dimethylbenz [a] Anthracene in Rat Hepatocytes In Vitro. *Cancer Lett.* **1998**, *128*, 155–160.

Puri; Rasayana, H. S. *Ayurvedic Herbs for Longevity and Rejuvenation*; CRC Press, 2002; pp 272–280.

Saija, A.; Scalese, M.; Lanza, M.; Marzillo, D.; Bonina, F.; Castelli, F. Flavonoids as Antioxidant Agents: Importance of Their Interaction with Biomembrane. *Free Rad. Biol. Med.* **1995**, *19*, 481.

Sen, P.; Maiti, P. C.; Puri, S.; Ray, A.; Audulov, N. A.; Valdman, A. V. Mechanism of Anti Stress Activity of *Ocimum sanctum* Linn, Eugenol, *Tinospora malabarica* in Experimental Animals. *Ind. J. Exp. Biol.* **1992**, *32*, 592–596.

Sharma, P. C.; Yelne, M. B.; Dennis, T. J. *Data Base on Medicinal Plants Used in Ayurveda*, Vol 2; CCRAS: Delhi, Reprint 2005; p 500.

Singh, N.; Verma, P.; Pandey, B. R.; Bhalla, M. Therapeutic Potential of *Ocimum sanctum* in Prevention and Treatment of Cancer. *Int. J. Pharma. Sci. Drug Res.* **2012**, *4* (2), 97–104.

Singh, N.; Hoette, Y.; Miller, R. *Tulsi 'The Mother Medicine of Nature'*, 2nd ed.; International Institute of Herbal Medicine: Lucknow, 2010; pp 28–47.

Singh, N. In the Symposium "Continuing Education Programme on Herbal Drug Research" Held at Institute of Nuclear Medicine and Allied Sciences; DRDO: Delhi, India on 3–7 October 2005.

Singh, S.; Majumdar, D. K.; Rehan, H. M. S. Evaluation of Anti-Inflammatory Potential of *Ocimum sanctum* (Holy Basil) and Its Possible Mechanism of Action. *J. Ethnopharmacol.* **1996**, *54*, 19–26.

Singh, V.; Amdekar, S.; Verma, O. *Ocimum tenuiflorum* (Tulsi): Biopharmacological Activities. *Pharmaco* **2010**, *1* (10).

Singh, D.; Chaudhuri, P. K. *Study of Some Reputed Indian Medicinal Plants and Chemical Modification of Major Bioactive Scaffold*, 2013; p 136.

Singh, D.; Chaudhuri, P. K.; Darokar, M. P. New Antiproliferative Tricyclic Sesquiterpenoid from the Leaves of *Ocimum sanctum*. Helv. *Chim. Acta* **2014**, *97*, 708–711.

Siva, M.; Shanmugam, K. R.; Shanmugam, B.; Venkata, S. G.; Ravi, S.; Sathyavelu, R. K.; Mallikarjuna, K. *Ocimum sanctum*: A Review on the Pharmacological Properties. *Int. J. Basic Clin. Pharmacol.* **2016**, *5* (3), 558–565.

Subramanian, M.; Chintalwar, G. J.; Chattopadhyay, S. Antioxidant and Radioprotective Properties of an *Ocimum sanctum* Polysaccharide. *Redox Rep* **2005**, *10*, 257–264.

Sugihara, N.; Arakawa, T.; Ohnishi, M.; Furuno, K. Anti and Pro-Oxidative Effects of Flavonoids on Metal-Induced Lipid Hydroperoxide-Dependent Lipid Peroxidation in Cultured Hepatocytes Are Loaded with Alpha-Linolenic Acid. *Free Radical Biol. Med.* **1999**, *27*, 1313–1323.

Summers, C. B.; Felton, G. W. Pro-Oxidant Effects of Phenolic Acids on the Generalist Herbivore *Helicoverpa zea* (Lepidoptera: Noctuidae): Potential Mode of Action for Phenolic Compounds in Plant Anti-Herbivore Chemistry. *Insect Biochem. Mol. Biol.* **1994**, *24*, 943–953.

Trevisan, M. T.; Vasconcelos, S. M. G.; Pfundstein, B.; Spiegelhalder, B.; Owen, R. W. Characterization of the Volatile Pattern and Antioxidant Capacity of Essential Oils from Different Species of the Genus Ocimum. *J. Agric. Food Chem.* **2006**, *54*, 4378–4382.

Vrinda, B.; Uma, D. P. Radiation Protection of Human Lymphocyte Chromosomes In Vitro by Orientin and Vicenin. *Mutat. Res.* **2001**, *498*, 39–46.

Warrier, P. K. In *Indian Medicinal Plants*; Longman O, Ed.; CBS Publication: New Delhi, 1995; p 168.

Yamanaka, N.; Oda, O.; Nagao, S. Prooxidant Activity of Caffeic Acid, Dietary Non-Flavonoid Phenolic Acid, on Cu2+-Induced Low-Density Lipoprotein Oxidation. *FEBS Lett.* **1997**, *405*, 186–190.

Yanpallewar, S. U.; Rai, S.; Kumar, M.; Acharya, S. B. Evaluation of Antioxidant and Neuroprotective Effect of *Ocimum sanctum* on Transient Cerebral Ischemia and Long-Term Cerebral Hypoperfusion. *Pharmacol. Biochem. Behav.* **2004**, *79*, 155–164.

Index